Fields of change
Progress in African archaeobotany

GRONINGEN ARCHAEOLOGICAL STUDIES
VOLUME 5

Editorial Board
Prof. dr. P.A.J. Attema
Dr. J. Bos
Prof.dr. R.T.J. Cappers
Prof. dr. L. Hacquebord
Dr. W. Prummel
Prof. dr. D.C.M. Raemaekers
Prof. dr. H.R. Reinders
Dr. S. Voutsaki

Groningen Institute of Archaeology
Poststraat 6
9712 ER Groningen
the Netherlands
gia@rug.nl

Website
www.gas.ub.rug.nl

Publisher's address
Barkhuis Publishing
Zuurstukken 37   9761 KP Eelde   the Netherlands
Tel. 0031 50 3080936 Fax 0031 50 3080934
info@barkhuis.nl   www.barkhuis.nl

Fields of change

Progress in African archaeobotany

EDITED BY

René Cappers

BARKHUIS &
GRONINGEN UNIVERSITY LIBRARY
GRONINGEN 2007

Cover design: Nynke Tiekstra, Noordwolde, the Netherlands
Illustration front cover: Agricultural fields in the Nile Valley (Ezbet Bazili, Egypt; April 2002).
   Photo: R.T.J. Cappers
Illustration back cover: Cow-drawn scoop-wheel (*Saqia*) at Aswan (Egypt; November 1996).
   Photo: R.T.J. Cappers

ISBN-13 9789077922309

Copyright © 2007 Groningen Institute of Archaeology, Groningen, the Netherlands

All rights reserved. No part of this publication or the information contained herein may be reproduced, stored in a retrieval system, or transmitted in any form or by any means, electronic, mechanical, by photocopying, recording or otherwise, without prior written permission from the Groningen Institute of Archaeology.

Although all care is taken to ensure the integrity and quality of this publication and the information herein, no reponsibility is assumed by the publishers nor the authors for any damage to property or persons as a result of operation or use of this publication and/or the information contained herein.

Table of contents

R.T.J. Cappers
Preface ... VII

R.M. Blench
The intertwined history of the silk-cotton and baobab ... 1

B. Eichhorn
Use of firewood resources in a hyperarid environment:
charcoal analysis from sites in the Skeleton Coast Park, Northern Namib Desert ... 21

A. Höhn
Where did all the trees go?
Changes of the woody vegetation in the Sahel of Burkina Faso during the last 2000 years ... 35

S.S. Murray
Medieval cotton and wheat finds in the middle Niger Delta (Mali) ... 43

S.S. Murray
Identifying African rice domestication in the middle Niger Delta (Mali) ... 53

M.A. Murray, D.Q. Fuller & C. Cappeza
Crop production on the Senegal River in the early First Millennium AD:
preliminary archaeobotanical results from Cubalel ... 63

D.Q. Fuller, K. Macdonald & R. Vernet
Early domesticated pearl millet in Dhar Nema (Mauritania):
evidence of crop processing waste as ceramic temper ... 71

J. Morales & T. Delgado
Figs and their importance in the prehistoric diet in Gran Canaria Island (Canary Isles) ... 77

A.M. Mercuri & E.A.A. Garcea
The impact of hunter/gatherers on the vegetation in the Central Sahara during the Early Holocene ... 87

P.L. Crawford
Beyond paper: use of plants of the Cyperaceae family in ancient Egypt ... 105

R. Hamdy
Plant remains from the intact garlands present at the Egyptian Museum in Cairo ... 115

R.T.J. Cappers, L. Sikking, J.C. Darnell & D. Darnell
Food supply along the Theban desert roads (Egypt):
the Gebel Roma[a], Wadi el-Huôl, and Gebel Qarn el-Gir caravansary deposits ... 127

C. Newton
Growing, gathering and offering: Predynastic plant economy at Adaïma (Upper Egypt) ... 139

A.J. Clapham & P.A. Rowley-Conwy
New discoveries at Qasr Ibrim, Lower Nubia   157

R.T.J. Cappers & R. Hamdy
Ancient Egyptian plant remains in the Agricultural Museum (Dokki, Cairo)   165

Preface

R.T.J. Cappers

Groningen Institute of Archaeology, University of Groningen, the Netherlands

The continent of Africa has played an important and independent role in the history of plant exploitation. Located in both the northern and southern hemisphere, Africa is characterized by different vegetation types, such as forests, woodlands, savannas, grasslands, deserts and semi-deserts, each with its own floristic elements.

As a result of climatic changes during the Holocene and the transition to agriculture, the vegetation in Africa has undergone major changes in the last 10,000 years. The study of plant macro remains and pollen as well as ethnobotanical and linguistic research can be used to improve our knowledge of the former vegetation development, the process of plant domestication and the introduction of economic plants from other continents by international trade.

The transition to agriculture in Africa is of comparatively recent date when compared with the emergence of agriculture in the Near East. This might be related to the different climatic conditions in North Africa during the early Holocene and the availability of a broad spectrum of edible plants in many parts of Africa. Nevertheless, our insight into this important shift in the food economy in Africa has only recently received serious attention among archaeobotanists. As a consequence, the domestication history of many African crops, such as Sorghum (*Sorghum bicolor*), is still under discussion and from others it has only recently been unravelled, namely the hyacinth-bean (*Lablab purpureus*) and the watermelon (*Citrullus lanatus*).

In May 1994, the 1st International Workshop on African Archaeobotany (IWAA) was organized in Mogilany (Poland) to enhance the archaeobotanical research in the African continent. This first workshop marked the beginning of a three-yearly meeting of archaeobotanists and specialists on African languages. Subsequent workshops have been organized at Leicester (England, 1993), Frankfurt (Germany, 2000), Groningen (the Netherlands, 2003) and London (England, 2006). The presentation of current research and the findings of laboratory sessions during these workshops offer a fruitful basis for integrating and updating our understanding of the history of the African vegetation. Ongoing research will, however, be necessary to refine our knowledge of the plant exploitation of certain areas and to fill in the major gaps in the data of other parts of the African continent.

The proceedings of the International Workshop on African Archaeobotany have provided us with a major insight into the vegetation development and plant exploitation in Africa. Papers presented at earlier workshops have been published by Stuchlik & Wasylikowa (1995), Van der Veen (1999) and Neumann et al. (2003). This book presents papers presented at the 4th International Workshop on African Archaeobotany, held in Groningen from 30th of June until the 2nd of July 2003. Several papers deal with the domestication history and related aspects of specific plants, including wheat (*Triticum*), rice (*Oryza*), pearl millet (*Pennisetum glaucum*), fig (*Ficus*), cotton (*Gossypium*), silk-cotton (*Ceiba pentandra*) and baobab (*Adansonia digitata*). Other contributions discuss the exploitation of woody vegetations, members of the sedge family (Cyperaceae) and the botanical composition of mummy garlands. Three papers present the subfossil plant remains from Egyptian sites: Pharaonic caravanserais along the Theban Desert Road, Predynastic Adaïma and Napatan to Islamic Qasr Ibrim. The last contribution presents an update inventory of the ancient plant remains present in the Agricultural Museum (Dokki, Cairo). The book covers a wide range of countries and includes Namibia, Burkina Faso, Mali, Senegal, Mauritania, Canary Isles, Libya and Egypt.

A special word of thanks goes to all the colleagues who gave useful comments on earlier versions of these papers, to Daphne Maring-Van der Pers, Marieke van der Wal, Erwin Bolhuis and Sander Tiebackx from the Groningen Institute of Archaeology, who were responsible for the copy-editing and the printing quality of the illustrations. The correction of some of the English texts was carried out by P. Bleeker.

References

Neumann, K., A. Butler & S. Kahlheber (eds.) 2003. *Food, fuel and fields. Progress in African Archaeobotany.* Heinrich Barth Institut, Köln.

Stuchlik, L. & K. Wasylikowa (eds.) 1995. *Acta Palaeobotanica* 35 (1), pp. 1-184.

Veen, M. van der (ed.) 1999. *The exploitation of plant resources in Ancient Africa.* Kluwer/Plenum, New York.

The intertwined history of the silk-cotton and baobab

R.M. Blench

Mallam Dendo, 8 Guets Road, Cambridge CB1 2AL, United Kingdom

Both the silk-cotton (*Ceiba pentandra*) and the baobab (*Adansonia digitata*) are widely associated with human settlement in West Africa and neither species is part of the 'natural' vegetation but owe their distribution to human activity. The paper reviews the ethnobotany of both species and the evidence for their dispersal in both prehistoric and historical times. The vernacular names seem to have been exchanged between species to another in areas where the trees co-exist. The paper analyses the biogeographical and linguistic evidence and suggests that the accepted models are at best problematic.

> "Quand elles ne sont pas méconnues, les fonctions de l'arbre dans les civilisations africaines sont généralement sous-estimées." Paul Pélissier (1980)

1 Introduction

How natural the vegetation of West-Central Africa is can be a matter for debate, but there is little doubt that human beings have been manipulating the distribution of various species of economic or ritual importance for some considerable time. Among the species whose Africa-wide distribution is very striking are the baobab (*Adansonia digitata*) and the silk-cotton (*Ceiba pentandra*). Neither are truly native to the continent; the baobab probably originated in Madagascar where all its relatives occur, while *Ceiba* apparently originated in the Americas. Both trees have economic uses, but it is unlikely that this is the primary reason for their spread; both are large, impressive and somewhat oddly-shaped and have been incorporated into ritual systems almost everywhere they occur.

This paper[1] explores the history and ethnobotany of the baobab and the silk-cotton and in particular looks at a phenomenon that might be call 'conceptual crossover'. Put simply, the silk-cotton apparently spread across the continent from West to East and the baobab the other way. At some highly speculative point in prehistory, the two trees must have 'met' one another, as they diffused in opposite directions. At this juncture, populations using them seem to have confused or at least conceptualised the two as similar, because both names and ideas about the trees are exchanged between species. The paper looks at the origin and spread of these species and then tabulates a sample of vernacular names in diverse African languages, to exemplify this process.

2 Origins and distribution

2.1 Baobab

Baobabs are members of the Bombacaceae, a pantropical family containing a number of well-known economically important plants like kapok, balsa and durian (Baum, 1995a; Baum, Small & Wendel, 1998). Six of the eight species of baobabs are restricted to western and southern Madagascar, a seventh is endemic to northwestern Australia, and the eighth is widespread in sub-Saharan Africa but now introduced by humans throughout the warm tropics (Armstrong 1983; Bowman 1997). The African baobab is the best known of the eight species. All baobabs are deciduous trees ranging in height from five to 30 meters with leaves with segments that radiate somewhat like the fingers of a hand, showy flowers, and large, many-seeded gourdlike fruits covered with a velvety thatch of hairs. The baobab is the subject of several popular reviews in addition to technical monographs (*e.g.* Codjia *et al.*, 2001; Sidibe & Williams, 2002; Bash, 2002; Pakenham, 2004; Wickens *in press*). Figure 1 shows a characteristic baobab in the Nuba mountains in Southern Kordofan.

In height, trunk shape, and girth, there is no such thing as a typical baobab, but the tremendous size of some African baobabs suggests that individual trees may be several thousand years old. Like many large tropical trees, the baobab has no reliable tree rings, and size does not necessarily indicate age because variation in water content of the trunk can cause large fluctuations. Adanson (1771), who first attempted age

Figure 1: Baobab in the Nuba mountains.

calculations, estimated that two trees on an island off Cape Verde were 5150 years old. Swart (1963) who radiocarbon-dated a tree 4.5 meters in diameter being cut down for the Kariba dam obtained a date of 1010 ± 100 BP and clearly trees can be older still (Von Breitenbach, 1985). Exactly how long the baobab has been spreading in Africa remains to be confirmed by archaeobotany, but some living examples may be several thousand years old. Kahlheber (pers. comm.) has found records of baobab in Senegal, Mali and Benin, with the oldest probably at the site of Arondo on the Senegal River AD 400-1000. Germer (1985) records baobab fruits in the Turin museum and in the Louvre in Paris of Egyptian provenance, but with no location or date. Cappers (pers. comm.) has found baobab seeds at Berenike on the Red Sea coast of Egypt, dated to the 4-5th century AD; these are likely to be imports, although whether from Yemen or further south in Africa is difficult to say.

The first recorded literary reference is by the Arab traveller Ibn Baṭṭūṭa in West Africa in AD 1352, who mentions the water-storage capacity of its massive trunk (Hamdun & King 1975: p. 30, 71). Leo Africanus also noted its presence.[2] The earliest description in the European literature is Scaliger in 1557, although the tree is recognisable in hindsight from some of the descriptions of fifteenth century navigators (Wickens, 1982: p. 175). The name baobab derives from the Arabic *bu hibab*, 'fruit with many seeds', and it is first referred to directly by Alpini (1591) who encountered its fruits for sale in Egypt. Alpini gives the name as *ba hobab*, hence the vowels of the common English name. Linnaeus gave the baobab its scientific name, Adansonia, in 1759, to honour the celebrated French botanist Michel Adanson (1727-1806), who lived in Senegal and who provided the first technical description and illustration of the tree.

The distribution of the baobab throughout the dry parts of Africa reflects both human activity and the depredations of elephants (Adam, 1962). Its medical and fibre uses make it essential to many communities and its shape carries numerous magico-religious associations (Owen, 1970). Its fruits represent a convenient source of food that is naturally conserved by the hard outer shell and pastoralists, in particular, often carry fruits with them. As they crack the shell and eat the seeds, trees spring up along former cattle-trails. Assogbadjo *et al.* (2005, 2006) report that baobabs show considerable genetic diversity between ecological zones in West Africa, suggesting that this reflects the east-west nature of trade routes as well as natural ecological zonation. Jaouen (1988) in a study of the vegetation of Mauritania, reports dead baobabs far into the arid zone, presumably also seeded by travellers. On the other hand, elephants are responsible both for destroying seedlings and damaging mature trees (Owen, 1974; Guy, 1982). Barnes (1980) argued that predation by elephants was gradually destroying the baobab populations in the Ruaha National Park, Tanzania. The literature suggesting that baobabs are endangered is venerable; Chevalier (1906: p. 486) considered it was disappearing[3], while Perrier de la Bathie (1953: p. 214) warned of the threat to the Malagasy baobabs. But a paradoxical consequence of the widespread elimination of wildlife, especially in West Africa, is that the baobab is spreading rapidly. Pullan (1974) noted that the baobab was a key species in the intensively farmed 'parkland' of West Africa. In the Hadejia-Nguru Wetlands in Northeast Nigeria, an area once rich in biodiversity but now stripped by predatory timber-cutting and poaching and starved of water by upstream extraction, the baobab has spread rapidly during the 1990s (Blench *et al.* 2003). Baobabs may thus indicate a depauperate fauna. More generally, as the indigenous forest is stripped away, and replaced by anthropic grassland, many dry-zone species, such as the tamarind and the fan-palm are moving further south. At present, the future of the Australian baobab seems secure. Survival of the African baobab and at least two Malagasy species (*A. rubrostipa* and *A. za*) also seems likely because of their comparatively widespread ranges and broad ecological tolerances. The long-term prognosis for the four Malagasy baobab species with more restricted distributions is a matter of concern because of continuing habitat destruction that goes along with population growth.

Figure 2 shows the African distribution of the baobab given by Wickens (1982). Wilson (1988: p. 200) notes that the baobab is found at more locations and at higher altitudes than this map suggests. Some of the sites on the map may be rather misleading; for example, the baobab was introduced into Gabon from Senegal as an ornamental in colonial times and thus its occurrence is both late and 'unnatural' (Walker, 1953; Raponda-Walker & Sillans, 1961: p. 104). Chevalier (1906: p. 492) considered that the baobab was only spontaneous in coastal areas, and its presence in the interior was due to human action; and that Islam had played an important role in this redistribution in recent centuries. He explained its absence in the heart of the continent simply by chronology, that it had simply not been spread there. Its Africa-wide distribution is quite strongly associated with river-systems, although the baobab is by no means a riverine species. This is probably a secondary effect, whereby pastoralists grazing in river valleys have increased its incidence. Armstrong (1983: p. 144) presents maps of the Malagasy species, all of which are confined to the west coast of Madagascar.

Burton-Page (1969) noted that the baobab was been spread around the Indian Ocean by Arab traders, and associates its presence in India with the export of Ḥabshis, or slaves from the Horn of Africa, who were brought to India in the early Middle Ages. The baobabs in Yemen, Oman, Zanzibar and Egypt may be attributed to the same source. Ironically, the African baobab has also been reintroduced in Madagascar, presumably its original centre of origin, by settlers from Zanzibar. In the twentieth century it was carried across the world as an ornamental. Armstrong (1983: p. 146) notes that the baobab was also carried to Sri Lanka, Java and the Philippines as a result of Indian Ocean trade. It was also taken to the Caribbean in the 1890s

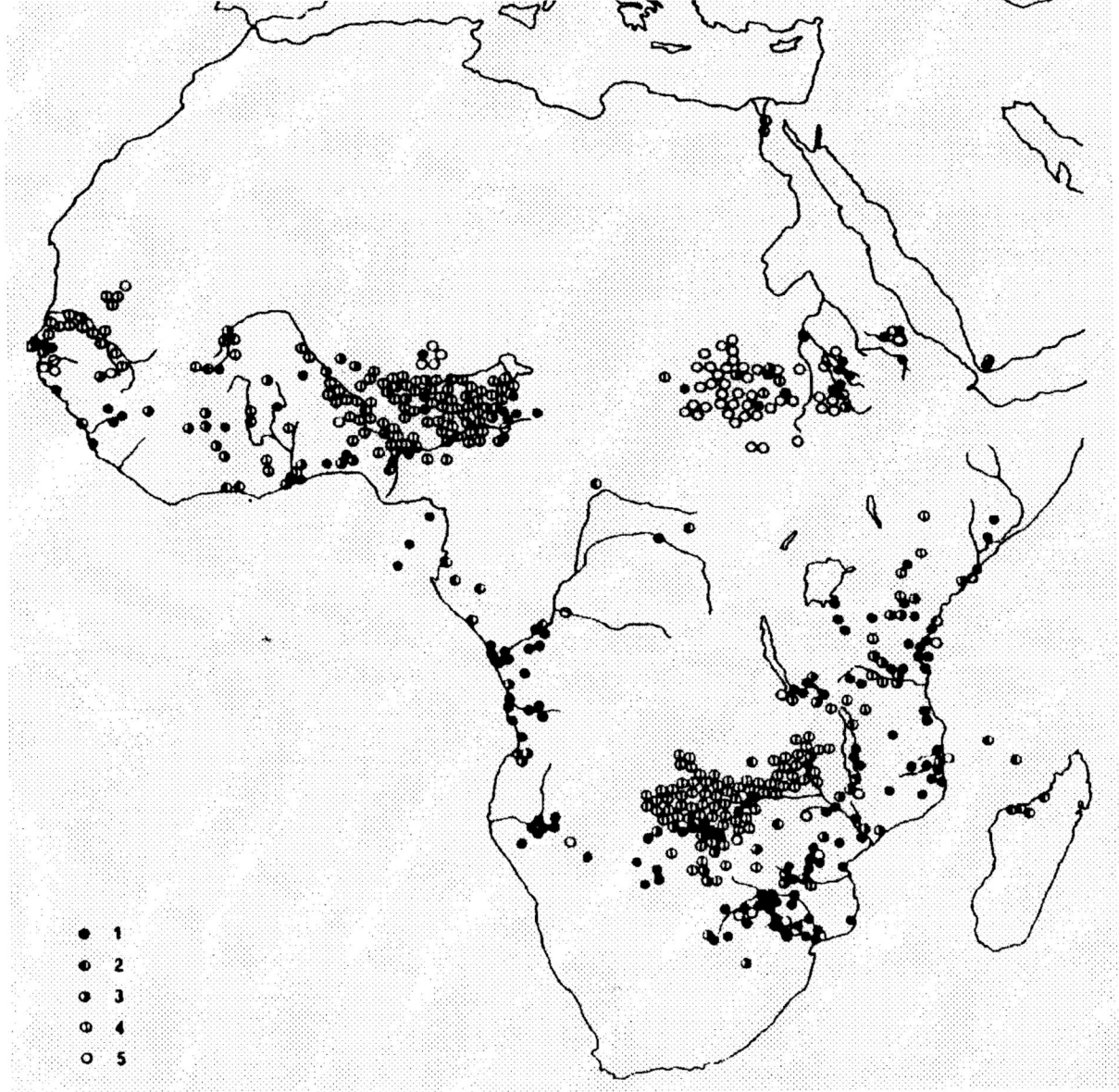

Figure 2: Africa-wide distribution of Adansonia digitata (Key: 1: Herbarium and flora records; 2: Cultivated or introduced; 3: Information from photographs; 4: Kew 'Baobab survey' data; 5: Data from travel literature. Source: Adapted from Wickens [1982]).

as an ornamental (Rashford 1996). Wickens (1982: p. 182) also discusses the distribution of baobabs in India. The tree has no Sanskrit name and appears to be associated with the area of Muslim influence; if so, then its introduction may be medieval. However, it is strongly associated with shrines and temples, which would suggest a greater antiquity.

Figure 3 shows the primary and secondary distributions of *A. digitata*. The Australian species, *A. gibbosa* [formerly *A. gregorii*], known as a *boab*, is something of a mystery, and it has been explained as a relic of Gondwanaland with plate tectonics accounting for its disjunct distribution. However, Armstrong (1977, 1983) argues that the 'Gondwanic' hypothesis is inadequate[4] and instead argues for a Tertiary era dispersal, presumably via floating seed-pods. Wickens (1982: p. 201) also concludes "There is still no satisfactory explanation for the presence of *A. gregorii* in Australia" and Bowman (1997:894) leaves the question open. It is notable however that *A. gibbosa* is part of the Malagasy section *Longitubae* which is pollinated by long-tongued hawk-moths (Baum 1995b). It has been speculated that the Australian boabs may have originated from seed pods carried for food carried either by coastal movements of early modern humans or even by Austronesian seafarers en route from Madagascar.[5]

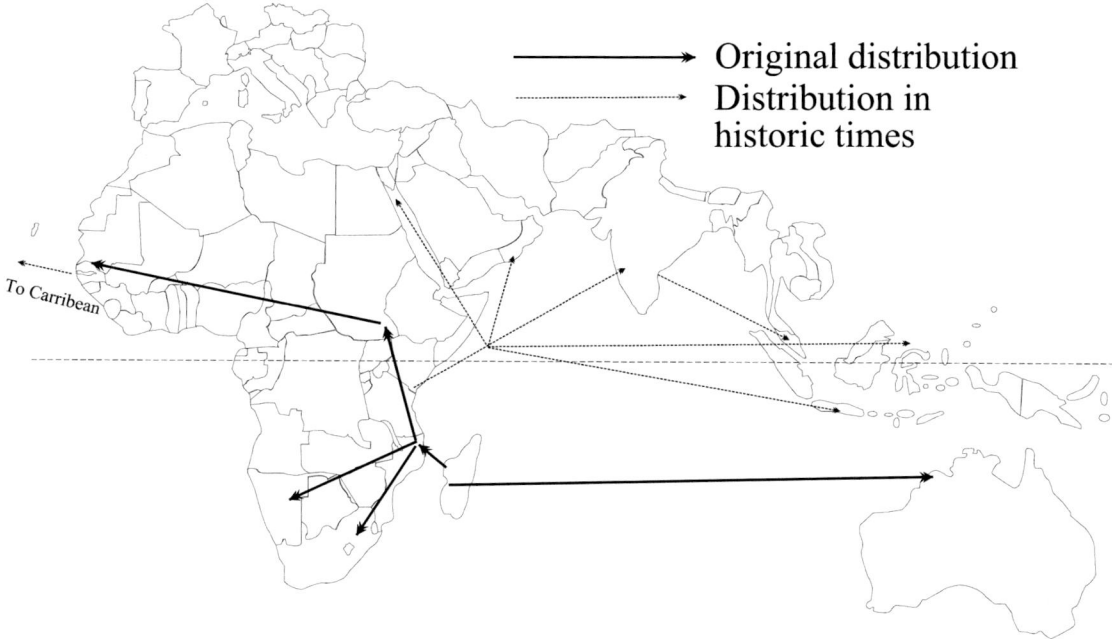

Figure 3: Worldwide distribution of Adansonia digitata.

2.2 Silk-cotton [kapok] *Ceiba pentandra* (L.) Gaertn. var. *pentandra* (1791)

The silk-cotton is 'a large or immense tree, the trunk with grey bark, buttressed, usually spiny, cylindrical and smooth far up, deciduous; branches in horizontal tiers; crown of leaves open; leaves of 5-9 leaflets palmately borne, each lanceolate, acuminate, glaucous beneath, entire or obscurely toothed, 7-18 cm long, 1-3.5 cm wide, distinctly petiolulate; petiole 7-20 cm long; flowers clustered on branchlets, cream-white or pale pink; 5 petals about 2.5 cm long; stamens in 5 bundles, anthers twisted; fruit oblong-ellipsoid, smooth, pendulous, 7-15 cm long, 5-celled, eventually dehiscent, the interior filled with soft long copious hairs ("kapok" or silk-cotton); seeds many, brown' (Stone, 1970). The silk-cotton is fire resistant and can sometimes become dominant in highly anthropic savannah landscapes (Swaine, 1992; Swaine *et al.*, 1992).

Chevalier (1931, 1937) was the first to point out that *Ceiba* is a Neotropical genus and the apparently ancient presence of *Ceiba* in Africa is anomalous. He counted nine species of *Ceiba*, eight of which are confined to the New World although the most recent revision suggests there are seventeen species (Gibbs & Semir, 2003). Chevalier thought that the seeds may have floated across the Atlantic attached to the floss, although this is apparently unlikely, because the seeds readily detach themselves from the floss (Burkill, 1985: p. 280). The distribution of the silk-cotton in the New World is highly anthropic (Gibbs & Semir, 2003, Fig. 6) and it plays an important role in many indigenous cosmologies. The very similar associations in West Africa suggest that it was intentionally brought across the Atlantic. Irvine (1961:191) assumed that "it was probably introduced by the Portuguese". There appear to be few archaeobotanical records for silk-cotton in sub-Saharan Africa, but Bedaux (1972) reports both *Ceiba* and the baobab on the Dogon escarpment between the 12th and 14th centuries, *i.e.* prior to any possible Portuguese introduction.[6] The antiquity of *Ceiba* in Africa would further seem to be confirmed by evidence for its presence further east in the Indian Ocean, for pictorial records may show that kapok had reached Java by AD 850 (Steinmann, 1934; see also Toxopeus, 1948). If so, then *Ceiba* would have had to travel across Africa from west to east and then spread around the Indian Ocean along the established maritime routes more than 1500 years ago. This also implies pre-Portuguese contacts between the east coast of South America and West Africa, a possibility usually scouted by prehistorians. However, Chevalier (1931) noted a number of species common to the east coast of South America and West Africa unlikely to have floated across on ocean currents. Indeed, there are some species that apparently went in the opposite direction; *Elaeis oleifera* is surprisingly close to the oil-palm, *Elaeis guineensis* (Henderson *et al.*, 1995: p. 165). Without more detailed distributional work, this must remain speculation.

The early presence of *Ceiba* in Southeast Asia via Indian Ocean dispersal is problematic in that *Ceiba*

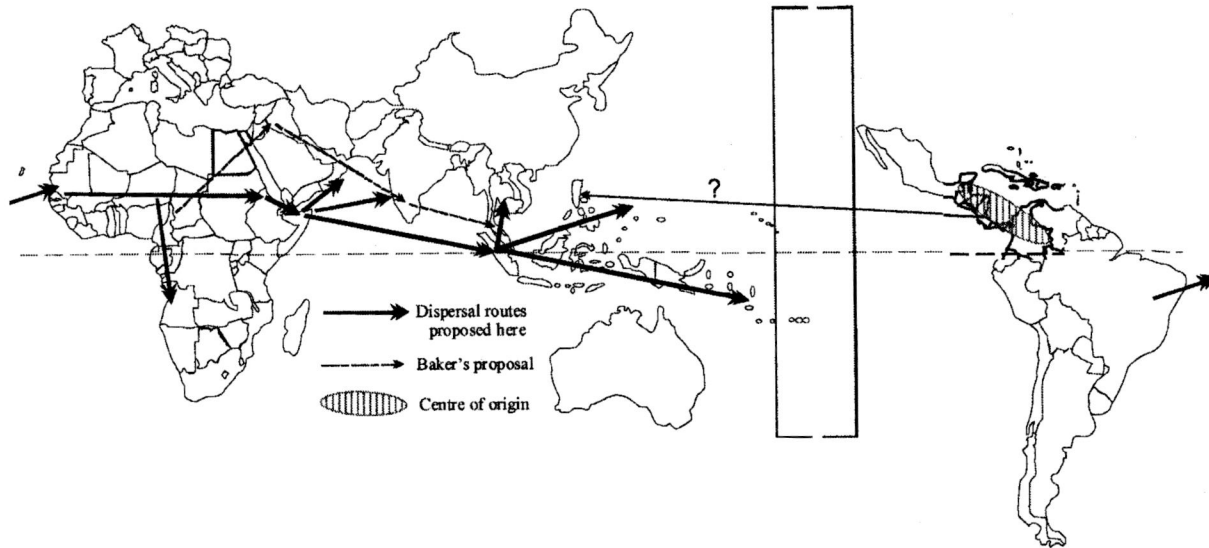

Figure 4: *Worldwide dispersal of* Ceiba pentandra.

hardly occurs in the floras of East Africa. Although Heine & Legère (1995: p. 221) note a Swahili name, *msufi*, they also observe that this tree was spread in Tanzania by the Germans. Noad & Birnie (1989: p. 37) in their tree flora of Kenya treat *Ceiba* as an exotic introduced from Central America. The solution canvassed by Baker (1965) is that cultivated kapok was introduced from West Africa to south-western Asia by the Arabs. In some ways this would explain the distributional gap, but since *Ceiba* is almost unknown in Southwest Asia and none of the vernacular names in any way indicate dispersal by Arabic-speakers, this is not easy to accept. A significant possibility is that *Ceiba* was brought from Central America to Manila by the Spanish and spread to some parts of the Pacific from the Philippines, as some of the vernacular names rather suggest (Appendix Table 3) and that Steinmann (1934) is simply wrong and has misidentified the species depicted, confusing it with one of the indigenous *Bombax* spp. In this case, the occurrences of *Ceiba* in Southeast Asia and the Pacific would all originate from the New World and thus be post-European contact. Figure 4 shows the worldwide dispersal of *Ceiba* canvassed in this paper (which imagines dispersal from the Horn of Africa) with Baker's alternative hypothesis marked separately.

3 Ethnobotany of the baobab and silk-cotton

Burkill (1985: pp. 270-274 & 278-283) provides a useful summary of the conventional ethnobotany of these two species as well as being a major source for vernacular names. Simpson (1995) and Sidibe & Williams (2002) represent recent overviews of the ecology and economic potential of the baobab.

3.1 Baobab

The baobab has many uses, but the collection of pods as a food source must undoubtedly account for part of the anthropic distribution. Fibre from the bark is used to make rope, baskets, cloth, musical instrument strings, and waterproof hats. Stripping the bark from the lower trunk of most trees usually leads to their death, but baobabs not only survive, they regenerate new bark. Fresh baobab leaves provide an edible vegetable similar to spinach which is also used medicinally to treat kidney and bladder disease, asthma, insect bites, and several other maladies. The hollowed-out trunk of a living tree can be used to store water. The fruits and seeds of several species are collected, while pollen from the African and Australian baobabs is mixed with water to make glue. Hobley (1922) describes the co-association of baobabs and ruined towns on the East African coast and attributes it to the collection of the seeds as a shampoo base. Palgrave (1983: p. 588) notes that the interior of old trees crumbles into a fibrous pulp and if they die suddenly, such as after a frost, the tree can burst spontaneously into flame.

The tag of an 'upside-down' tree features in a number of purported traditional stories. Wickens (1982) recounts one in which the Creator is said to have initially planted the baobab in the rainforests of the Congo Basin, but the tree complained that the dampness made its trunk swell. So the Creator moved it to the high slopes of the Ruwenzori range, the

Table 1. The silk cotton and its conceptual relatives in Iten.

Singular	Plural		
ɛkum	nikum	silk-cotton tree	Ceiba pentandra
ɛkum èdaà nìrèè	nikum èdaà nìrèè	red-flowering silk-cotton tree; lit. 'silk-cotton of the [neighbouring] Kwakwi people'	Bombax buonoponenze
ɛkum nè'warâng	nikum nè'warâng	false baobab	Adenium obesum

Mountains of the Moon. But the baobab continued to grumble about the humidity. Angered by the incessant wailing, the Creator took the swollen trunk and tossed it into a dry part of Africa. The tree landed upside down with its roots in the air.

Baobabs are often considered the abode of witches or spirits. Luxureau (1994: p. 73) depicts an enormous split baobab in Maradi, Niger where women place money in order for it to assist them to realise their wishes. The Dompo people of Western Ghana claim to have arisen from a horseman who emerged from a split baobab that is still alive a short walk from their settlement. Alternatively, baobabs may be emblematic of the coherence of a community. In the Volta Region of Ghana, the health of specific baobabs is associated with the community; should they die, the community may break up. Livingstone (1857) mentions a baobab near the Mozambique coast with a room-sized central cavity enough to shelter an entire family. In Australia, where the baobab is known as a 'boab', the tree had similar uses. In the early Kimberley pioneering days, boabs were often temporarily used to contain prisoners. Grates were fitted to the openings, the prisoners put inside and the grate locked. Twenty kilometres from Wyndham is the Prison Tree, which still bears bolts and studs from its service as a prison.

3.2 Silk-cotton

The silk-cotton is only marginally a food-plant; its young leaves and seeds are eaten. But it has probably been spread by humans for the floss; used to stuff pillows, as tinder for fire-making and many other uses. In West Africa, there are two varieties an armed (thorny) and an unarmed cultivars. The armed *Ceiba* is commonly used for fences, as it keeps out goats and the young leaves can be regularly harvested. Raponda-Walker & Sillans (1961: p. 106) note that in Gabon the unarmed type is only found in the forest.[7] It has been brought into plantations in Java for the kapok, although this trade is in decline due to synthetics. In addition, its important spiritual associations make it a prime candidate to spread to new village sites. The growth of chieftaincy institutions in Northern Ghana, for example, encouraged the spread of the silk-cotton because the pillows for chiefs must be filled with kapok. Hauenstein (1997) describes the complex role played by the silk-cotton in the peoples of central Cote d'Ivoire. Irvine (1961: p. 191) notes that it is 'one of the most sacred trees of West Africa', which would be very surprising if it had really been introduced by the Portuguese.

4 Conceptualisation and vernacular terminology of baobab and silk-cotton

4.1 Vernacular terminology in Africa

Both the baobab and the silk-cotton have conceptual twins. For the baobab it is the false baobab or desert rose, *Adenium obesum,* which it resembles at the level of the trunk, although *Adenium* is much shorter and has brilliant red flowers. For the silk-cotton, it is the red-flowered silk-cotton, *Bombax buonoponenze,* a similar tree with, as advertised, red flowers. The ecological ranges of these analogues largely coincide, although the false baobab is usually confined to more semi-arid regions. Table 1 shows the basic term for *Ceiba* in Iten, a language spoken SW of the Jos Plateau in Nigeria. The baobab does not grow in this area and is known only as a Hausa loan. However, both the red-flowering silk-cotton and the false baobab are treated as relatives of the silk-cotton.

The terminology for baobabs is highly ramified in many African languages, which is indicative of its ancient establishment in many parts of the continent. As an example, Table 2 shows the terminology for the baobab and its parts in Hausa, spoken in Nigeria and Niger. The origin of many of these terms is unknown.

4.2 Analysis of vernacular names

The vernacular names for silk-cotton and baobab in Africa are tabulated in Appendix Table 1 and Appendix Table 2. The data is arranged by language phylum and within that by family. For more detail on the internal classification and terminology of African language phyla, Heine & Nurse (2000) is a recent overview. I have assigned numbers to the widespread roots for silk-cotton in the column marked 'R', indicating lexemes that should probably be grouped together.

Table 2. Baobab terminology in Hausa.

Term	Dialect	Gloss	Comment
Ɓákkô		baobab	<Fulfulde?
Bambu		baobab	
Dunku		baobab	
Gàatsííkà	Kano	young baobab	
Gullutu	Kabi	baobab	
Gùntsúú		baobab seeds	
Gwàrgwámíí		baobab seeds	
Kubali	Katsina	baobab	
Kulambali	Katsina	baobab	
Kúmbàlíí	Katsina	baobab flower buds	
Kúúkà *pl.* kúúkóókíí		baobab	< Songhay?
Kwámè		baobab fruit	
Mùrnàà	Sokoto	baobab	

Unnumbered roots seem to be local innovations. Sources are marked in the final column, although I have tried to use commonly available collections of terms rather than citing each individual document in order to keep the bibliography to a manageable length. Skinner (1996: p. 151) discusses some etymologies for baobab. Rashford (1994) has a discussion of the association of monkeys with baobab names, but this is entirely confined to European-related languages; no African language makes this connection.

4.2.1 Baobab

Vernacular names for baobab are manifestly more diverse than those for *Ceiba*. If the distributional argument given in paragraph 2.1 is accepted, the interpretation is that the baobab only began to spread in Africa *after* the main outlines of its language phyla were established and it was diffusing east to west, *i.e.* against the direction of the expansion of Niger-Congo. As a consequence, when farming groups expanded eastwards they encountered the baobab as an unfamiliar tree and were forced to construct a name, perhaps borrowing it from resident hunter-gatherers or comparing it to a species they already knew. This is particularly the case in some names in the Plateau languages of Nigeria and further west in Mali and Cote d'Ivoire, where the terms are clearly borrowed from established names for *Ceiba*.

Despite a reconstructed Bantu form appearing in the Tervuren Lexical Reconstruction database[8], the Bantu could not have carried the baobab with them as they expanded eastwards across the tropical forest, because it will not grow in the humid areas of Cameroun, Gabon and Congo where the Bantu are thought to have originated. They must have developed new terms once they encountered it, on emerging into the savanna. There are clearly two competing Bantu roots in Eastern and Southern Africa, "#mbuyu" and "#muramba" or similar. It seems possible the nearly identical forms in Swahili and Tonga might be loans rather than true genetic cognates. Nonetheless, the forms in South African Bantu languages are clearly cognate with "#mbuyu" and it is possible that the baobab was carried southwards with the Bantu expansion. An intriguing name is the Ovambo "omu-kura" which appears to resemble some West African forms. It is conceivable that the name was carried with the Bantu groups who expanded southwards along the coast of West Africa, although the humid forest that occurs there today no longer supports baobabs.

Newman (1977: p. 22) reconstructed 'baobab' as "★kuka" for Proto-Chadic, and observed "Kanuri *kuwa* is undoubtedly a borrowing from Chadic". In view of the widespread presence of this root in Nilo-Saharan, this is very unlikely and the direction of borrowing is more likely the reverse. The original source of Hausa "kúúkà" may have been a borrowing from the West Kainji languages, as there is considerable diversity of forms, pointing to an original "#kukpa" or "#kugba". Alternatively, Hausa could have borrowed from Kanuri/Manga "kúwà", since the tone-pattern is identical. But there is no doubt that roots of the "#kuka" form are widespread due to secondary borrowing from Hausa. The reason may be that the practice of drying the leaves, powdering them and selling them as soup ingredients has definitely been widely spread along the Hausa trade routes, even if they did not initiate this practice. In Nupe, for example, the baobab tree itself is "muci", an old Nupoid root. However, the Nupe name for the leaves, "kúka", is a relatively recent borrowing from Hausa. Forms such as Daffo are probably later borrowings from Hausa, but

Gashua Bade and Bagirmi probably derive their forms directly from Kanuri/Kanembu. If this was an old Nilo-Saharan term then it was borrowed *into* proto-Kainji, perhaps from Songhay (Dendi is today a direct neighbour of West Kainji languages). Skinner (1996) gives some rather sparse Chadic lookalikes usually referring only to tree species. The Tuareg names are very puzzling, as they are not obviously derived from other languages, and yet the baobab would not have been familiar in North Africa. Probably these are transferred from another tree species. The Malagasy names for *A. digitata* are almost all borrowed from the vernacular names for native *Adansonia* species, except in one case when Sakalava borrows from French.

4.2.2 Silk-cotton
Ceiba pentandra terms are more obviously patterned than those for baobab. The numbered roots all have interesting distributions, although II, "#kum-", is by far the most widespread.

I. Derived from a proto-form something like "*bàntVŋ" for Central Mande languages, it also appears in a clutch of Atlantic languages, including Wolof and Fulfulde. The extensive migrations of Fulɓe pastoralists have given it currency much further east in West Africa, but it is probably most useful to think of it as characteristic of the western zone.

II. Derived from a proto-form something like "#kum-" and widespread across West Africa, also apparently spreading in to the Nilo-Saharan languages. This characteristic root suggests an early date for *Ceiba* and a link with the diffusion of Niger-Congo languages. Its widespread occurrence in Gabonese languages (see Raponda-Walker & Sillans, 1961: p. 106 for a complete list) would seem to make it a candidate for Proto-Bantu, although the East Africanist bias of Bantu reconstructions has so far excluded it.

III. Confined to the East Mande languages and Atlantic languages that appear to have borrowed from them (although Baga Koba is hardly in contact with these languages today).

IV. Derived from a proto-form something like "#-nyã̀ĩ-", this root is confined to the Kwa languages, but is dispersed across the family, suggesting considerable time-depth (Kwa itself being very internally divided).

V. This root is the most complex, since it appears in Nilo-Saharan, Afroasiatic and Bantu. Probably originating either in West Chadic or Nilo-Saharan, it underwent a significant change of vowel from high front /i/ to high back /u/ and was then loaned into Arabic as "rûm" and perhaps thence into Somali as "dum". Arabic may also be the source for various Central African languages, such as Gula "rum".

VI. This root is entirely confined to the Ijoid languages spoken in the Niger Delta in Nigeria. Kay Williamson proposes that the forms can be reconstructed to proto-Ịjọ which would independently indicate considerable antiquity for the tree in this region.

The presence of a distinctive root in Senufic, "#ʃeŋe", which probably reconstructs to proto-Senufic, is good evidence for the antiquity of the silk-cotton in West Africa. The fact that it does not appear to be a loan and does not appear outside Senufic suggests some time-depth.

One of the complex etymologies yet to be fully understood is the name *pemba* or *pamba*, which occurs in NE Madagascar. This is an area under strong Swahili influence, but this is not the Swahili name, which is *msufi* and may derive from the island of Pemba. Boiteau *et al.* (1997) note that *Ceiba* on Madagascar, while an exotic, appears to be subspontaneous all along the coast.

4.3 *Conceptual crossover*
Baobabs and silk-cotton trees may both have been introduced into the African continent by humans, and have certainly been spread by them. Both are tall, impressive trees with numerous ethnobotanical uses and both attract belief complexes. The silk-cotton appears to have spread in the same direction as the expansion of Niger-Congo languages (west to east across the continent) and for this reason, can be reconstructed to quite a high level in that phylum. The lexemes for silk-cotton are strongly associated with those for 'death' and 'corpse', and with various initiation cults. The terms for baobab are more diverse and fall less easily into patterns. This may be because the baobab spread across Africa from east to west, against the direction of flow of the expansion of Niger-Congo. At some place in Central Africa, perhaps between Nigeria and Cameroun, the two species began to co-occur, and in particular, the baobab began to take on names originally applied to the silk-cotton. Prost (1964: p. 423), discussing the vernacular names in Gur languages observes '*Fromager n'est pas toujours distingué de kapokier*'. Throughout this whole region the names and attributes of the two species form a complex networks of borrowings.

5 Conclusion

The history of the baobab and the silk-cotton remain little-understood, despite their economic importance, and there are various reasons for thinking that the published hypotheses do not account for all the facts. Both trees have considerable economic and spiritual importance where they are linked to village communities and it is reasonable to assume they were translocated by human action, perhaps long prior to agriculture. Vernacular names for these species certainly suggest considerable antiquity in Africa; and indeed significant exchanges of associated ideas and terminology. Presumably further work on the genetics of specimens growing in widely separated locations could contribute to unravelling some of these problems.

6 Notes

1. This paper was first given at the 4th International Workshop for African Archaeobotany Groningen, 30 June – 2 July 2003 and has since been substantially revised. I would like to thank participants at the workshop for their comments, and particular thanks to René Cappers, Dorian Fuller, Stephanie Kahlheber and Katharina Neumann for additional unpublished materials. Gerald Wickens is publishing a book on the baobab and has kindly commented at length on the present text. I have also made use of unpublished linguistic data kindly supplied by Richard Gravina, Russell Schuh, Valentin Vydrine, Kay Williamson and Guillaume Segerer to whom my thanks are due. The most recent revision was in May 2006.
2. At least, according to Wickens (1982: p. 175). However, Dietrich Rauchenberger (pers. comm.), Leo's most recent editor, finds no such reference and this may be an error.
3. "Quoi qu'il en soit, les Adansonia sont bien des arbres en voie de disparition"
4. "It is my view that to assume the persistence of Adansonia in sub-Saharan Africa, Madagascar and north-west Australia since before the fragmentation of the southern super-continent initiates more problems than it solves" (Armstrong 1983: p. 143).
5. This not as bizarre as it might first appear. Linguistic and botanical evidence for contact between Austronesians and the north coast of Australia is quite abundant. Against this, A. gregorii, has a high chromosome number (2n=96) in contrast to the lower numbers found in the Malagasy Adansonia spp.
6. Cappers (pers. comm.) has extracted references to Ceiba from his personal database giving dates for Nqoma am Tsodilo in Botswana at AD 9-1000 and Matlhapaneng in Botswana at AD 7-1000 which seem credible.
7. The data is too sketchy to make a strong case, but it seems possible that the contrast between the two types may represent two different introductions; the unarmed ssp. being ancient and the armed, 'cultivated' types of Portuguese vintage.
8. This can be consulted on-line at http://linguistics.africamuseum.be/blr/BLR_Home.html.

7. References

Abraham, R.C., 1949. *Dictionary of the Hausa language.* London, University of London Press.

Abraham, R. C., 1958. *Dictionary of Modern Yoruba.* London, University of London Press.

Adam, J.G., 1962. Le baobab (*Adansonia digitata* L.). *Notes africaines,* 94, pp. 33-44.

Adami, P., 1981. *Lexique Bediondo-Français.* Sarh, Collège Charles Lwanga.

Adanson, M., 1771. Description d'un arbre d'un nouveau genre appelé Baobab observé au Sénégal. *Histoire Academie Royale Scientifique* 1791, pp. 77-85, 218-243.

Alpini, P., 1591. *De medicina Aegyptiorum libri quatuor: in quibus multa cum de vario mittendi sanguinis vsu de'q; inustionibus, & alijs chyrurgicis operationibus, tum de quamplurimis medicamentis apud Aegyptios frequentioribus, elucescunt.* Venetiis, Apud Franciscum de Franciscis Senensem.

Andrews, F.W., 1953. *Vernacular names of plants as described in "Flowering plants of the Anglo-Egyptian Sudan Vol. II".* Sudan, McCorquodale & Co.

Armstrong, P., 1977. Baobabs: remnant of Gondwanaland? *New Scientist* 73, pp. 212-213.

Armstrong, P., 1983. The disjunct distribution of the genus *Adansonia. National Geographical Journal of India* 29, pp. 142-163.

Assogbadjo, A.E., B. Sinsin, C. Codjia, P. Jean & P. Van Damme, 2005. Ecological diversity and pulp, seed and kernel production of the baobab (*Adansonia digitata*) in Benin. *Belgian Journal of Botany* 138 (1), pp. 47-56.

Assogbadjo, A.E., T. Kyndt, B. Sinsin, G. Gheysen & P. Van Damme, 2006. Patterns of genetic and morphometric diversity in Baobab (*Adansonia digitata*) populations across different climatic zones of Benin (West Africa). *Annals of Botany* 97 (5), pp. 819-830.

Bailleul, C., 1996. *Dictionnaire Bambara-Français.* Bamako, Donniya.

Baker, H.G., 1965. The evolution of the cultivated Kapok tree: a probable West African product. In: D. Brokensha (ed.), *Ecology and Economic Development in Tropical Africa.* Berkeley, University of California Press, pp. 185-216.

Barnes, R.F.W., 1980. The decline of the baobab tree in the Ruaha National Park, Tanzania. *African journal of Ecology* 18, pp. 243-252.

Barnes, R.F.W., K.L. Barnes, E.B. Kapela, 1994. The long-term impact of elephant browsing on baobab trees at Msembe, Ruaha National Park, Tanzania. *African journal of ecology* 32(3), pp. 177-184.

Bash, B., 2002. *Tree of Life: The World of the African Baobab*. New York, Gibbs Smith.

Baum, D.A., 1995a. A systematic revision of *Adansonia* Bombacaceae. *Annals of the Missouri Botanical Gardens*, 82 (3), pp. 440-470.

Baum, D.A., 1995b. The comparative pollination and floral biology of baobabs (*Adansonia*-Bombacaceae). *Annals of the Missouri Botanical Garden* 82, pp. 322-348.

Baum, D.A., R.L. Small & J.F. Wendel, 1998. Biogeography and floral evolution of Baobabs (*Adansonia*, Bombacaceae) as inferred from multiple data sets. *Systematic Biology*, 47(2), pp. 181-207.

Bedaux, R.M.A., 1972. Tellem, reconnaissance archéologique d'une culture de l'ouest africain au moyen âge: Recherches architectoniques. *Journal de la Société des Africanistes* 42 (2), pp. 103-185.

Blench, R.M. *et al.*, 2003. Access rights and conflict over common pool resources in the Hadejia-Nguru wetlands. Unpublished report to the JEWEL project, Nigeria.

Boiteau, P., M. Boiteau, L. Allorge-Boiteau & C. Paulsen, 1997. *Dictionnaire des noms Malgaches de végétaux*. Grenoble, Éditions Alzieu.

Bowman, D.M.,1997. Observations on the demography of the Australian boab (*Adansonia gibbosa*) in the north-west of the Northern Territory, Australia. *Australian Journal of Botany* 45(5), pp. 893-90.

Breitenbach, F. Von, 1985. Aantekeninge oor die groeitempo van aangeplante kremetartbome (*Adansonia digitata*) en opmerkinge ten opsigte van lewenstyd, groeifases en genetiese variasie van die spesie. *Journal of Dendrology* 5, pp. 1-21.

Burkill, H.M., 1985. *The Useful Plants of West Tropical Africa, Families A-D*. Kew, Royal Botanic Gardens.

Burton-Page, J., 1969. The problem of the introduction of *Adansonia digitata* into Africa. In: P.J. Ucko & G.W. Dimbleby (eds.), *Domestication and exploitation of plants and animals*. London, Duckworth, pp. 331-335

Calame-Griaule, G., 1968. *Dictionnaire Dogon: dialecte Tôrò: Langue et Civilisation*. Langues et Littératures de L'Afrique Noire IV. Paris, Klincksieck.

Chevalier, A., 1906. Les Baobabs (*Adansonia*) de l'Afrique continentale. *Bulletin de la Société Botanique de France* 53, pp. 480-496.

Chevalier, A., 1931. Le rôle de l'Homme dans la dispersion des plantes tropicales: échanges d'espèces entre l'Afrique tropicale et l'Amerique du Sud. *Revue Internationale d'Agriculture Tropicale et Botanique appliquée* 120, pp. 631-650.

Chevalier, A., 1937. Arbres à Kapok et Fromagers. *Revue Internationale d'Agriculture Tropicale et Botanique appliquée* 188, pp. 245-268.

Chevalier, A., 1949. Nouvelles observations sur les arbres à Kapok de l'Ouest Africain. *Revue Internationale d'Agriculture Tropicale et Botanique appliquée* 321 (2), pp. 377-385.

Codjia, J.T.C., B. Fonton-Kiki, A.E. Assogbadjo & M.R.M. Ekue, 2001. *Le baobab (Adansonia digitata), une espèce à usage multiple au Bénin*. Cotonou, Bénin, CECODI/CBDD/Veco/SNV/FSA.

Colombel, V. De, 1995. Noms de plantes: classification, reconstruction et histoire à partir des noms de six cents plantes en dix langues tchadiques des monts du Mandara. In: D. Leger & R. Leger (eds.), *Studia Chadica et Hamitosemitica*. Köln, Rüdiger Köppe, pp. 229-251.

Cosper, R., 1999. *A Wordlist of Eight South Bauchi (West Chadic) Languages: Boghom, Buli, Dott, Geji, Jimi, Polci, Sayanci and Zul*. Munich, Lincom Europa.

Ducroz, J-M & M-C. Charles, 1978. *Lexique Soŋey (Songay) Français*. Paris, Harmattan.

Germer, R., 1985. *Flora des Pharaonischen Agypten*. Mainz am Rhein, Von Zabern.

Gibbs, P. & J. Semir, 2003. A taxonomic revision of the Genus *Ceiba* Mill. (Bombacaceae). *Anales del Jardín Botánico de Madrid* 60 (2), pp. 259-300.

Guy, G.L., 1982. Baobabs and elephants. *African Journal of Ecology* 20, pp. 215-220.

Hamdun, S. & N. King, 1975. *Ibn Battuta in Black Africa*. London, Rex Collings.

Hauenstein, A., 1997. La dendrolâtrie en Côte d'Ivoire (Dendrolatry in the Ivory Coast). *Ethnographie* 93 (121), pp. 89-99.

Heine, B. & K. Legère, 1995. *Swahili plants*. Köln, Rüdiger Köppe.

Heine, B. & D. Nurse (eds.), 2000. *African languages: an introduction*. Cambridge, Cambridge University Press.

Henderson, A.., G. Galeano & R. Bernal, 1995. *Field guide to the palms of the Americas*. Princeton, Princeton University Press.

Hérault, G., 1983. *Atlas des Langues Kwa de Côte d'Ivoire, Tôme 2*. Abidjan, ACCT and ILA.

Hobley, C.W., 1922. On baobabs and ruins. *Journal of the East African and Uganda Natural history society* 17, pp.75-77.

Irvine, F.R., 1961. *Woody Plants of Ghana with special reference to their uses*. London, Oxford University Press.

Jaouen, X., 1988. *Arbres, arbustes et buissons de Mauritanie*. Nouakchott, Centre Culturel Français.

Julien De Pommerol, P., 1999. *Dictionnaire arabe tchadien*. Paris, Karthala.

Le Mbaindo, D.M. & J. Fedry, 1979. *Lexique ngàmbáy-Français, Français-ngàmbáy*. Sarh, Collège Charles Lwanga.

Livingstone, D., 1857. *Missionary travels and researches in South Africa*. London, John Murray.

Luxureau, A., 1994. Usages, représentations, évolutions de la biodiversité végétale chez les Haoussa du Niger. *JATBA*, 36(2), pp. 67-85.

Malgras, D., 1994. *Arbres et arbustes guérisseurs des savanes maliennes.* Paris, ACCT-Karthala.

Manessy, G., 1975, *Les langues Oti-Volta.* Paris, SELAF.

Marchese, L., 1983. *Atlas Linguistique Kru.* Abidjan, ACCT and ILA.

Maundu, P.M., G.W. Ngugi & C.H.S. Kabuye, 1999. *Traditional food plants of Kenya.* Nairobi, National Museums of Kenya.

Mous, M. & R. Kießling, 2004. *Reconstruction of proto-West Rift.* Köln, Rüdiger Köppe.

Newman, P., 1977. *Chadic Classification and Reconstructions.* Afroasiatic Linguistics 5/1.

Ngila, B., 2000. *Expérience végétale bolia.* Köln, Rüdiger Köppe Verlag.

Noad, T. & A. Birnie, 1989. *Trees of Kenya.* Nairobi:, Noad & Birnie.

Nougayrol, P., 1999. *Les parlers gula, Centrafrique, Soudan, Tchad. Grammaire et lexique.* Paris, Éditions CNRS.

Owen, J., 1970. The medico-social and cultural significance of *Adansonia digitata* (baobab) in African communities. *African Notes* 6, pp. 24-36.

Owen, J. 1974. A contribution to the ecology of the African baobab. *Savanna* 3, pp. 1-12.

Pakenham, T., 2004. *The Remarkable Baobab.* London, Weidenfeld & Nicolson.

Palgrave, K.C., 1983. *Trees of Southern Africa.* Cape Town, Struik Publishers.

Pelissier, P., 1980. L'arbre en l'Afrique tropicale: la fonction et la signe. *Cahiers ORSTOM, séries Science Humaine* 13 (1), pp. 127-130.

Perrier De La Bathie, H., 1953. Les Adansonia de Madagascar et leur utilisation. *Revue Internationale d'Agriculture Tropicale et Botanique appliquée* 367 (8), pp. 211-214.

Pipe-Wolferstan, K., 1988. *Traditional food plants. A resource book for promoting the exploitation and consumption of food plants in arid, semi-arid and sub-humid lands of Eastern Africa.* Food and Agriculture Organisation [FAO]. Rome, Food and Agriculture Organization of the United Nations.

Prost, R.P.A., 1953. *Les langues Mandé-Sud du groupe Mana-Busa.* Dakar, IFAN.

Prost, R.P.A., 1964. *Contribution à l'étude des langues voltaïques.* Dakar, IFAN.

Pullan, R.A., 1974. Farmed parkland in West Africa. *Savanna* 3, pp. 119-151.

Raponda-Walker A. & R. Sillans, 1961. *Les plantes utiles du Gabon.* Paris, Le Chevalier.

Rashford, J., 1994. Africa's Baobab tree: Why monkey names? *Journal of Ethnobiology* 14, pp. 173-183.

Rashford J., 1996. Distribution, history and use of the African Baobab in Barbados. *Proceedings of the Thirty-first Annual Meeting of the Caribbean Food Crops Society, St Michael, Barbados, 10 -14 July 1995*, pp. 50-61. Kingshill, St Croix, US Virgin Islands, CFCS.

Reh, M., 1999. *Anywa-English and English-Anywa Dictionary.* Köln, Rüdiger Köppe.

Rongier, J., 1995. *Dictionnaire français-éwé.* Paris, ACCT-Karthala.

Sidibe, M. & J.T. Williams, 2002. *Baobab: Adansonia digitata L. Fruits for the Future 4.* Southampton, International Centre for Under-Utilised Crops.

Sillans, R., 1958. *Les savanes de L'Afrique centrale.* Paris, Paul Lechevalier.

Simpson, M., 1995. *The ecology of the baobab (Adansonia digitata L.) -a literature review.* PhD thesis, School of Agricultural and forest sciences, University of Wales, Bangor, UK.

Skinner, N.A., 1996. *Hausa comparative dictionary.* Köln, Rüdiger Köppe Verlag.

Steinmann, A., 1934. The oldest pictures of the kapok tree in Java. *Tropisch Natuur* 23, pp. 110-113.

Stone, B., 1970. The flora of Guam. *Micronesica* 6, p. 418.

Swaine, M. D., 1992. Characteristics of Dry Forest in West Africa and the Influence of Fire. *Journal of Vegetation Science* 3 (3), pp. 365-374.

Swaine, M.D., W. D. Hawthorne & T. K. Orgle, 1992. The Effects of Fire Exclusion on Savanna Vegetation at Kpong, Ghana. *Biotropica* 24 (2), pp. 166-172.

Swart, E.R., 1963. Age of the baobab tree. *Nature* 198, pp. 708-9.

Thornell, C., 2004. Wild plant names in the Mpiemo language. *Africa & Asia* 4, pp. 57-80.

Toxopeus, H.J., 1948. On the origin of the kapok tree, *Ceiba pentandra. Nederlandse Alg. Proefstat Landbouw* 56, p. 19.

Walker, A., 1953. Le Baobab au Gabon. *Revue Internationale d'Agriculture Tropicale et Botanique appliquée* 365 (6), pp. 174-175.

Weibegue, C. & P. Palayer, 1982. *Lexique Lele-Français.* Sarh, Collège Charles Lwanga.

Wickens, G.E., 1982. The baobab -Africa's upside-down tree. *Kew Bulletin* 37, pp. 172-209.

Wickens, G.E. in press. *Baobabs: Pachycauls of Africa, Madagascar and Australia.* Berlin, Springer.

Wild, H., 1972. *A Rhodesian Botanical dictionary of African and English Plant names.* Salisbury, Government Printer.

Wilson, R.T., 1988. Vital statistics of the baobab (*Adansonia digitata*). *African journal of ecology* 26, pp. 197-206.

Websites

Baobab

Websites for baobab are numerous and often repeat the same information so I have selected the three most informative ones.

http://www.icuc-iwmi.org/resources.htm
http://www.museums.org.za/bio/plants/malvaceae/adansonia_digitata.htm

http://florawww.eeb.uconn.edu/acc_num/200100525.html

Silk-cotton

http://florawww.eeb.uconn.edu/acc_num/198500310.html.
http://www.hear.org/pier/species/ceiba_pentandra.htm
http://www.uog.edu/cals/site/POG/ceiba.html.
http://www.tropilab.com/ceiba-pen.html.
http://www.tis-gdv.de/tis_e/ware/fasern/kapok/kapok.htm#informationen.

8 Appendix Datasheets

Appendix Tables 1 and 2 compile vernacular names for baobab and silk-cotton in sub-Saharan Africa. I have tried to use the most linguistically accurate transcription possible. In a language group where the same root occurs many times I have cited representative forms rather than every recorded form as the intention is to uncover geographical and linguistic patterns. RMB in the source column indicates the data is from my own fieldwork. Existing summary compilations of vernacular names for baobab in Africa are Burkill (1985) and Sidibe & Williams (2002: p. 11).

Abbreviations

AA = Afroasiatic
AN = Austronesian
KS = Khoesan
NC = Niger-Congo
NS = Nilo-Saharan
P = Phylum
R = Numbered root
RMB = own data

Appendix Table 1. Vernacular names of the baobab Adansonia digitata in Africa.

P	Family	Language	attestation	R	Comment	Source
NC	Dogonic	Dogon Toro	ɔ́rɔ	? I	and many other Dogon languages	Calame-Griaule (1968)
	Kordofanian	Heiban	kwor	I	and many other Nuba languages	Andrews (1953)
		Abri	kwugwor	I		Andrews (1953)
		Tira	θɔr	I		RMB
	Mande	Bambara	(n)sìra	IV		Bailleul (1998)
		Soninke	kide			Burkill (1985)
		Susu	kiri			Burkill (1985)
		Kono	sela	IV		Burkill (1985)
		Mende	gbowulo	II		Burkill (1985)
		Bobo-Fing	pii	II	*cf.* Bobo-Fing *Ceiba*	Malgras (1994)
		Guro	bèlé	II	*cf.* root I for *Ceiba* esp. Nwan	Burkill (1985)
	Atlantic	Balanta	laté			Burkill (1985)
		Bassari	a-màk			Burkill (1985)
		Bedik	ga-mak			Burkill (1985)
		Diola Fogny	babaq			Burkill (1985)
		Fulfulde	bokki			Burkill (1985)
		Serer	bak			Burkill (1985)
		Konyagi	a-mbu	II		Burkill (1985)
		Mankanya	bedôal	II		Burkill (1985)
		Wolof	gui		*cf.* root I for *Ceiba* esp. Toura	Burkill (1985)
		Bijogo	u-áto			Burkill (1985)
	Kru	Guéré	go pl. gwê		*cf.* root I for *Ceiba* esp. Toura	Burkill (1985)
		Wobe	gblé-tu		*cf.* root I for *Ceiba* esp. Mwa	Burkill (1985)
	Gur	Bieri	tebu/tora			Manessy (1975)
		Tayari	ñor-ga/-əri			Manessy (1975)
		Nawdm	todde/tuura			Manessy (1975)
		Dagbane	tú-á/-hé			Manessy (1975)
		Moore	tɛɛga/teese			Manessy (1975)
		Dagara	twoo/tooru			Manessy (1975)

Appendix Table 1. (Cont.)

P	Family	Language	attestation	R	Comment	Source
		Bwa	'iya			Malgras (1994)
	Senufic	Minyanka	zige	III		Burkill (1985)
		Senufo	zeŋe	III		Burkill (1985)
		Supyire	zhengè	III		Burkill (1985)
	Kwa	Ewe	àdìdó			Rongier (1995)
		Ga	sààlò	IV		Irvine (1961)
		Dangme	salɛtʃo	IV		Burkill (1985)
		Baule	fromdo			Burkill (1985)
		Twi	ɔdadɛ			Irvine (1961)
			and ɔtɔtɔwaa			Irvine (1961)
		Guan	totɔ			Irvine (1961)
		Brong	kɛlau	I		Irvine (1961)
WBC						
	Kainji	Lopa	kufwə	I	*cf.* Hausa	RMB
		cLela	k-kubu	I	*cf.* Hausa	RMB
		Ror	u-kuk	I	*cf.* Hausa	RMB
		sSaare	u-kup	I	*cf.* Hausa	RMB
		Rogo	u-ub	I	*cf.* Hausa	RMB
	Plateau	Kuki	upə	I	*cf.* root II 'silk-cotton'	RMB
		Təsu	kúkúrú	I		RMB
		Hasha	ikum		*cf.* root II 'silk-cotton'	RMB
		Berom	kugul leng			RMB
		Jijili	ulici			RMB
	Bantu	PB	#bùjú			BLR3[1]
		PB	#dámbà			BLR3
		Kamba	mwamba			FAO (1988)
		Swahili	mbuyu			FAO (1988)
		Embu, Meru	muramba			Maundu *et al.* (1999)
		Taita	mlamba			Maundu *et al.* (1999)
		Chewa	mlambe			FAO (1988)
		Yao	mlonje			FAO (1988)
		Ndebele	umkhomo			FAO (1988)
		Tonga	mubuyu		*cf.* Swahili	FAO (1988)
		Hlengwa	muwu			FAO (1988)
		North Sotho	motsoo			*cf.* Website 2.
		Ovambo	omukura			*cf.* Website 2.
		Tsonga	ximuwu			*cf.* Website 2.
		Tswana	movana			*cf.* Website 2.
		Venda	muvhuyu			*cf.* Website 2.
		Zulu	isimuku			*cf.* Website 2.
NS		Songhay	kò pl. kòà	I	*cf.* Hausa	Ducroz & Charles (1978)
		Dendi	kɔɔ	I	*cf.* Hausa	Burkill (1985)
		Kanuri	kúwa	I	*cf.* Hausa	Burkill (1985)
		Maasai	ol-mesera			Maundu *et al.* (1999)
		Samburu	lamai			Maundu *et al.* (1999)
		Nuer	kusha			Andrews (1953)
		Dinka	dungwol			FAO (1988)
AA	Semitic	Chad Arabic	hamray *pl.* hamar			Julien de Pommerol (1999)
		Chad Arabic	kalakûkay *pl.* kalakûka	I	? <Hausa	Julien de Pommerol (1999)
		Chad Arabic	tabalday *pl.* tabaldi			Julien de Pommerol (1999)
		Amharic	bamba			FAO (1988)

Appendix Table 1. (Cont.)

P	Family	Language	attestation	R	Comment	Source
		Tigre	duma		*cf.* Somali 'silk-cotton'	FAO (1988)
	Cushitic	Orma	yak			Maundu *et al.* (1999)
		Somali	yag			Maundu *et al.* (1999)
		Goroa	dakaa'umó			Mous & Kießling (2004)
		Alagwa	dakaa'imoo			Mous & Kießling (2004)
		Burunge	daka'u			Mous & Kießling (2004)
	Berber	Tadghaq	tăkudust			Kossmann (pers. comm.)
		Iwellemmeden	tadyəmt			Kossmann (pers. comm.)
	C. Chadic	Bacama	kawtə	I		Newman (1977)
		Bana	kwɔ́kwə̀	I		Gravina (pers. comm.)
		Daba	kàkāw	I	name of fruit	Gravina (pers. comm.)
		Giziga	mbaatay			Gravina (pers. comm.)
		Hdi	ka'u	I		Gravina (pers. comm.)
		Logone Kotoko	kuka		< Hausa	Gravina (pers. comm.)
		Mada	kokormbana			Gravina (pers. comm.)
		Makary Kotoko	kalkuka	I	? < Hausa	Gravina (pers. comm.)
		Malgwə	kwakwa	I		Gravina (pers. comm.)
		Muyang	ăkrām			Gravina (pers. comm.)
		Podoko	huhuwá	I		Gravina (pers. comm.)
		Tera	kukwa	I	? < Hausa	Newman (1977)
		Zulgo	mátàkwambúrzùm			Gravina (pers. comm.)
	W. Chadic	Hausa	kúúkà *pl.* kúúkóókíí	I	<Kanuri?	Abraham (1949)
		Bole	dəmbər			Schuh (pers. comm.)
		Ngizim	kuku	I	<Kanuri?	Newman (1977)
		Karekare	kuci	I	<Kanuri?	Newman (1977)
		Gashua Bade	kukwáu *pl.* kùkun	I	? < Hausa	Schuh (pers. comm.)
		Duwai	kuko	I	? < Hausa	Schuh (pers. comm.)
		Miya	kushi	I	<Kanuri?	Schuh (pers. comm.)
		Guruntum	kwàslà			Cosper (1999)
		Jimi	girum			Cosper (1999)
		Tal	bòkwo		? < Fulfulde	Cosper (1999)
		Boghom	mbùɣdi			Cosper (1999)
		Mangas	bokò		? < Fulfulde	Cosper (1999)
		Saya	dót			Cosper (1999)
		Jimi	girim			Cosper (1999)
		Polci	pə́t roon			Cosper (1999)
		Zul	bəlime			Cosper (1999)
		Buli	lúùn			Cosper (1999)
		Dot	róon			Cosper (1999)
		Geji	daahooli			Cosper (1999)
		Guus	duɓul			Cosper (1999)
AN		Sakalava	baobaba		< French	Boiteau *et al.* (1997)
			bontona, vontona		< name for *Andansonia madagascarensis*	Boiteau *et al.* (1997)
			boringy		< name for *A. fony*	Boiteau *et al.* (1997)
			sefo			Boiteau *et al.* (1997)
			za		< name for *A. za*	Boiteau *et al.* (1997)

[1] The reconstructions are drawn from the Tervuren 'Bantu Lexical Reconstructions 3' website. As the citations show, evidence for these reconstructions is strictly confined to Eastern Africa.

Appendix Table 2. Vernacular names of the silk-cotton Ceiba pentandra in Africa.

P	Family	Language	attestation	R	Comment	Source
NC	Dogonic	Dogon Toro	jǔ			Calame-Griaule (1968)
	Ijoid	Kalaḅarị	sıkákáá	VI		Williamson (pers. comm.)
		Oiyakiri	asısayá	VI		Williamson (pers. comm.)
		Kolokuma	ìsàgháí́	VI		Williamson (pers. comm.)
		Oruma	sìyáí́	VI		Williamson (pers. comm.)
		proto-Ijoid	*ı-sìkákà, *a-sìsákà	VI		Williamson (pers. comm.)
	Mande	Bambara	bàna(n)	I		Bailleul (1998)
		Mandinka	bàntaŋ	I		Vydrine (pers. comm.)
		Xasonka	bàntiŋ	I		Vydrine (pers. comm.)
		Vai	ɓàndá	I		Vydrine (pers. comm.)
		Lele	bándà	I		Vydrine (pers. comm.)
		Koranko	bándã̀	I		Vydrine (pers. comm.)
		Jogo	bȑá	I		Vydrine (pers. comm.)
		Nwan	bre	I		Prost (1953)
			*bàntvŋ	I		Vydrine (pers. comm.)
		Toura	gwéè	II		Vydrine (pers. comm.)
		Wan	kwɛē	II		Vydrine (pers. comm.)
		Mende	ngúwɔ	II		Vydrine (pers. comm.)
		Loko	ŋguuhɔ	II		Vydrine (pers. comm.)
		Looma	gúò	II		Vydrine (pers. comm.)
		Kpelle	wuyɛ	II		Vydrine (pers. comm.)
		Dan	gwē̄	II		Vydrine (pers. comm.)
			*gwúwē̄	II		Vydrine (pers. comm.)
		Bisa	hor	III		Prost (1953)
		Mwa	gbure	II		Prost (1953)
		Beng	poro	III		Prost (1953)
		Yaure	fere	III		Prost (1953)
		San	ko(no)		?	Prost (1953)
		Bobo-Fing	peda	III	?	Prost (1953)
		Bobo-Fing	pirii	III	*cf.* Bobo-Fing 'baobab'	Malgras (1994)
		Busa	gbe	II		Burkill (1985)
	Atlantic	Baga Koba	porõ	III	<E Mande	Segerer (pers. comm.)
		Balanta	rumbum			Segerer (pers. comm.)
		Banyun	kidem			Segerer (pers. comm.)
		Basari	a-ndin			Segerer (pers. comm.)
		Bedik	gi-ndii			Segerer (pers. comm.)
		Biafada	bregwe	?II		Segerer (pers. comm.)
		Bijogo	cobbe			Segerer (pers. comm.)
		Diola Flup	bosanobo		= canoe	Segerer (pers. comm.)
		Diola Fogny	busanay			Segerer (pers. comm.)
		Diola Kwataay	étufay			Segerer (pers. comm.)
		Fulfulde Senegal	batigehi	I	*cf.* Xasonka	Segerer (pers. comm.)
		Fulfulde Gambia	bantehi	I	*cf.* Xasonka	Segerer (pers. comm.)
		Serer	m-buday	I		Segerer (pers. comm.)
		Wolof	betene, bentenki	I	*cf.* Xasonka	Segerer (pers. comm.)
		Mandyak	pentya	I		Segerer (pers. comm.)
		Mankanya	pentene	I		Segerer (pers. comm.)
		Pepel	mecene, ntene	I		Segerer (pers. comm.)
		Kissi	banda	I	*cf.* Mande	Segerer (pers. comm.)
		Konyagi	a-man			Segerer (pers. comm.)

Appendix Table 2. (Cont.)

P	Family	Language	attestation	R	Comment	Source
		Non	len			Segerer (pers. comm.)
	Kru	Krao	jwe			Marchese (1983)
		Tepo	jò			Marchese (1983)
		Bete	gɔ̄ɔ̄	II		Marchese (1983)
		Neyo	vīdā			Marchese (1983)
		Godie	gbādā			Marchese (1983)
		Koyo	vādā			Marchese (1983)
	Gur	Bieri	hun-ga/-si	II		Manessy (1975)
		Tayari	ku-m/-na	II		Manessy (1975)
		Nawdm	gom-be/-ti	II		Manessy (1975)
		Dagbane	gu-ŋwa/-nse	II		Manessy (1975)
		Moore	gù-ŋga /-msi	II		Manessy (1975)
		Bwa	tyaa			Malgras (1994)
		Moba	gbang	II		Irvine (1961)
		Baatonun	guma	II		Burkill (1985)
	Senufic	Minyanka	ʃeŋe			Malgras (1994)
		Senufo	ʃiŋe			Malgras (1994)
		Supyire	sììŋè			RMB
	Ubangian	Banda	kopu			Sillans (1958)
		Langbase	kepu			Sillans (1958)
		Gbaya	gela			Sillans (1958)
	Kwa	Ewe	vŭtí			Rongier (1995)
		Ga	ònyãĩ	IV		Irvine (1961)
		Abbey	kpè, òbà			Hérault (1983)
		Abidji	lókpá			Hérault (1983)
		Abouré	èɲĩ̀ɱì	IV		Hérault (1983)
		Abron	jĩnã́			Hérault (1983)
		Adyukru	lĕkp			Hérault (1983)
		Alladian	ecɔ́ tè			Hérault (1983)
		Baule	ɲɲɛ̀	IV		Hérault (1983)
		Twi	onyãã	IV		Irvine (1961)
		Nzema	èɲĩ̀	IV		Hérault (1983)
		Brong	ekile			Irvine (1961)
			danta			Irvine (1961)
		Ega	ɔ̀vè			Hérault (1983)
		Eotile	èɲè	IV		Hérault (1983)
		Gonja	kàlèlə̀			Burkill (1985)
	WBC	Yoruba	ɛgún	II		Abraham (1958)
		also	àràbà			Abraham (1958)
		Yoruba Ife	vuti		< Ewe?	Burkill (1985)
		Igala	agwu	II		Burkill (1985)
		Işekiri	egungun	II		Burkill (1985)
		Isoko	ahe			Burkill (1985)
		Urhobo	óháhèn			Burkill (1985)
		Igbo	ákpū̄	II		Burkill (1985)
		Nupe	kúci	II		Burkill (1985)
		Yala	igu	II		Burkill (1985)
	EBC	Kulu	gù-kúúmú	II		RMB
	Plateau	Berom	kugul	II		RMB
		Iten	ɛkum pl. nikum	II		RMB

Appendix Table 2. (Cont.)

P	Family	Language	attestation	R	Comment	Source
		Cara	fum *pl.* akum	II		RMB
		Izere	kâkúm	II		RMB
		Hyam of Nok	cum	II		RMB
		Ayu	íkúm	II		RMB
		Ningye	kum	II		RMB
		Toro	kumu	II		RMB
		Tǝsu	kúmú	II		RMB
		Ake	ifɔŋ			RMB
		Eggon	ebzi akum	II		RMB
		Jijili	ukumu	II		RMB
		Jili	kúkúmú	II		RMB
	Cross	Abua	ù-mùùm *pl.* àrù-			Burkill (1985)
	River	Anaang	úkúm	II		Burkill (1985)
		Ibibio	úkím	II		Burkill (1985)
	Bendi	Bokyi	bokum	II		Burkill (1985)
	Tivoid	Tiv	vàmbè			Burkill (1985)
	Bantu	Duala, Kele	bŭma	II		Burkill (1985)
		Isongo	buma	II		Sillans (1958)
		Mpongwe cluster	oguma	II		Raponda-Walker & Sillans (1961)
		Mpiemo	dumɔ		< Arabic?	Thornell (2004)
		Fang	oduma			Raponda-Walker & Sillans (1961)
		Vili	mukuma	II		Raponda-Walker & Sillans (1961)
		Eshira	mufuma	II		Raponda-Walker & Sillans (1961)
		Bolia	bo-hɔngɔ́			Ngila (2000)
		Swahili	mbuyu			Maundu et al. (1999)
		Nyanja	mpilila			FAO (1988)
		Tumbuka	myali			FAO (1988)
		Chewa	usufu		< Swahili?	FAO (1988)
		Nkonde	mutunda			FAO (1988)
NS	Songhay	Songhay	bántàm	I	< Mande	Ducroz & Charles (1978)
		Zarma	fórgò	III?		Ducroz & Charles (1978)
		Dendi	bantan		? < Mandinka	Burkill (1985)
	Saharan	Kanuri	tôm	V		Burkill (1985)
		Ngambay	kura	II?	*cf.* Songhay?	Le Mbaindo & Fedry (1979)
	E. Sudanic	Bejondo	kunœ	II?		Adami (1981)
		Anywa	dhégò *pl.* dhék		? *cf.* Dinka 'baobab'	Reh (1999)
		Bagirmi	tumu	V	?< Kanuri	De Colombel (1995)
		Gula	rum	V	< Arabic?	Nougayrol (1999)
AA	Semitic	Chad Arabic	rûm	V	? < Hausa	Julien de Pommerol (1999)
	Cushitic	Somali	dum	V	? < Arabic	FAO (1988)
	W. Chadic	Hausa	ríímíí *pl.* ríímààyéé	V		Abraham (1949)
		Guus	mbəràán			Caron (pers. comm.)
		Gashua Bade	līmi *pl.* līmaksat	V	< Hausa	Schuh (pers. comm.)
		Duwai	rimi *pl.* rīmi	V	< Hausa	Schuh (pers. comm.)
	C. Chadic	Mora, Muktele etc.	təwme	V	?< Kanuri	De Colombel (1995)
		Uldeme	tiwme		?< Kanuri	Gravina (pers. comm.)
		Bana	tìpə̀			Gravina (pers. comm.)
		Logone Kotoko	laɓe			Gravina (pers. comm.)
		Mafa	kwərmbala			De Colombel (1995)
	E. Chadic	Lele	mànynà			Weibegué & Palayer (1982)

Appendix Table 2. (Cont.)

P	Family	Language	attestation	R	Comment	Source
AN	Malagasy	Sakalava	hazomorengy			Boiteau et al. (1997)
		Sakalava	kaboaka		< French 'kapok'	Boiteau et al. (1997)
		Sakalava	landahazobe			Boiteau et al. (1997)
		Betsimisaraka	laoaty			Boiteau et al. (1997)
		Tankara	pamba, pemba		*cf.* Amharic [!] 'baobab' but possibly from the island of Pemba	Boiteau et al. (1997)

Appendix Table 3. Indo-Pacific names for *Ceiba pentandra*.

Place/language	Vernacular name	Comment
Lao	ngiou^2 ban^2	
Vietnamese	gòn	
American Samoa, Samoa, Niue, Tonga	vavae	
Chuuk	koton	< English
Northern Mariana Islands, Guam	*algodon de Manila, atgodon di Manila, algidon, atgidon de anila* [i.e. cotton of Manila]	< Spanish
French Polynesia	vavai	
Cook Islands Mangaia, Aitutaki?	vavai mama'u, vavai maori	
Fiji	vauvau ni vavalangi, semar	
Kosrae	kuhtin, cutin	< English
Marshall Islands	koatoa, atagodon, bulik, kotin	< English
Belau	kalngebard, kalngebárd, kerrekar ngebard	
Pohnpei	cottin, koatun, koatoa	< English
Saipan	arughuschel	
Yap	batte ni gan' ken	

Use of firewood resources in a hyperarid environment: charcoal analysis from sites in the Skeleton Coast Park, Northern Namib Desert

B. Eichhorn
Forschungsstelle Afrika, Universität zu Köln, Germany

In contrast to rock shelter sites in the interior of north-western Namibia which allow for long-term charcoal analyses, recently investigated surface sites in the Northern Namib Desert were occupied for only short periods in the late Holocene. During this phase, climatic conditions in the coastal desert most probably resembled those presently prevailing. Anthracological analysis from these sites provides palaeoethnobotanical information on the use of the scarce woody vegetation for fuel. Results generally indicate that even in this hyperarid environment, availability and accessibility were of major importance for firewood choice. However, the anthracological proof of some woody species, presently occurring at larger distances from the coastal sites, may either indicate mobility of the former inhabitants, fluctuations of ephemeral rivers in the hinterland or collecting of marine driftwood.

1 Introduction

While charcoal analysis has generally played an important role in expanding our knowledge of the vegetation history of southern Africa (see Eichhorn & Jürgens, 2002, table 1), few charcoal assemblages have been analysed from the western part of this subcontinent. The investigations of February (1992), Cartwright & Parkington (1997), Cowling et al. (1999) and Parkington et al. (2000) have dealt with the palaeoecology of the western Cape, situated in the winter rainfall zone of the Republic of South Africa. However, from

Table 1: Interpretation of anthracological results from rock shelters in eastern and central Kaokoland (characteristic species composition is valid for site Omungunda N 99/1).

Period	Pleistocene/Holocene transition phase	Early Holocene	Mid- to Late Holocene	Late Holocene
Age of charcoal deposits	around 12.000 BP	around 8000 BP	around 5300 BP	since about 2000 BP
Vegetation type	probably contracted woody vegetation in depressions and dry river beds	dry savanna or shrubland	savanna, little human influence	savanna, signs of degradation
characteristic species	*Acacia* spp. Capparaceae	*Colophospermum mopane* *Terminalia prunioides*	*Colophospermum mopane* *Terminalia prunioides* *Combretum apiculatum* *Spirostachys africana* *Lonchocarpus nelsii*	*Colophospermum mopane* *Terminalia prunioides* *Combretum apiculatum* *Spirostachys africana* *Lonchocarpus nelsii* *Bridelia tenuifolia* *Flueggea virosa*
climatic interpretation	cooler and more arid than today	more arid than today	generally similar to today	generally similar to today

the summer rainfall part of the Namib Desert further north, no anthracological analysis was published before the start of the investigations carried out within the collaborative research centre ACACIA in 1995. One single publication, which was mainly of methodological character, dealt with the analysis of uncharred wood fragments recovered from a rock shelter in the Central Namib (Sandelowsky, 1977). Besides this, the only other reference made to such remains appeared in a compilation of botanical macroremains from a site at the Spitzkoppe inselbergs on the interior Namib edge and mentioned the presence of *Combretum* spp. wood chips (Kinahan, 1990). This situation is complicated by the fact that terrestrial sediments with sufficent pollen preservation for palynological analysis are scarce in arid regions; thus knowledge on the late Quaternary vegetation history of the region was fragmentary and partially had to be deduced from results from adjacent areas (Van Zinderen Bakker, 1984 a,b; Van Zinderen Bakker & Müller, 1987; Scott et al., 1991; Scott, 1996; Scott et al., 1997). Although recent marine pollen analyses have expanded our knowledge of the south-western African vegetation history on a large supra-regional scale during this phase (Shi & Dupont, 1997; Shi et al., 1998, 2000; Shi et al., 2001; Dupont & Wyputta, 2003), this work has not really aided research with the focus on the regional to local development of vegetation in north-western Namibia.

Within the interdisciplinary research project ACACIA (SFB 389) at the University of Cologne, the former Kaokoland, a landscape now forming part of the Kunene Region in north-western Namibia, has been one regional focus for research. ACACIA is an acronym for „Arid Climate, Adaptation and Cultural Innovation in Africa", and the project focuses on the Holocene cultural and environmental history of arid regions in north-eastern and south-western Africa, with the final aim to compare the development in both regions. In order to obtain information on the Holocene vegetation and climate history, charcoal assemblages from several rock shelters in the mopane savannas of eastern and central Kaokoland (savannas being characterised by the dominance of *Colophospermum mopane* (J. Kirk ex Benth.) J. Kirk ex J. Léonard, Fabaceae-Caesalpinioideae) have been analysed (Albrecht et al., 2001; Eichhorn, 2002; Eichhorn & Jürgens, 2002; Vogelsang et al., 2002; Eichhorn & Jürgens, 2003). A short summary of the most important results is given in table 1. By the means of charcoal analysis, knowledge on the regional vegetation history since the Pleistocene/Holocene transition could thus be enlarged, although a continuous reconstruction is still not practicable.

Anthracological analyses from the most western part of the region, belonging to the Northern Namib Desert in the narrow sense, have thus far been missing for two reasons: On the one hand, access to the region within the Skeleton Coast Park is restricted due to its status as a protected natural reserve with limited touristic access (Jacobson et al., 1995: p. 92; Schoemann, 2000: p.p. 131-145). On the other hand, rock shelters that might yield long-term sequences due to their continuous or repeated use are extremely scarce in the Northern Namib landscape, which is flat except for dunes and rare inselbergs. The few discovered rock shelters on the interior edge of the desert have contained only a small amount of sediment. As a consequence, charcoal pieces of different ages lay close together, inhibiting a reliable palaeoecological interpretation of analyses (Eichhorn, 2002: pp. 304-307; unpublished data).

During archaeological surveys and excavations in 2002 and 2003, several open air sites were investigated in the coastal desert area of the Skeleton Coast Park. Subsequently, anthrocological analysis was conducted on material from four of these sites which were radiocarbon dated to between approximately 200 and 1200 BP (see chapter Radiocarbon dating). Due to this relatively young age, the charcoal analysis from these sites cannot provide us with long-term information on climate and vegetation history of the region. Instead, the palaeoethnobotanical information can be used to reconstruct the use of rare firewood resources in a hyperarid environment and possibly provide us with some clues as to the seasonal movements of the former inhabitants between inland areas and the coast. The probability of such seasonal movements in southern Africa has been discussed vividly in the archaeological literature (e.g. Parkington & Poggenpoel, 1971; Wendt, 1972; Sandelowsky, 1977, 1983; Carr et al., 1978; Wadley, 1979, 1984; Avery, 1984; Richter, 1984, 1991: pp. 239-247; Sealy & Van der Merwe, 1986; Plug, 1998; Smith et al., 2000: pp. 10-25). A comparable mobility in Kaokoland, probably using dry beds of ephemeral rivers as linear oases which provided food and subterranean water, is corroborated by the similarity of archaeological finds at coastal and interior sites (Vogelsang, 2003, see Kinahan & Kinahan, 1984).

It is highly likely that only minor climatic oscillations affected the coastal part of the Northern Namib during the late Holocene (e.g. Heine, 1998a, 1998b, 2002; Hüser et al., 1998; Blümel et al., 2000; Eichhorn, 2002: pp. 19-22). This facilitates a palaeoethnobotanical interpretation, as - with ecological condions resembling the present - particularities in the charcoal spectra can be mainly attributed to human behaviour. However, the special ecological conditions prevailing in large ephemeral river systems makes an interpretation of the results of the anthracological analysis from sites in their proximity more difficult.

1.1 Environmental setting

Landscape and vegetation types of Kaokoland are subject to extreme variability as one progresses from the east to the west, and this on a transect of less than 300 km (Viljoen, 1980; Becker, 2000: pp. 39-42.; Brunotte et al., 2002; Sander, 2002). In eastern and central Kaokoland, the featureless main surface and the mountain area of the marginal swell, dissected by valleys and broad basins, support relatively dense tree savannas and savannas. The western part of the mountain area of the marginal swell is commonly referred to as an escarpment (figure 1), although, according to Brunotte et al. (2002), there is no clear boundary to the coastal plain to justify the use of this term. However, in contrast to the eastern part, it is characterised by dry mountain savannas and shrubland. The coastal plain is commonly divided into two parts: the flat and gravely pro-Namib plain on the eastern edge of the Namib Desert supports predominantly ephemeral grasslands, whereas the Northern Namib Desert with its rare inselbergs, gravel plains, dunes and pans is either virtually bare of vegetation or partially covered by a dwarf shrub vegetation.

Besides less important orographic and geological factors, the major cause for this phenological change is climate. Due to its position near the tropic of Capricorn, Kaokoland is under the influence of the Subtropical High Pressure Zone, and exhibits an arid, strongly seasonal climate. Convective rainfall is linked to the movement of the Southern Intertropical Convergence Zone entering the region in a north-eastern direction in the summer (Hutchinson, 1995). In the winter months the stability of the high pressure cell over Botswana (Botswana Anticyclone) prevents rainfall. The main factor determining the distribution of vegetation units is the strong rainfall gradient from about 350mm in the north-east to almost zero in the south-west (Viljoen, 1980; Becker, 2000; Becker & Jürgens, 2000, 2002; Mendelsohn et al., 2002: p. 84). The moist tropical air masses, which might result in rainfall over the coastal plains, are blocked by sea winds blowing inland. Furthermore, the air warms and dries out when blowing over the escarpment and then sinking down (Mendelsohn et al., 2002: p. 74). Only a narrow coastal belt of the Namib Desert regularly receives fog precipitation, reducing evapotranspiration and providing a small amount of additional moisture for plant growth (Nagel, 1959; Olivier, 1995; Hachfeld & Jürgens, 2000). The occurrence of coastal fog is a result of the air-cooling effect of the Benguela Current and cold upwelling zones in combination with mainly western and south-western winds. The cool coastal waters also result in lower mean temperatures in the western part of the desert and in a lower daily and yearly temperature amplitude in comparison to inland areas (Sander & Becker, 2002). Frost is virtually absent from the coastal area.

Whereas Giess (1971, 1998) preliminarily distinguished two large vegetation zones, namely the mopane savanna in the east and the Northern Namib in the west, a more detailed map of the Kaokoland vegetation was compiled by Viljoen (1980). The distribution of vegetation units mainly follows the climatic gradient (figure 1). Plant cover values, species number and evenness decrease significantly from the north-east to the south-west (Becker, 2000: pp. 82-96; Becker & Jürgens, 2000, 2002). Ephemeral *Stipagrostis*-grasslands and different types of dry dwarf shrub vegetation are typical for western Kaokoland. According to Viljoen (1980), the main part of the Skeleton Coast Park belongs to the *Acanthosicyos horridus-Zygophyllum stapfii-Hermannia gariepina*-desert vegetation unit of the Northern Namib, comprising four subunits (table 2). Additional floristic and vegetation ecological information on the area is available in Giess (1968), Malan & Owen-Smith (1974), Craven & Marais (1986), Jacobsen & Moss (1987), Tarr & Tarr (1989), Jürgens et al. (1997), and Craven (2002).

The large, westward flowing ephemeral rivers that dissect the Namib Desert have often been characterised as linear oases (Seely et al., 1979-81; Seely, 1987:10; Loutit, 1991; Jacobson et al., 1995: p. 39; Barnard, 1998: p. 48, p. 62; Schoemann, 2000: p. 49;). They are fed by rainfall which occurs mainly in the eastern higher rainfall catchment areas, and support dense riparian forests in their upper and middle reaches. In the lower reaches, where subterranean water level is lower and overground flow is rare, occurring after heavy rainfalls in the hinterland, the riverine vegetation becomes increasingly sparse. Within the research area, the Hoanib, Hoarusib, and Khumib Rivers regularly or at least occasionally reach the Atlantic Ocean, whereas smaller rivers like the Ondodonjengo and Nadas are blocked by dunes and end in salty pans (Jacobsen & Moss, 1987; Loutit, 1991, Schoemann, 2000: pp. 49-63; Jacobson et al., 1995: pp. 121-145; Brunotte et al., 2002). However, after exceptional rainfall events, e.g. during the rainy season of 1995, new river channels can open up and break through as far as the coast.

Today, the vegetation in eastern and central Kaokoland is strongly influenced by nomadic and, increasingly, by sedentary livestock breeding (Sander et al., 1998; Bollig & Schulte, 1999; Becker, 2000: pp. 255-301; Becker & Jürgens, 2000, 2002; Schulte, 2001, 2002; Brunotte et al., 2002). This is, however, not

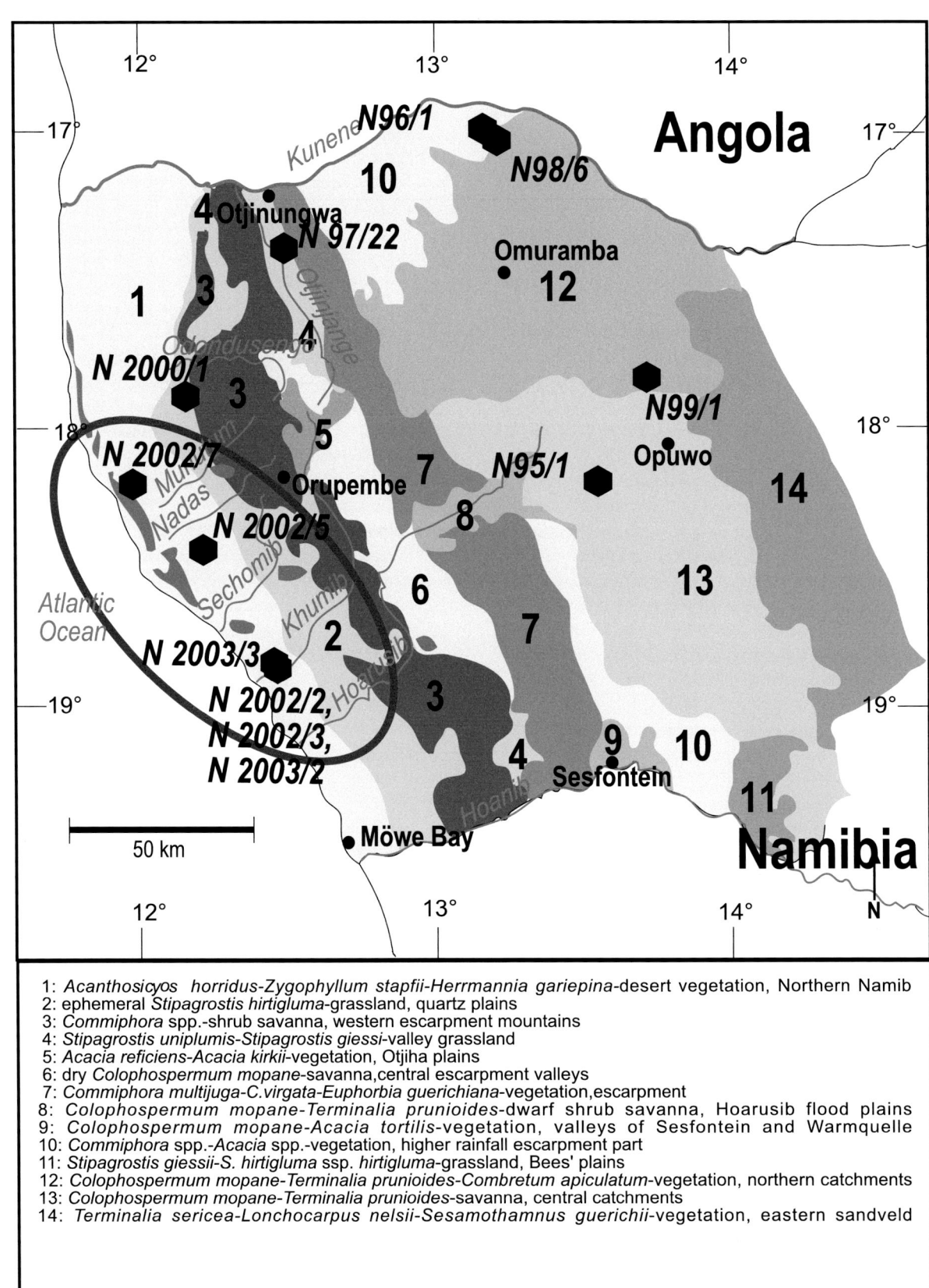

Figure 1: Vegetation map of Kaokoland (from Viljoen 1980, modified). Anthracologically investigated sites are indicated and the oval marks the Skeleton Coast Park sites.

Table 2: The four subunits of the "Acanthosicyos horridus-Zygophyllum stapfii-Hermannia gariepina-desert vegetation unit" (after Viljoen 1980, modified, scientific plant names modified according to Craven (ed.), 1999).

vegetation subunit	site characteristics	typical plant species
Salsola spp.-Zygophyllum spp.-association	dry river beds allow the establishment of typical plant species, subsequently small dunes are built up around the plants	Salsola arborea Salsola nollothensis Zygophyllum clavatum Zygophyllum stapfii Ectadium rotundifolium
Hermannia gariepina-Ectadium rotundifolium-association	gravel plains, small sand hummocks are built up around the plants	Hermannia gariepina Ectadium rotundifolium Indigofera cunenensis Crotalaria leubnitziana Stipagrostis ramulosa Petalidium angustitubum Stipagrostis hermannii Arthraerua leubnitziae
Acanthosicyos horridus-Cladoraphis cype-roides-association	dune fields	Acanthosicyos horridus Cladoraphis cyperoides Merremia multisecta
Centropodia glauca-Crotalaria leubnitziana-association	small graniticic inselbergs	Centropodia glauca Crotalaria leubnitziana Othonna lasiocarpa Marcelliopsis denudata Merremia mulisecta Barleria solitaria Lithops ruschiorum

the case for the protected Skeleton Coast Park area. Despite of its extreme aridity, the area might have been used in the past by pastoralists when pasture was available. This was the case either after rare rainfall events or in dry river beds favouring year-round plant growth. From the Walfish Bay region further south in the central Namib, historical reports as well as archaeological investigations have yielded evidence that even in this proximity to the coast, cattle and small stock occurred and were bartered there (J.H.A. Kinahan, 1990, 1992, 2000; J. Kinahan, 1991). However, as archaeological surveys and preliminary archaeozoological results indicate, subsistence near the coast of Kaokoland was probably based predominantly on hunting-gathering (Ralf Vogelsang, personal comm. 2003; Vogelsang, 2003; archaeozoological analysis by Joris Peters is in progress). In addition to the use of marine resources such as shells, fish, seals, seabirds and stranded whales, hunting and probably the collecting of fruits of the !nara melon (Acanthosicyos horridus, Cucurbitaceae) completed the diet. The importance for subsistence in the Namib of this staple food, yielding highly nutritious pips and a fruit pulp that can be eaten either fresh or processed into a long-lasting dried mass, has been demonstrated by ethnobotanical studies as well as by archaeobotanical finds (Budack, 1977; Dentlinger, 1977; Sandelowsky, 1977, 1983; Richter, 1991: p. 153; Van den Eynden et al., 1992; Van Damme, 1998; Van Wyk & Gericke, 2000: p. 34; Eichhorn, unpublished data).

1.2 Site characteristics

In spite of the hyperaridity and hostility of the Kaokoland coastal desert, numerous archaeological sites occur in close vicinity to present tracks. Leaving these tracks is prohibited, but most probably a similar site density occurs elsewhere (Vogelsang 2003). Two main categories of sites were encountered, shell middens (figure 2) and stone circle sites. Some are obvious settlement sites whereas more frequent groupings of half circles may suggest an alternative interpretation of these sites as hunting shelters. In one case, whale bones formed part of the stone circle, a phenomenon restricted to the coastal area (figure 3). The shell midden sites, which consist of large amounts of shell remains associated with fire places, potsherds and stone tools, probably served mainly for the exploitation of this single marine resource. In contrast to this, the only stone circle thus far investigated in detail shows a more diverse spectrum of archaeozoological remains, including fish vertebrae as well as bones of seabirds

Figure 2: A shell midden site in the Skeleton Coast Park (photo R. Vogelsang).

and sea mammals, ostrich eggshell fragments and beads (Vogelsang 2003; R. Vogelsang, pers. communication 2003). The picture of a 'bushmen' family in front of their whalebone hut structure near the lower Orange River in 1779 gives an idea of the assumed original state of whalebone-stone circle sites during their occupation (figure 4).

As the availability of water is extremely limited in the Namib Desert environment, site distribution is restricted to the vicinity of pans and dry river beds where water is occasionally present overground after heavy rainfall in the catchment area, and all-year-round in the form of subterranean water. However, the relative position to water resources is not similar at each of the sites investigated anthracologically. Site complex N 2003/1, comprising several sites in close proximity (N 2002/2, N 2002/3, N 2003/2), is situated near the Khumib River mouth, i.e. in vicinity of a large ephemeral river regularly flooding and occasionally reaching the coast (Jacobson et al., 1995: p. 123). The other sites are located close to small pans with a high groundwater table. In addition, large coastal pans occur near site N 2002/7 (figure 1).

1.3 Radiocarbon dating

One radiocarbon date obtained from charcoal is available for site N 2002/5 (KN 5565: 400±50 BP). Three AMS radiocarbon dates were obtained at site N 2002/7, one from charred organic material adhering to a potsherd (KIA 18993: 1175±25 BP) and two from charcoal (KIA 21034: 1235±35 BP; KIA 21035: 1150±30 BP).

Three AMS radiocarbon dates are available for charcoal originating from the site complex N 2003/1: two for the shell midden N 2002/3 (KIA 21036: 100±25 BP; KIA 21037: 205±25 BP) and one for a fireplace with a large associated whale bone, N 2002/2 (KN 5566: 120±45 BP). Up to present, no radiocarbon dates are available for the fireplace N 2003/2 which was covered by a small dune. Another radiocarbon date is available for a stone circle site (N 2003/3) in close vicinity of the site complex (KIA 21033: 165±20 BP). Unfortunately, all these measurements fall within a radiocarbon plateau. Therefore, a more precise classification than the period between AD 1650 to AD 1955 is not possible. The only diagnostic potsherd is a lug of a Khoi pot from

Figure 3: A stone/whalebone circle site in the Skeleton Coast Park (photo R. Vogelsang).

the site complex, pointing to an age of at least 100 to 200 years (R. Vogelsang, pers. communication 2003).

2 Methods

During the excavations, all charcoal pieces larger than 2.5 mm were picked out by hand from the sieves used to filter the sediment removed from the excavation.

At site N 2002/5 one stone circle was excavated. Excavation size was 3m x 3m, all sediment was dry-sieved. Two ash concentrations were visible under the loose sand cover. One of them was interpreted as a fireplace and the other as a possible refuse heap. Charcoal was almost exclusively limited to these concentrations.

At site N 2002/7 the 5-20 cm thick, loose sand cover from within and around the stone circles was removed and dry-sieved. The settlement horizon was excavated in its entirety. One of the visible find concentrations was a fireplace covered with a stone slab.

At the site complex N 2003/1 a test excavation of 1m x 1m in artificial spits of 5 cm was conducted in the shell midden N 2002/3. Finds were only present on the surface and within the upper 5 cm of the sediment (charcoal analysis was conducted but its results not presented here due to the small number of charcoal pieces). At site N 2003/2, a small dune was cut and an underlying fireplace uncovered. It was entirely excavated (excavation size about 1m x 1m), and again all sediment was dry-sieved. The fireplace with the associated whale bone N 2002/2 was also excavated. The loose sand cover was only thin in this case, and again all sediment was dry-sieved.

In the laboratory, standard charcoal analytical methods were used. Pieces were fractured by hand in transversal plane and, with help of a razor blade, in longitudinal-tangential and longitudinal-radial planes. The three diagnostic planes were examined with the means of incident light microscopy and, in doubtful cases, using a Scanning Electron Microscope. Identification followed IAWA standards (Wheeler *et al.*, 1989) and was undertaken with the help of the wood anatomical catalogue in Eichhorn (2002) and the reference collection of charred Namibian woods, deposited in the Forschungsstelle Afrika/UFG of the University of Cologne.

Figure 4: A 'bushmen' family living in the lower Orange River valley in 1779 (from Smith et al., 2000, p. 16, picture 3.5; original source: Gordon Atlas, Rijksprentenkabinet, Rijksmuseum, Amsterdam). Note the fireplace, the high number of ostrich eggshell water containers, shell accumulations in the foreground, whale vertebrae and whale bones used as seats.

3 Anthracological results and interpretation

In the case of site N 2002/5 (figure 5), people maintained their fire solely with the coastal desert dwarf shrub vegetation, typical for the smaller river washes (table 2), using the woody base of leaf-succulent, perennial *Zygophyllum* species. Grass was also used, probably when lighting the fire (Eichhorn, 2002: p. 319). There is no indication that the inhabitants moved far from the site, or indeed needed to do so, in order to cover their firewood requirements.

Charcoal analyses from N 2002/7 are shown in figure 6. Similar to N 2002/5, people used mainly the woody parts of *Zygophyllum* species as fuel. Chenopodiaceae, typical for localities with a high salinity, i.e. with pans, brackish waterholes or dry river beds (Giess 1968), are also present in the charcoal samples. Amaranthaceae may be represented by *Calicorema capitata*, a dwarf shrub typical for ephemeral grassland communities of the pro-Namib plain. This species penetrates into the desert in small river washes. Another suffrutescent Amaranthaceae, *Arthraerua leubnitziae*, occurs sporadically near the coast, mainly in the Central Namib, but at least as far north as Cape Frio (table 2, Giess 1968). *Marcelliopsis denudata* occurs on Northern Namib granitic inselbergs (table 2). All these taxa were potentially accessible in the vicinity of the settlement. This is most probably not the case for *Colophospermum mopane*, a savanna species which reaches the fringes of the Namib desert in dry river beds. Only three charcoal pieces of this species were determined from site

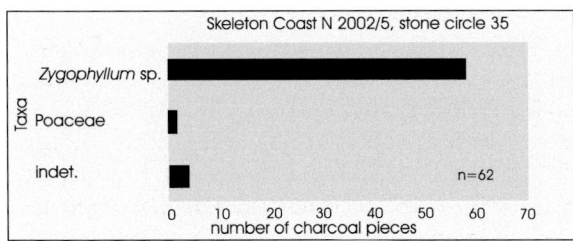

Figure 5: Anthracological diagram, site N 2002/5.

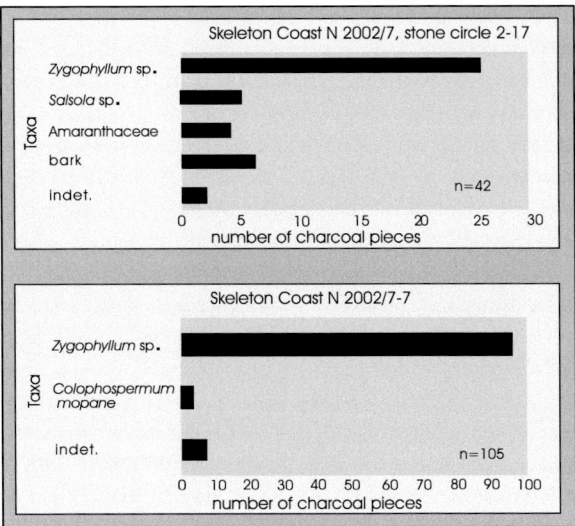

Figure 6: Anthracological diagram, site N 2002/7.

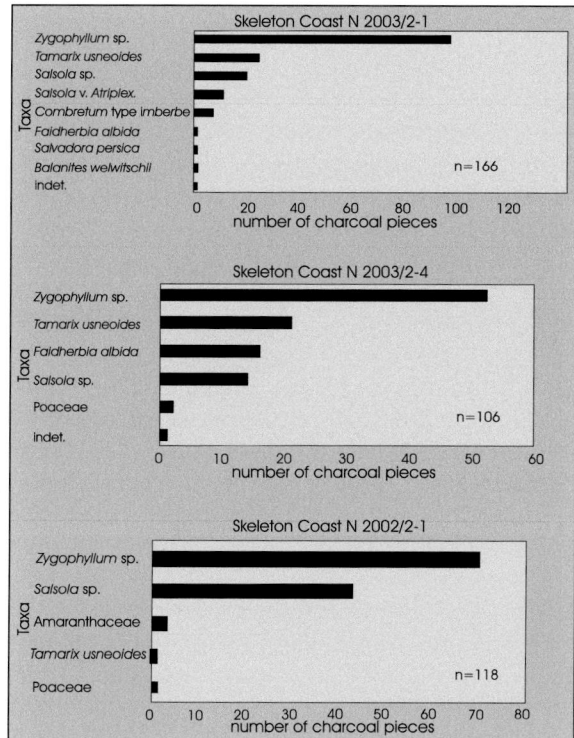

Figure 7: Anthracological diagram, site complex N 2003/1 (sites N 2003/2-1, N 2003/2-4, N 2002/2-1).

N 2002/7. Due to its high quality, the wood of this tree is highly favoured as fuel (Malan & Owen-Smith, 1974; National Academy of Sciences, 1980; Prior & Cutler, 1992; Van Wyk & Gericke, 2000: p. 283, p. 286; Eichhorn, 2002: pp. 316-330; Eichhorn & Jürgens, 2002). At present, it has a religious significance in that it is the only wood used to maintain the holy fires of the Ovahimba people in the Kunene Region (Malan & Owen-Smith, 1974). The wood from certain species with a particular cultural significance may be a reason for importing it over several kilometers, either from dry river beds, inselbergs or inland savannas (compare Archer, 1990; Shackleton & Prins, 1992). However, it cannot be entirely ruled out that single mopane individuals were available at favourable sites in closer proximity (see Giess, 1998: p. 32).

The charcoal spectra from site complex N 2003/1 (in detail site N 2002/2, surface 1, N 2002/3 and site N 2003/2 with the surfaces 1 and 4) near the mouth of the Khumib River, exhibit quite different results with a higher diversity of determined taxa (figure 7). Similarly to the former sites, *Zygophyllum* species were most extensively used for firewood. Charcoal of Chenopodiaceae and of *Tamarix usneoides*, a species which is common in Namibia at sites with high salinity, e.g. in the lower reaches and near the mouths of ephemeral rivers, is present in the assemblage too.

Most other taxa are riverine trees and shrubs, mainly typical for the middle and upper reaches further away from the coast. These depend on more regular flooding and a higher subterranean water table. There are several possible reasons for their occurrence in the charcoal spectra:

1. During strong floods of ephemeral rivers, wood of riverine plant species is transported over great distances. Even the transport of entire cast up trees has been reported from western Namibia (Vogel & Rust, 1987; Jacobson *et al.* 1995: pp. 39-47, p.p. 118-120). Thus, the collecting of driftwood in dry river beds is an easy means to supplement the scarce fuel resources of the coastal desert.

2. The vegetation of dry river beds is an important food source for roaming animals in the Kunene Region, especially during the dry season and during dry years when grass cover is low or absent (Malan & Owen-Smith; 1974; Viljoen, 1980; 1982; Loutit, 1991; Jacobson *et al.*, 1995: pp. 48-51). When hunting game or collecting ostrich eggs, people probably followed them upstream collecting fuelwood as they went.

3. Firewood or wood for construction was collected in the interior during seasonal inland-coast movements and imported from a larger distance.

The latter assumption is the most improbable one, on the one hand, because the transport of wood over greater distances requires more time and energy, and on the other hand, because typical, highly-favoured inland species such as *Terminalia prunioides* are absent from the assemblage. Ethnobotanical interviews with the present inhabitants of Kaokoland have shown that even at the margin of the Namib desert, where natural plant cover is low, or in densely populated areas, where woody resources have become sparse due to over-exploitation, people do not move further than 4-5 km to collect firewood (Eichhorn, 2002: pp. 320-328).

However, the proof of *Faidherbia albida* in the charcoal assemblage needs consideration. In Namibia, this tree occurs in many larger ephemeral river systems, provided mean annual temperature is high enough and relatively strong flooding supports seed dispersal and the rare establishment of the drought-sensitive seedlings (Giess, 1971, 1998; Jacobson *et al.*, 1995: pp. 39-42). But, according to Viljoen (1980: p. 186, p. 198), this species is absent from the Khumib River's gallery forest. In lists of common riparian species of the Khumib, neither Schoemann (2000: p. 54) nor Jacobson *et al.* (1995: p. 123) mention *Faidherbia albida*. Investigations in the Khuiseb river bed in central Namibia have demonstrated that phytoecological standard methods may fail to seize all woody species in an ephemeral river system (Seely *et al.*, 1979-81). Therefore, the possibility that rare individuals of *Faidherbia albida* occur in the Khumib and were not recorded cannot be ruled out entirely. Other possible explanations for *Faidherbia albida* charcoal at the Khumib sites are based on ephemeral river ecology. Changes in the composition of riparian forests of an ephemeral river system may reflect climatic fluctuations in the hinterland. The flooding and subterranean water table of rivers crossing the Namib depend on precipitation rates and events much further east or north-east in the region of their headwaters. The dependence on the availability of over- and underground water is clearly demonstrated by the fact that large stands of *Faidherbia albida* in the Huab River in central Namibia have become extinct due to farm dams being built further upstream in the river catchment (Barnard, 1998: p. 68). The demise of old *Faidherbia albida* trees in the lower Khuiseb River in the 1980s was correlated with a downward trend in flow amplitude (Ward & Breen, 1983).

Therefore, on the one hand, anthropogenic influence or drought periods in the hinterland can lead to the death of tree populations in the lower reaches of a river. On the other hand, rare events of extreme flooding which have uprooted and killed whole tree populations, finally depositing them downstream or even in the sea, have been reported on repeated occasions. Floods in 1982 removed mature *Faidherbia albida* trees near Purros in the Hoarusib River bed (Loutit, 1991). It is possible that the Khumib River once supported a population of *Faidherbia albida* trees in the past which is now extinct due to undetermined reasons.

If this species is and was in fact absent from the river system, the collecting of driftwood in - or mobility along - the Khumib are not suitable explanations for the proof of its charcoal in the samples from site 2003/3. Movements of people to other river systems supporting stands of *Faidherbia albida*, like the Hoarusib or Hoanib further south, are one possible explanation. Although not extremely favoured as fuel by present inhabitants of Kaokoland, *Faidherbia albida* plays a role in one of the traditional methods of kindling fire using flintstone. The spark is caught in a small container, filled with ground charcoal of selected trees, one of them being *Faidherbia albida* (Eichhorn, 2002: p. 319). A selective use of a woody species as in this case might indeed represent a further explanation for an increased transport distance of its wood and deposition in a charcoal assemblage, a factor otherwise difficult to explain.

Finally, although plenty of driftwood was probably available in the dry river close to the site, the accidental collecting of driftwood on the Atlantic beach has to be considered as an additional 'allochthonous' source of firewood. Marine driftwood can be transported over large distances and may thus explain the archaeobotanical proof of woody species far from their present distribution areas (Dickson, 1992; Alix, 1998; Vermeeren, 1999). Off the north-west Namibian coast, surface flow of the Atlantic is directed northwards (Wedepohl et al., 2000; Lydie Dupont, personal comm. 2003). Driftwood which was deposited in the sea by an ephemeral river further south may have been washed ashore near site complex N 2003/1.

4 Conclusion

Once more, availability and accessibility have been proven to be of major importance in the choice of woody species as fuel (compare Heizer, 1963; Prior & Tuohy, 1987; Smart & Hoffmann, 1988; February, 1992; Neumann, 1999; Van Wyk & Gericke, 2000: p. 283; Eichhorn, 2002: pp. 316-330; Zapata-Pena et al., 2003), even in the hyperarid environment of the coastal Namib, where trees only very seldomly occur. This is shown by the dominance of *Zygophyllum* charcoal in all investigated Skeleton Coast sites. In addition to the utilisation of the desert dwarf shrub vegetation, drift wood was most probably collected in ephemeral river beds when available. This is consistent with studies from other regions bare of trees, and where driftwood use for fuel or other purposes is indicated (e.g. Heizer, 1963; Alix, 1998; Adams & Hedberg 2002).

However, the presence of charcoal from several species growing at some distance from their present distribution area, might either point to the mobility of people, to ecological fluctuations in the hinterland of ephemeral rivers or to the use of marine driftwood. Anthracological analysis does not allow us to decide in favour of one or the other of these possibilities. The mere presence of stone circle sites in the hyperarid Skeleton Coast Park has led Hüser et al. (1998) and Blümel et al. (2000) to the assumption that during their occupation climatic conditions in the coastal desert must have been more favourable than at present. However, neither the presence of charcoal from *Colophospermum mopane* at site N 2002/7 nor *Faidherbia albida* and other riparian species at site N 2002/7 should be interpreted as indicators of climatic fluctuations in the Northern Namib Desert proper: The dominance of *Zygophyllum* sp. charcoal clearly indicates that the 'zonal' woody vegetation in the vicinity of all three sites was similar to the one presently prevailing.

5 Acknowledgements

I am very much indebted in the Deutsche Forschungsgemeinschaft for funding of archaeobotanical research within the ACACIA project. I am also grateful to all members of the National Botanical Research Institute and to Dr. Holger Kolberg of the Ministry of Environment and Tourism, (both Windhoek, Namibia) for friendly support and for granting the necessary research permits.

Special thanks are due to Dr. Ralf Vogelsang who conducted the excavations and Lee Clare for assistance with the English language.

I wish to thank Dr. Lydie Dupont, Prof. Dr. Norbert Jürgens, Dr. Rudolf Kuper, Priv.-Doz. Dr. Katharina Neumann and Prof. Dr. Jürgen Richter for support and friendly cooperation. Many thanks are also due to Dr. Marlies Klee, Dr. Werner Schuck and Dr. Ursula Tegtmeier.

6 References

Adams, E.C. & C. Hedberg, 2002. Driftwood use at Homol'ovi and implications for interpreting the archaeological record. *KIVA* 67 (4), pp. 363-384.

Albrecht, M., Berke, H., Eichhorn, B., Frank, T., Kuper, R., Prill, S., Vogelsang, R. & S. Wenzel, 2001. Oruwanje 95/1: a Late Holocene stratigraphy in north-western Namibia. *Cimbebasia* 17, pp. 1-19.

Alix, C., 1998. Provenances et circulation des bois en milieu arctique: quels choix pour les Thuléens. *Revue d'Archaeométrie* 22, pp. 11-22.

Archer, F., 1990. Planning with people – ethnobotany and African uses of plants in Namaqualand (South Africa). *Mitteilungen aus dem Institut für Allgemeine Botanik Hamburg* 23 b, pp. 959-972.

Avery, G., 1984. Late Holocene avian remains from Wortel, Walvis Bay, SWA/Namibia, and some observations on seasonality and Topnaar Hottentot prehistory. *Madoqua* 14 (1), pp. 63-70.

Barnard, P. (ed.), 1998. *Biological diversity in Namibia: a country study*. Namibian National Biodiversity Task Force, Windhoek.

Becker, T., 2000. *Muster der Vegetation und ihre Determinanten in einem desertifikationsgefährdeten Raum im Nordwesten Namibias (Kaokoland)*. PhD-thesis, University of Cologne, Cologne. Elektronische Dissertationen: URN: urn:nbn:de:hbz:38-4786, URL: http://kups.ub.uni-koeln.de/volltexte/2003/478/.

Becker, T. & N. Jürgens, 2000. Vegetation along climate gradients in Kaokoland, North-West Namibia. *Phytocoenologia* 30, pp. 543-565.

Becker, T. & N. Jürgens, 2002. Vegetationsökologische Untersuchungen im Kaokoland, Nord-West Namibia In: Bollig, M., Brunotte, E. & T. Becker (eds.), *Interdisziplinäre Perspektiven zu Kultur- und Landschaftswandel im ariden und semiariden Nordwest Namibia*. (=Kölner Geographische Arbeiten 77). Geographisches Institut der Universität zu Köln, Cologne, pp. 81-100.

Blümel, W.D., Hüser, K. & B. Eitel, 2000. Landschaftsveränderungen in der Namib. Klimawandel oder Variabilität? *Geographische Rundschau* 5, pp. 17-23.

Bollig, M. & A. Schulte, 1999. Environmental change and pastoral perceptions: Degradation and indigenous knowledge in two African pastoral communities. *Human Ecology* 27, pp. 493-514.

Brunotte, E., Sander, H. & J. Frangen, 2002. Human-induced environmental changes in areas favorable and unfavorable for land-use in Kaokoland, Namibia. *Die Erde*, pp. 133-154.

Budack, K.F.R., 1977. The Aonin or Topnaar of the lower !Kuiseb valley and the sea. *Khoisan Linguistic Studies* 3, pp. 1-42.

Carr, M.J., Carr, A.C. & L. Jacobson, 1978. Hut remains and related features from the Zerrissene Mountain area: their distribution, typology and ecology. *Cimbebasia* (B) 2 (11), pp. 237-258.

Cartwright, C. & J. Parkington, 1997. The wood charcoal assemblages from Eland's Bay Cave, southwestern Cape: Principles, procedures and preliminary interpretation. *South African Archaeological Bulletin* 52, pp. 59-72.

Cowling, R.M., Cartwright, C.R. & J.E. Parkington, 1999. Fossil wood charcoal assemblages from Eland's Bay Cave, South Africa: implications for Late Quaternary vegetation and climates in the winter-rainfall Fynbos biome. *Journal of Biogeography* 26, pp. 367-378.

Craven, P. (ed.), 1999. *A checklist of Namibian plant species* (Southern African Botanical Diversity Network Report 7). SABONET, Windhoek.

Craven, P., 2002. Plant species diversity in the Kaokoveld, Namibia. In: Bollig, M., Brunotte, E. & T. Becker (eds.), *Interdisziplinäre Perspektiven zu Kultur- und Landschaftswandel im ariden und semiariden Nordwest Namibia* (=Kölner Geographische Arbeiten 77). Geographisches Institut der Universität zu Köln, Cologne, pp. 75-80.

Craven, P. & C. Marais, 1986. *Namib Flora*. Gamsberg Macmillan, Windhoek (reprint 1998).

Dentlinger, U. 1977. *The !nara plant in the Topnaar Hottentot culture of Namibia: ethnobotanical clues to an 8000 year old tradition*. (=Munger Africana Library Notes 38). California Institute of Technology, Pasadena.

Dickson, J.H., 1992. North American driftwood, especially Picea (spruce), from archaeological sites in the Hebrides and northern Isles of Scotland. *Review of Palaeobotany and Palynology* 73, pp. 49.

Dupont, L.M. & U. Wyputta, 2003. Reconstructing pathways of aeolian pollen transport to the marine sediments along the coast line of SW Africa. *Quaternary Science Reviews* 22, pp. 157-174.

Eichhorn, B., 2002. *Anthrakologische Untersuchungen zur Vegetationsgeschichte des Kaokolandes, Nordwest-Namibia*. PhD.-thesis, Mathematisch-Naturwissenschaftliche Fakultät, University of Cologne, Cologne. Elektronische Dissertationen: URN:nbn:de:hbz:38-11784; URL: http://kups.ub.uni-koeln.de/volltexte/2004/1178/.

Eichhorn, B. & N. Jürgens, 2002. Vegetationsgeschichte und Nutzung pflanzlicher Ressourcen im Kaokoland - Stabilität oder Wandel? In: Bollig, M., Brunotte, E. & T. Becker (eds.), *Interdisziplinäre Perspektiven zu Kultur- und Landschaftswandel im ariden und semiariden Nordwest Namibia* (=Kölner Geographische Arbeiten 77). Geographisches Institut der Universität zu Köln, Cologne, pp. 119-135.

Eichhorn, B. & N. Jürgens, 2003. The contribution of charcoal analysis to the late Pleistocene and Holocene vegetation history of north-western Namibia. In: Neumann, K., Butler, A. & S. Kahlheber (eds.), *Progress in African Archaeobotany*. (=Africa Praehistorica 15), Heinrich-Barth-Institut, Cologne, pp.151-162.

February, E., 1992. Archaeological charcoals as indicators of vegetation change and human fuel choice in the late Holocene at Eland's Bay, western Cape Province, South Africa. *Journal of Archaeological Science* 19: pp. 347-354.

Giess, W., 1968. A short note on the vegetation of the Namib coastal area from Swakopmund to Cape Frio. *Dinteria* 1, pp. 13-29.

Giess, W., 1971. Eine vorläufige Vegetationskarte von Namibia. *Dinteria* 4, pp. 1-114.

Giess, W., 1998. Eine vorläufige Vegetationskarte von Namibia. *Dinteria* 4 (2., enlarged edition), pp. 1-112.

Hachfeld, B. & N. Jürgens, 2000. Climate patterns and their impact on the vegetation in a fog driven desert: The Central Namib Desert in Namibia. *Phytocoenologia* 30, pp. 567-589.

Heine, K., 1998a. Climate change over the past 135,000 years in the Namib Desert (Namibia) derived from proxy data. *Palaeoecology of Africa* 25, 1998, pp. 171-198.

Heine, K. 1998b. Late Quaternary climatic changes in the Central Namib Desert, Namibia. In: Alsharhan, A.S., K.W. Glennie, Whittle, G.L. & C.G.S.C. Kendall (eds.), *Quaternary deserts and climatic change*. Balkema, Rotterdam, pp. 293-304.

Heine, K., 2002. Sahara and Namib/Kalahari during the late Quaternary – inter-hemispheric contrasts and comparisons. *Zeitschrift für Geomorphologie N.F.*, Supplement-Bd. 126, pp. 1-29.

Heizer, R.F., 1963. Domestic fuel in primitive society. *Journal of of the Royal Anthropological Institute of Great Britain and Ireland* 93, pp. 186-194.

Hüser, K., Blümel, W.D. & B. Eitel, 1998. Landschafts- und Klimageschichte des südwestlichen Afrika. *Geographische Rundschau* 50, pp. 238-244.

Hutchinson, P., 1995. The climatology of Namibia and its relevance to the draught situation. In: Moorsom, R., Franz, J. & M. Mupotola (eds.), *Coping with aridity*. Brandes & Apsel/NEPRU, Frankfurt/Windhoek, pp. 17-37.

Jacobsen, N.H.G. & H. Moss, 1987. A contribution to the flora of the Northern Namib. *Dinteria* 19, pp. 27-68.

Jacobson, P.J., Jacobson, K.M. & M.K. Seely, 1995. *Ephemeral rivers and their catchments. Sustaining people and development in western Namibia*. Desert Research Foundation of Namibia, Windhoek.

Jürgens, N., Burke, A., Seely, M.K. & K.M. Jacobson, 1997. Desert. In: Cowling, R.M., Richardson, D.M. & S.M. Pierce (eds), *Vegetation of Southern Africa*. Cambridge University Press, Cambridge, pp. 189-214.

Kinahan, J., 1990. Four thousand years at the Spitzkoppe: changes in settlement and landuse on the edge of the Namib Desert. *Cimbebasia* 12, pp.1-14.

Kinahan, J., 1991. *Pastoral nomads of the Central Namib Desert. The people history forgot*. New Namibia Books, Windhoek.

Kinahan, J.H.A., 1990. The impenatrable shield: HMS Nautilus and and the Namib coast in the late 18th century. *Cimbebasia* 12, pp. 23-61.

Kinahan, J.H.A., 1992. *By command of their Lordships: The exploration of the Namibian coast by the Royal Navy, 1795-1895*. Namibia Archaeological Trust, Windhoek.

Kinahan, J.H.A., 2000. *Cattle for beads. The archaeology of historical contact and trade on the Namib coast*. (=Studies in African Archaeology 17). University of Uppsala, Department of Archaeology and Ancient History, Uppsala.

Kinahan, J. & J.H.A. Kinahan, 1984. Holocene subsistence and settlement on the Namib coast: the example of the Ugab River mouth. *Cimbebasia* (B) 4 (6), pp. 59-72.

Loutit, R., 1991. Western flowing ephemeral rivers and their importance to wetlands in Namibia. *Madoqua* 17, pp.135-140.

Malan, J.S. & G.L. Owen-Smith, 1974. The ethnobotany of Kaokoland. *Cimbebasia* (B) 2, pp.151-178.

Mendelsohn, J., Jarvis, A., Roberts, C. & T. Robinson, 2002. *Atlas of Namibia*. Philip, Cape Town.

Nagel, J.F., 1959. Fog precipitation at Swakopmund. *Newsletter of the Weather Bureau of the Union of South Africa*, pp. 1-8.

National Academy of Sciences, 1980. *Firewood crops*. National Academy of Sciences, Washington 1980.

Neumann, K., 1999. Charcoal from West African savanna sites: Questions of identification and interpretation. In: Van der Veen, M. (ed.), *The exploitation of plant ressources in ancient Africa*. Kluver Academic/Plenum Publishers, New York, pp. 205-219.

Olivier, J., 1995. Spatial distribution of fog in the Namib. *Journal of Arid Environments* 29, pp. 129-138.

Parkington, J., Cartwright, C., Cowling, R.M., Baxter, A. & M. Meadows, 2000. Palaeovegetation at the Last Glacial Maximum in the western Cape, South Africa: wood charcoal and pollen evidence from Elands Bay Cave. *South African Journal of Science* 96, pp. 543-546.

Parkington, J. & C. Poggenpoel, 1971. Excavations at De Hangen, 1968. *South African Archaeological Bulletin* 26, pp. 3-36.

Plug, I., 1998. Some evidence for seasonality amongst Later Stone Age hunter-gatherers in southern Africa. *Environmental Archaeology* 3, 1998, pp. 103-107.

Prior, J. & D. Cutler, 1992. Trees to fuel Africa's fires. *New Scientist* 135, pp. 35-39.

Prior, J. & J. Tuohy, 1987. Fuel for Africa's fires. *New Scientist* 115, pp. 48-51.

Richter, J., 1984. Messum 1: A later Stone Age pattern of mobility in the Namib Desert. *Cimbebasia* (B) 4 (1), pp. 1-12.

Richter, J., 1991. *Studien zur Urgeschichte Namibias* (=Africa Praehistorica 3). Heinrich-Barth-Institut, Cologne.

Sandelowsky, B., 1977. Mirabib – an archaeological study in the Namib. *Madoqua* 10, pp. 221-283.

Sandelowsky, B. 1983., Archaeology in Namibia. *American Scientist* 71, pp. 606-615.

Sander, H. 2002., Zur naturräumlichen Gliederung des nordöstlichen Kaokolandes. In: Bollig, M., Brunotte, E. & T. Becker (eds.), *Interdisziplinäre Perspektiven zu Kultur- und Landschaftswandel im ariden und semiariden Nordwest Namibia* (=Kölner Geographische Arbeiten 77). Geographisches Institut der Universität zu Köln, Cologne, pp. 101-118.

Sander, H. & T. Becker, 2002. Klimatologie des Kaokolandes. In: Bollig, M., Brunotte, E. & T. Becker (eds.), *Interdisziplinäre Perspektiven zu Kultur- und Landschaftswandel im ariden und semiariden Nordwest Namibia* (=Kölner Geographische Arbeiten 77). Geographisches Institut der Universität zu Köln, Cologne, pp. 57-68.

Sander, H., Bollig, M. & A. Schulte, 1998. Himba paradise lost – Stability, degradation and pastoralist management of the Omuhonga Basin (Namibia). *Die Erde* 129, 1998, pp. 301-315.

Schoemann, A., 2000. *Skeleton Coast*. Struik Publishers, Cape Town.

Schulte, A., 2001. *Weideökologie des Kakolandes. Struktur und Dynamik einer Mopane-Savanne unter pastoralnomadischer Nutzung*. PhD-thesis, University of Cologne, Cologne. Elektronische Dissertationen: URN: urn:nbn:de:hbz:38-112312740, URL: http://kups.ub.uni-koeln.de/volltexte/2003/432/.

Schulte, A., 2002. Stabilität oder Zerstörung? Veränderungen der Vegetation des Kaokolandes unter pastoralnomadischer Nutzung. In: Bollig, M., Brunotte, E. & T. Becker (eds.), *Interdisziplinäre Perspektiven zu Kultur- und Landschaftswandel im ariden und semiariden Nordwest Namibia* (=Kölner Geographische Arbeiten 77). Geographisches Institut der Universität zu Köln, Cologne, pp 101-118.

Scott, L., 1996. Palynology of hyrax middens: 2000 years of palaeoenvironmental history in Namibia. *Quaternary International* 33, pp. 73-79.

Scott, L., Anderson, H.M. & J.M. Anderson 1997. Vegetation history. In: Cowling, R.M., Richardson, D.M. & S.M. Pierce (eds.), *Vegetation of southern Africa*, Cambridge 1997, pp. 62-84.

Scott, L., De Wet, J.S. & L.C. Vogel, 1991. Holocene environmental changes in Namibia inferred from pollen analysis of swamp and lake deposits. *The Holocene* 1, pp. 8-13.

Sealy, J.C. & N.J. Van der Merwe, 1986. Isotope assessment and the Seasonal-Mobility-Hypothesis in the Southwestern Cape of South Africa. *Current anthropology* 27, pp. 135-150.

Seely, M., 1987. *Die Namib*. Shell Oil SWA Limited, Windhoek.

Seely, M.K., Buskirk, W.H., Hamilton, W.J.III, & J.E.W. Dixon, 1979-81. Lower Kuiseb River perennial vegetation survey. *Journal SWA Wissenschaftliche Gesellschaft* 34/35, pp. 57-86.

Shackleton, C.M. & F. Prins 1992, Charcoal analysis and the principle of least effort – a conceptual model. *Journal of Archaeological Science* 19, pp. 631-637.

Shi, N. & L.M. Dupont, 1997. Vegetation and climatic history of southwest Africa: a marine palynological record of the last 300,000 years. *Vegetation History and Archaeobotany* 6, pp. 117-131.

Shi, N., Dupont, L.M., Beug, H.-J. & R. Schneider, 1998. Vegetation and climate changes during the last 21,000 years in S.W. Africa based on a marine pollen record. *Vegetation History and Archaeobotany* 7, pp. 127-140.

Shi, N., Dupont, L.M., Beug, H.-J. & R. Schneider, 2000. Correlation between vegetation in southwestern Africa and oceanic upwelling in the past 21,000 years. *Quaternary Research* 54, pp. 72-80.

Shi, S., Schneider, R., Beug, H.-J. & L.M. Dupont, 2001. Southeast trade wind variations during the last 135 kyr: evidence from pollen spectra in eastern South Atlantic sediments. *Earth and Planetary Science Letters* 187, pp. 311-321.

Smart, T.S. & E.S. Hoffman, 1988. Environmental interpretation of archaeological charcoal. In: Popper, V.S. & C.A. Hastorf (eds.), *Current palaeoethnobotany*. University of Chicago Press, Chicago, pp. 167-205.

Smith, A., Malherbe, C., Guenther, M. & P. Berens, 2000. *The Bushmen of southern Africa. A forgaging society in transition*. Philip, Cape Town.

Tarr, P.W. & J.G. Tarr 1989. Veld dynamics and utilisation of vegetation by herbivores on the Ganias Flats, Skeleton Coast Park, SWA/Namibia. *Madoqua* 16 (1), pp. 15-22.

Van Damme, P. 1998. Wild plants as food security in Namibia and Senegal. In: Bruins, H.J. & H. Lithwick (eds.), *The arid frontier*, pp. 229-247.

Van den Eynden, V., Vernemmen, P. & P. Van Damme, 1992. *The ethnobotany of the Topnaar*. Universiteit Gent, Gent.

Van Wyk, B.-E. & N. Gericke, 2000. *People's plants. A guide to useful plants of southern Africa*. Briza Publications, Pretoria 2000.

Van Zinderen Bakker, E.M., 1984a. Aridity along the Namibian coast. *Palaeoecology of Africa* 16, pp. 149-160.

Van Zinderen Bakker, E.M., 1984b. A late- and post-Glacial pollen record from the Namib Desert. *Palaeoecology of Africa* 16, pp. 421-435.

Van Zinderen Bakker, E.M. & M. Müller, 1987. Pollen studies in the Namib Desert. *Pollen et Spores* 29, pp. 185-205.

Vermeeren, C., 1999. The use of imported and local wood species at the Roman port of Berenike, Red Sea coast, Egypt. In: Van der Veen, M. (ed.), *The exploitation of plant ressources in ancient Africa*. Kluver Academic/Plenum Publishers, New York, pp. 199-204.

Viljoen, P.J., 1980. *Veldtipes, verspreiding van die grooter soogdiere, en enkele aspekte van die ekologie van die Kaokoveld*. Unpublished M.Sc.-thesis, Pretoria.

Viljoen, P.J., 1982. The distribution and population status of the larger mammals in Kaokoland, South West Africa/Namibia. *Cimbebasia* (A) 7, pp. 5-33.

Vogel, J.C. & U. Rust, 1987. Environmental changes in the Kaokoland Namib Desert during the present millenium. *Madoqua* 15, pp. 5-16.

Vogelsang, R., 2003. Die Skeleton Coast in Namibia – ein Mythos verliert seine Schrecken. *Heinrich-Barth-Kurier* 11/2003, pp. 4-7.

Vogelsang, R., Eichhorn, B. & J. Richter, 2002. Holocene human occupation and vegetation history in northern Namibia. *Die Erde* 133, pp. 113-132.

Wadley, L., 1979. Big Elephant Shelter and ist role in the Holocene prehistory of central South West Africa. *Cimbebasia* (B) 3 (1), pp.1-76.

Wadley, L., 1984. On the move: a look at prehistoric food scheduling in central Namibia. *Cimbebasia* (B) 4 (4), pp.41-50.

Ward, J.D. & C.M. Breen, 1983. Drought stress and the demise of *Acacia albida* along the lower Khuiseb River, Central Namib Desert: preliminary findings. *South African Journal of Science* 79, pp. 444-447.

Wedepohl, P.M., Lutjeharms, J.R.E & J.M. Meeuwis, 2000. Surface drift in the south-east Atlantic Ocean. *South African Journal of Marine Science* 22, pp. 71-79.

Wendt, W.E., 1972. Preliminary report on an archaeological research programme in South West Africa. *Cimbebasia* (B) 2 (1), pp. 2-61.

Wheeler, E.A., Baas, P. & P.E. Gasson (eds.), 1989. IAWA list of microscopic features for hardwood identification by an IAWA committee. *IAWA bulletin new series* 10, pp. 219-332.

Zapata-Pena, L., Pena-Chocarro, L., Ibanez Estevez, J.J. & J.E. Gonzalez Urquijo, 2003. Ethnoarchaeology in the Moroccan Jebala (Western Rif): wood and dung as fuel. In: Neumann, K., Butler, A. & S. Kahlheber (eds.), *Progress in African Archaeobotany*. (=Africa Praehistorica 15), Heinrich-Barth-Institut, Cologne, pp. 163-175.

Where did all the trees go? Changes of the woody vegetation in the Sahel of Burkina Faso during the last 2000 years

A. HÖHN

Seminar für Vor- und Frühgeschichte, JW Goethe-Universität, Frankfurt, Germany

1 Introduction

Recent investigations have shown that the woody vegetation of the Sahel in Burkina Faso is under constant anthropogenic pressure (*e.g.* Ganaba & Guinko, 1995; Albert *et al.*, 2004). In particular the extraction of firewood has a strong impact on the woody vegetation and villagers currently note the disappearance of, especially, Sudanian and Sub-Sahelian species from the region (Ganaba & Guinko, 1995; Müller & Wittig, 2002). Is this decrease in diversity only a recent phenomenon or did it begin before scientific monitoring started? Anthracological investigations of material from archaeological sites enable us to reconstruct the woody vegetation and land-use in former times and to detect changes in the woody flora of the region. The analysis of several charcoal assemblages, dated from the first to the 14th century AD, reveals that the woody flora was more diverse than today. The questions arising are when and why the diversity decreased.

2 The region

The charcoal assemblages come from archaeological sites in the region of the Mare d'Oursi in Northern Burkina Faso (figure 1), which belongs to the Sahelian zone. The climate is semiarid – a long dry season alternates with a short rainy season from July to September, and the rainfall pattern is characterised by high variability in time and space. The mean annual rainfall from 1971 to 2000, including the last drought period in the 1970s and early 1980s, amounts to about 400 mm. Looking just at the 1990s, a period with higher rainfalls, the annual mean rises to about 500 mm (Some, 2002).

Dunes of Quaternary origin cross the region in several long, northeast – southwest oriented belts (figure 1), which are up to ten kilometres wide (Albert *et al.*, 1997) During the rainy season temporary lakes, called *mares*, develop in large depressions along these belts. After rainy seasons with high precipitations, large *mares*, like the Mare d'Oursi, may contain water even through the dry season. Rivers often run along the northern rim of the dunes and eventually drain into the Niger. The plain on which the dunes have accumulated, is mostly eroded to the crystalline base. Some inselbergs rise above this plain.

The settlements, past and present, are concentrated on the dune belts near permanent or semi-permanent water bodies (figure 1). The *mares* and rivers supply drinking water and the relatively lush riverine vegetation bears many resources of wood and wild fruits. Moreover, cultivation is only possible on the sandy soils of the dunes, which have a relatively high water storage capability (Vogelsang *et al.*, 1999).

Cultivation takes place in agroforestry systems. They are defined as land-use units combining the cultivation of crops and/or livestock breeding with the systematic exploitation of woody plants growing on the same plot, thus ensuring the supply of a high diversity of food plants. The staple crop is pearl millet (*Pennisetum glaucum*), grown during the rainy season. The fruit trees, interspersed within the fields, are mainly *Faidherbia albida* and *Balanites aegyptiaca*. Other woody species, such as *Guiera senegalensis*, are stimulated by shifting cultivation. Such easily burgeoning species sprout prolifically after cutting and thrive on fallow lands. The fallows are often used to collect firewood (Slingerland & Wiersum, 2001).

Phytogeographically the region belongs to the Sahel. The typical vegetation is described as wooded grassland (White, 1983) or Mimosaceae savanna (Le Houérou, 1989). Acacias dominate within the woody vegetation. The flora is poor in species and endemites. To the south the 600 mm isohyete marks the northern limit of the neighbouring Sudanian zone. Species with Sudanian distribution are for instance *Anogeissus leio-*

36 A. HÖHN

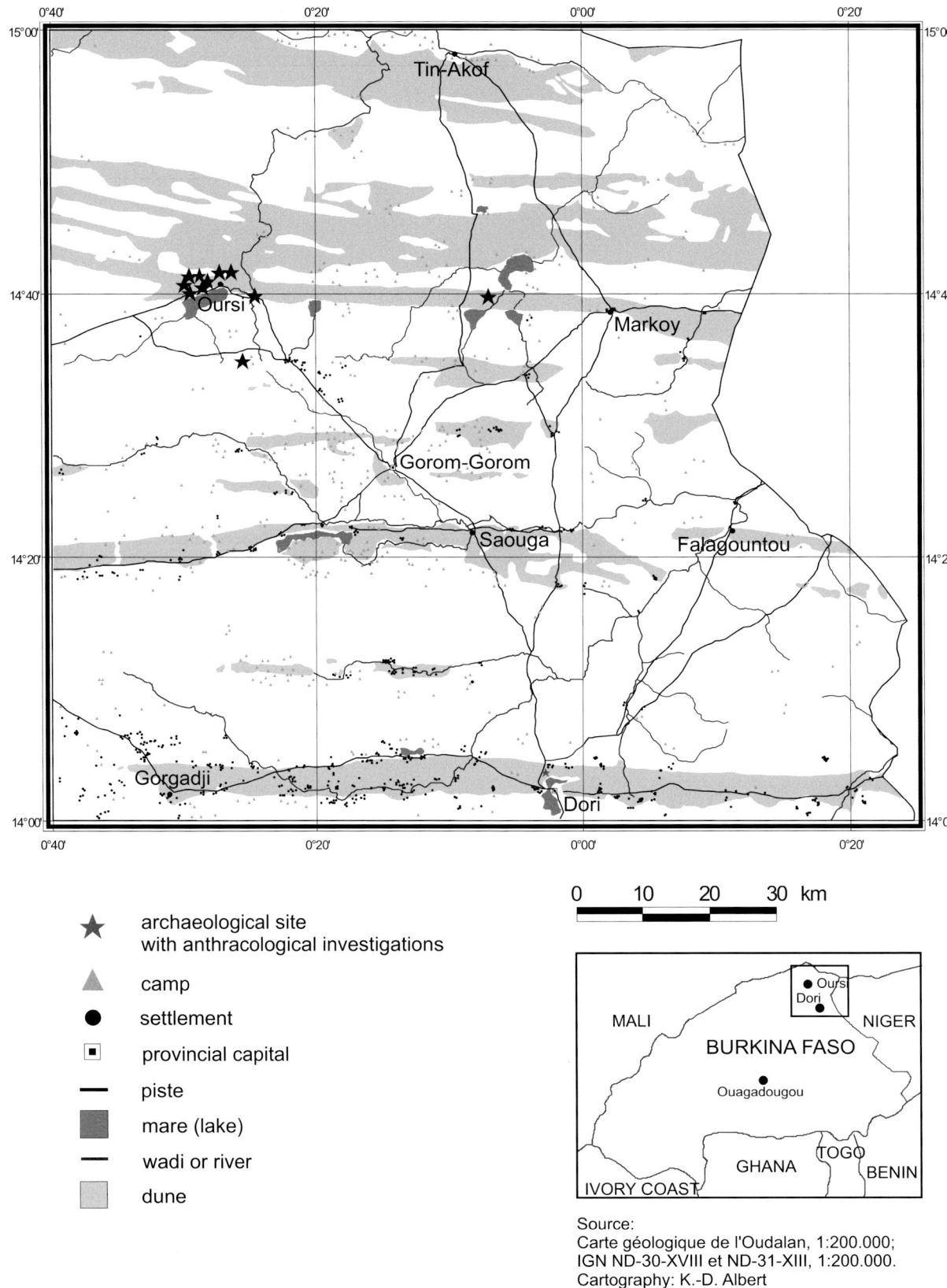

Figure 1: The environment and location of archaeological sites and modern settlements (Vogelsang et al. 1999, modified).

carpus, Annona senegalensis and *Tamarindus indica*. The distribution of these species extends into the Sahel, where, however, they are bound to hydrologically favourable sites (Le Houérou, 1989).

3 The charcoal assemblages

The investigated charcoal assemblages are obtained from iron age settlement mounds in the vicinity of the Mare d'Oursi (figure 1). The deposition of settlement mound debris began in the first centuries AD and ended in the 14th century AD. Based on ceramic evidence, Von Czerniewicz (2002) subdivides the iron age in northern Burkina Faso into three phases (early, middle and late), each lasting 400 to 500 years.

Anthracological material from ten archaeological sites (figure 1) was examined, coming from small test excavations of one cubic meter, with only few samples, as well as from large excavations of several cubic meters (BF 97/13 and 94/45), where many samples were taken and analysed. Due to the different preservation conditions, sample size varied from one fragment to 501 fragments. Samples with more than 200 fragments rendered reliable results concerning the taxonomic composition, with at least 20 charcoal types recorded. The highest diversity was furnished by a sample of 235 fragments, where 25 types were found. Altogether 35 charcoal types were distinguished.

4 Results and interpretation

The differences between the charcoal assemblages can mainly be attributed to anthropogenic changes of the woody vegetation in the course of the iron age. Agriculture and stockbreeding led to the development of a cultural landscape (*Kulturlandschaft*) at least in the vicinity of the settlements. Combretaceae, regarded as fallow species, have high proportions in the charcoal assemblages dating around AD 1000, while *Acacia* sp. dominates in the first centuries AD. High amounts of *Guiera senegalensis* in some sites are interpreted as a result of intensified livestock breeding (Höhn, 2005).

The comparison of the iron age charcoal flora with the woody flora of the region today is used to elucidate the development of the woody vegetation after the end of the iron age. For that purpose the species possibly belonging to the 35 charcoal types found in the iron age assemblages were compared to the woody species found by Müller (2003) in the same region. The charcoal types were put into three groups:
1. taxa of the modern vegetation showing natural regrowth,
2. taxa which still grow in the region but show no natural regrowth,
3. taxa missing in the region today. Nine charcoal types belong to the last group (table 1).

Most of them are listed as typical Sudanian species by Le Houérou (1989), White (1983) or Arbonnier (2000), but many can grow in hydrologically favorable habitats like the riverine vegetation of rivers or *mares* or in dune depressions. Only *Detarium microcarpum* and *Vitellaria paradoxa* show considerably higher precipitation needs than the region receives today and cannot endure temporary inundation.

According to Thies (1995) and Maydell (1990) the other species of the third group do not have distinctly higher needs of precipitation. Still, *Parinari curatellifolia* and *Strychnos spinosa* have probably been extinct in the region some time before the first half of the twentieth century. According to Aubréville (1950) their northern limit of distribution was already then lying clearly to the south of the region (figure 2). Others, like *Terminalia avicennioides* must have vanished only recently, since its distribution still extended into the region about 50 years ago.

Climatic as well as anthropogenic changes must be taken into account, when trying to explain the loss of woody species during the last 500 years.

5 Climatically induced changes

During the iron age the region probably received precipitations which are typical for the northern sudanian zone or the sahelo-sudanian transition zone. Instead of currently 400 to 500 mm/a at least 600 mm/a, rather about 700 to 750 mm of annual rain, should have fallen. This estimation is mainly based on the presence of two species, *Detarium microcarpum* and *Vitellaria paradoxa*, in the charcoal assemblages. They need more rain than the region receives today and do not grow on temporarily inundated sites (Arbonnier, 2000). The other taxa present in the charcoal assemblages but missing in the region today could have been present in the region in hydrologically favourable sites (table 1, column 5, empty circles) and connot be used for climatic reconstruction.

Detarium microcarpum today occurs in the northern sudanian zone. It is recorded in fallows and dry forests of areas with about 750 mm/a of precipitation (Hahn-Hadjali, 1998; Rietkerk et al., 1998; Neumann & Müller-Haude, 1999). Thus, the finds of *Detarium microcarpum* indicate higher rainfalls than the north of Burkina Faso receives today, even though the 1,000 mm/a (table 1) given as the minimum rainfall by Thies (1995) seem too high. The shrub or tree is typical for

Table 1: Comparison of the iron age charcoal flora with the regional woody flora of today. Information about floristic category according to Aubréville (1950), Le Houérou (1989) and White (1983), about regional vegetation and natural regrowth according to Müller (2003), information concerning modern habitat according to Müller (2003) and Arbonnier (2000) and concerning minimum precipitation according to Maydell (1990) and Thies (1995).

Charcoal types	floristic category	part of regional vegetation today	natural regrowth in the region	at mares, and rivers,	minimum annual rainfall (mm)
Acacia sp.	.	•	•	•	50
Balanites aegyptiaca	.	•	•	.	100
Bauhinia/Piliostigma	Sa	•	•	•	300
Boscia sp.	.	•	•	.	100
Combretum aculeatum/paniculatum	.	•	•	.	300
Diospyros mespiliformis	Su	•	•	•	500
Faidherbia albida	.	•	•	.	50
Guiera senegalensis	Su	•	•	.	400
Annona senegalensis/Hexalobus monopetalus	Su	•	.	.	600
Anogeissus leiocarpus	Su	•	.	•	200
Celtis integrifolia	Su	•	.	•	500
Combretum glutinosum	Su	•	.	.	200
Combretum micranthum	?	•	.	.	300
Commiphora africana/pendunculata	Sa	•	.	.	< 100
Flueggea virosa/Hymenocardia acida	.	•	.	•	400
Grewia bicolor/flavescens	.	•	.	.	200
Grewia venusta/villosa	.	•	.	.	300
Maerua angolensis/crassifolia	.	•	.	.	100
cf. Parkia biglobosa	Su	•	.	•	500
cf. Prosopis africana	Su	•	.	•	500
Pterocarpus lucens	Sa	•	.	.	250
Rubiaceae Typ I	Su	•	.	•	500
Rubiaceae Typ II	Su	•	.	•	300
cf. Sclerocarya birrea	Su	•	.	.	200
cf. Tamarindus indica	Su	•	.	.	400
Ziziphus sp.	Su	•	.	•	150
Capparis decidua	.	.	.	.	100
Detarium microcarpum	Su	.	.	.	1000
Lannea sp.	Su	.	.	°	600
Parinari curatellifolia	Su	.	.	°	400
Strychnos innocua	.	.	.	°	?
Strychnos spinosa	.	.	.	°	500
Terminalia avicennioides/macroptera	Su	.	.	°	500
Vitellaria paradoxa	Su	.	.	.	600
Ximenia americana	Su	.	.	°	500

fallows of the Sudanian zone (Hahn-Hadjali, 1998) because it resprouts well from the stump. It is well suited and easily collected as firewood (Sieglstetter, 2002). Tree savannas of the Sudanian zone, like the protected Forêt classée de Nazinon in Burkina Faso, where *D. microcarpum* is one of the dominant woody species (Rietkerk et al., 1998), seem to be the natural habitat.

Vitellaria paradoxa does not grow in areas with less than 600 mm of annual rainfall (Hall et al., 1996). The tree is typical for and common in the agroforestry systems of the Sudanian zone. Probable natural habitats of *Vitellaria paradoxa* are tree savannas and maybe even dry forests (Neumann & Müller-Haude, 1999), as well as the stony soils of elevations, where the tree also grows spontaneously (Küppers, 1996).

While charcoal fragments of *Vitellaria paradoxa*

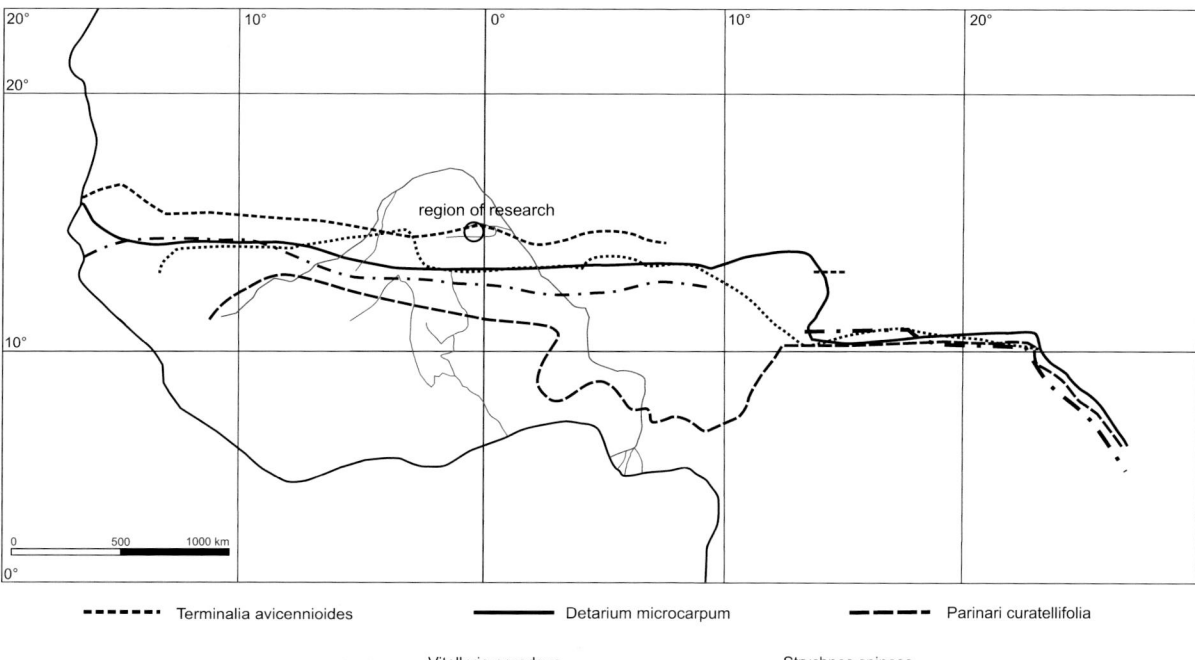

Figure 2: Northern border of distribution of selected species according to Aubréville (1950).

are present throughout the iron age, *Detarium microcarpum* appears only from around AD 500 onwards. Fruit shells, however, from *Detarium* sp. have already been found in a late stone age site, dating to 1000 to 1300 BC (Kahlheber, 2003). The later appearance of *Detarium microcarpum* in the charcoal record might be connected with the intensification of agriculture during the iron age (Kahlheber et al., 2001, Höhn et al.,2004). In the course of intensification the species could have penetrated from its natural habitat into the fallows, where it was only then collected as firewood in higher numbers. Palynologically the *D. microcarpum* has not been proved for the area, but that may be due to bad pollen preservation (Ballouche & Neumann, 1995). Sapotaceae pollen, probably from *Vitellaria paradoxa*, is present only in a pollen profile from the Mare de Kissi, about 40 km to the east of the Mare d'Oursi, around AD 1400 (Ballouche, 2001).

Since *Detarium microcarpum* and *Vitellaria paradoxa* are still present in the youngest layers of the settlement mounds, the higher precipitation regime must have prevailed until about AD 1300 to 1400 and climatic changes resulting in the modern conditions could have taken place only at the end or after the iron age period.

6 Anthropogenically induced changes

Other taxa present in the charcoal record but missing in today's vegetation, like *Terminalia avicennioides/macroptera*, *Parinari curatellifolia* and *Ximenia americana*, are rather victims of anthropogenic pressure than of climatic changes. With 400 to 500 mm of minimum precipitation per year they do not need more water than the region receives today and they have similar needs as species that are still present in the region (table 1). Except for *Strychnos innocua*, these taxa also potentially grow in riverine sites (table 1, column 5, empty circles). Thus they could have survived gradual aridification and drier periods in hydrologically favorable habitats like, for instance, *Prosopis africana* and *Tamarindus indica* did. These species need 500 respectively 400 mm of annual precipitation and today are found next to water courses. From ethnographic studies it is known that especially the riverine vegetation is heavily exploited for firewood in the region (Ganaba et al., 1998). It seems reasonable to assume that the extinct species were wiped out by man as a result of overexploitation.

The case of *Prosopis africana* could serve as a model for the degradation of the woody flora in the past. The last known tree of *Prosopis africana* in the region stands in a dune depression with a temporary watercourse about 40 km to the north of Oursi. The wood of the species is valued for the production of high quality charcoal for iron smelting and as very good firewood. Having been heavily collected in the past, anthropogenic selection has probably almost finished off this species, which is anyhow at the brink of its existence in the Sahel of Burkina Faso.

7 Conclusion

Palaeoecological data indicate that from 7000 BP onwards the Sahara and the Sahel have become increasingly drier (*e.g.* Ritchie *et al.*, 1985; Haynes *et al.*, 1989; Lezine *et al.*, 1990; Schulz, 1991; Guo *et al.*, 2000; Hoelzmann *et al.*, 2004). In Nigeria a strictly Sahelian vegetation comparable to that of today was established already around 1300 BC (Salzmann & Waller, 1998). For the Sahel of Burkina Faso similar results were obtained by pollen analysis. The increase of arboreal pollen, especially Combretaceae in the pollen diagram of Oursi, around 1000 BC, was interpreted as anthropogenically induced (Ballouche & Neumann, 1995). Archaeological and archaeobotanical investigations, however, have shown, that human impact could have begun to rise significantly only in the first centuries AD and that the change in the pollen diagram of Oursi around 1000 BC rather should be interpreted in terms of a drier climate (Kahlheber *et al.*, 2001).

The results of charcoal analyses show that during the iron age in the Sahel of Burkina Faso the composition of the vegetation was clearly different from today. A larger number of Sudanian species remained at least until AD 1300. *Detarium microcarpum* and *Vitellaria paradoxa* point to higher annual precipitations as one reason for this diversity. This conclusion is supported by the existence of cisterns in the neighbouring Gourma region of Mali dating to AD 1000. They are thought to have been effective only at a precipitation of at least 800 mm per year (Reichelt, 1977). Most of the Sudanian taxa, however, could have prevailed the gradual desiccation and could be part of contemporary Sahelian vegetation. These species must have fallen victim to human impact. It can be stated that some Sudanian elements in the charcoal assemblages testify for a former moister climate, while others can be interpreted as evidence of a hypothetical natural vegetation which could be more diverse, if there was less human impact. The example of *Prosopis africana*, as well as the lack of regrowth in many taxa (table 1, column 4), show that the loss of Sudanian species in the modern Sahelian vegetation is an ongoing process.

The comparison of the current woody flora of the Oursi-region with archaeological charcoal assemblages yields information about vegetation changes, and the question why some tree species are missing today can be addressed. However, to answer the question, when the different species had vanished from the region, we need archaeological sites, filling the gap between AD 1400 and the beginning of scientific monitoring.

8 Acknowledgements

The federal ministry of education and research (BMBF) has financed this study as part of the BIOTA-project "Phytodiversity in the Sahel and Sudan Zone of West Africa – Development and Evaluation". For critically reading the manuscript, I would like to thank PD Dr. Katharina Neumann and Stefanie Kahlheber. Dick Byer smoothed out the English. However, for any shortcomings that may remain in the paper the author is fully responsible.

9 References

Albert. K.D., Andres, W. & A. Lang, 1997. Palaeodunes in NE Burkina Faso; pedo- and morphogenesis in a chronological framework provided by luminescence dating. *Z. Geomorph. N.F* 41/2, pp. 167-182.

Albert, K.D., Müller, J., Ries, J.B. & I. Marzolff, 2004. Aktuelle Landdegradation in der Sahelzone Burkina Fasos. In: K.D. Albert, D. Löhr & K. Neumann (eds.), *Mensch und Natur in Westafrika*. Weinheim, Wiley-VCH.

Arbonnier, M., 2000. *Arbres, arbustes, et lianes des zones sèches d'Afrique de l'Ouest*. Paris, CIRAD, Montpellier.

Aubréville, A., 1950. *Flore Forestière Soudano-Guinéene*. Soc. D'Edit. Géogr. Mar. Col., Paris.

Ballouche, A., 2001. Un diagramme pollinique de la Mare de Kissi (Oudalan, Burkina Faso). Nouveux elements pour l'histoire anthropique de la vegetation sahelienne. In: S Kahlheber & K. Neumann (eds.), *Man and Environment in the West African Sahel – an Interdisciplinary Approach*. Berichte des Sonderforschungsbereichs 268/17, pp. 129-135.

Ballouche, A. & K. Neumann, 1995. A new contribution to the Holocene vegetation history of the West African Sahel: pollen from Oursi/Burkina Faso and charcoal from three sites in NE-Nigeria. *Vegetation History & Archaeobotany* 4/1, pp. 31-39.

Czerniewicz, M. von, 2002. *Studien zur Chronologie der Eisenzeit in der Sahel-Zone von Burkina Faso/Westafrika*. PhD thesis, J.W. Goethe-Universität Frankfurt am Main. [http://publikationen.ub.uni-frankfurt.de/volltexte/2004/373]

Ganaba, S. & S. Guinko, 1995. Etat actuel et dynamique du peuplement ligneux de la région de la Mare d'Oursi (Burkina Faso). *Etudes flor. Vég. Burkina Faso* 2, pp. 3-14.

Ganaba, S., Ouadba J.M. & O. Bognounou, 1998. Les ligneux à usage de bois d'énergie en region sahélienne du Burkina Faso: préférences des groupes ethniques. *Sécheresse* 4, pp. 261-268.

Guo, Z., Petit-Maire, N. & S. Kröpelin, 2000. Holocene non-orbital climatic events in present-day arid areas of northern Africa and China. *Global and Planetary Change* 26, pp. 97-103.

Hahn-Hadjali, K., 1998. Les groupements végétaux des savanes du sud-est du Burkina Faso (Afrique de l'Ouest). *Etudes flor. Vég. Burkina Faso* 3, pp. 3-79.

Hall, J.B., Aebischer, D.B., Tomlinson, H.F., Osei-Amaning, E. & J.R. Hinde, 1996. *Vitellaria paradoxa. A monograph*. Bangor, University of Wales.

Haynes C.C., Eyles, C.H., Pavlish, L.A., Ritchie, J.C. & M. Rybak, 1989. Holocene palaeoecology of the Eastern Sahara: Selima Oasis. *Quarternary Science Reviews* 8, pp. 109-136.

Höhn, A., 2005: Zur eisenzeitlichen Entwicklung der Kulturlandschaft im Sahel von Burkina Faso. Untersuchungen von archäologischen Holzkohlen. Unpublished PhD thesis, J.W. Goethe-Universität Frankfurt am Main.

Höhn, A., Kahlheber, S. & M. Hallier-von Czerniewicz, 2004. Den frühen Bauern auf der Spur – Siedlungs- und Vegetationsgeschichte der Region Oursi (Burkina Faso). In: K.D. Albert, D. Löhr & K. Neumann (eds.), *Mensch und Natur in Westafrika*. Weinheim, Wiley-VCH, pp. 221-255.

Hoelzmann, P., Gasse, F., Dupont, L.M., Salzmann, U., Staubwasser, M., Leuschner, D.C. & F. Sirocko, 2004. Palaeoenvironmental changes in the arid and subarid belt (Sahara-Sahel-Arabian peninsula) from 150 Kyr to present. In: R.W. Batterbee, F. Gasse & C.E. Stickley, *Past Climate Variability through Europe and Africa*. Springer, Dordrecht.

Kahlheber, S., 2003. *Perlhirse und Baobab – Archäobotanische Untersuchungen im Norden Burkina Fasos*. PhD thesis, J.W. Goethe-Universität Frankfurt am Main. [http://publikationen.ub.uni-frankfurt.de/volltexte/2005/561]

Kahlheber, S., Albert, K.D. & A. Höhn, 2001. A contribution to the palaeoenvironment of the archaeological site Oursi in North Burkina Faso. In: S. Kahlheber & K. Neumann (eds.), *Man and Environment in the West African Sahel – an Interdisciplinary Approach*. Berichte des Sonderforschungsbereichs 268/17, pp. 145-159.

Küppers, K., 1996. Die Vegetation der Chaîne de Gobnangou. Unpublished PhD thesis, J.W. Goethe-Universität Frankfurt am Main.

Le Houérou, H.N., 1989. *The Grazing Land Ecosystems of the African Sahel* (Ecological Studies 75)., Springer, Berlin, Heidelberg, New York.

Lezine, A.M., Casanova, J. & M. Hillaire, 1990. Across an early Holocene humid phase in Western Sahara. Pollen and isotope stratigraphy. *Geology* 18, pp. 264-267.

Maydell, H.-J., 1990. *Trees and Shrubs of the Sahel: their Characteristics and Uses*. Margraf, Weikersheim.

Müller, J., 2003. Zur Vegetationsökologie der Savannenlandschaft im Sahel Burkina Fasos. Unpublished PhD-thesis, Frankfurt am Main, J.W. Goethe-Universität.

Müller, J. & R. Wittig, 2002. L'état actuel du peuplement ligneux et la perception de sa dynamique par la population dans le Sahel burkinabé – presenté à l'exemple de Tintaboora et de Kollangal Alyaakum. *Etudes sur la flore et la végétation du Burkina Faso et des pays avoisinants* 6, pp. 19-30.

Neumann, K. & P. Müller-Haude, 1999. Forêts sèches au sud-ouest du Burkina Faso: végétation – sols – action de l'homme. *Phytocoenologia* 29, pp. 53-85.

Reichelt, R., 1977. Sur les aménagements hydrauliques anciens et récent dans le Gourma, Sahel tropical, République du Mali. *Sci. Geol. Bull.* 30/1, pp. 19-31.

Rietkerk, M., Blijdorp, R. & M. Slingerland, 1998. Cutting and resprouting of Detarium microcarpum and herbaceous forage ability in a semiarid environment in Burkina Faso. *Agroforestry Systems* 41, pp. 201-211.

Ritchie, J.C., Eyles, C.H. & C.V. Haynes, 1985. Sediment and pollen evidence for an early to mid-Holocene humid period in the eastern Sahara. *Nature* 314, pp. 352-354.

Salzmann, U. & M. Waller, 1998. The holocene vegetation history of the Nigerian Sahel based on multiple pollen profiles. *Review of Palaeobotany & Palynology* 100, pp. 39-72.

Schulz, E., 1991. Holocene environments in the Central Sahara. *Hydrobiologica* 214, pp. 359-365.

Sieglstetter, R., 2002. *Wie die Haare der Erde. Vegetationsökologische und soziokulturelle Untersuchungen zur Savannenvegetation der Südsudanzone Westafrikas und ihrer Nutzung und Wahrnehmung durch die ländliche Bevölkerung am Beispiel der Region Atakora im Nordwesten Benins*. PhD-thesis, Frankfurt am Main, J.W. Goethe-Universität. [http://publikationen.ub.uni-frankfurt.de/volltexte/2003/266].

Slingerland, M. & F. Wiersum, 2001. Wood production in Sahelian villages. In: L. Stroosnijder & T van Rheenen (eds.), *Agro-Silvo-Pastoral Land Use in Sahelian Villages*. Catena, Reiskirchen, pp. 275-286.

Some, L., 2002. Analysis of climate variability in Burkina Faso. http://www.econ.ox.ac.uk/CSAEadmin/conferences/2002-uPaGISSA/papers/some-csae2002.pdf.

Thies, E., 1995. *Principaux Ligneux (Agro-)Forestiers de la Guinée. Zone des Transition*. Roßdorf, TZ Verlag.

Vogelsang, R., Albert, K.D. & S. Kahlheber, 1999. Le sable savant: les cordons dunaires sahéliens au Burkina Faso comme archive archéologique et paléoécologique du Holocène. *Sahara* 11/1999, pp. 51-68.

White, F., 1983. *The Vegetation of Africa. A descriptive memoir to accompany the Unesco/AETFAT/UNSO vegetation map of Africa*. Paris, Unesco.

Medieval cotton and wheat finds in the Middle Niger Delta (Mali)

S.S. Murray

Anthropology Department, University of Wisconsin, Madison, United States of America

At Dia, in the Middle Niger Delta (Mali), archaeobotanical research recovered 560 whole *Gossypium* (cotton) seeds and more than 1,000 seed fragments, as well as four grains identified to *Triticum turgidum* ssp. *durum* or *T. aestivum* (durum and bread wheat). These finds came from deposits widely separated across the sites and dating through the 8th to 15th centuries AD. Due to the charred nature of the cotton seeds, identification to species was not possible, however because many of the seeds are pre-Columbian in date they likely belong to Old World species *G. arboreum* or *G. herbaceum*. Islamic texts state that in the Sahel, thriving markets existed for cotton cloth and that it was a principal commodity of trade between towns to the north and south, and later, west to the Atlantic. This period may also coincide with the presence in Dia of the Diakhanké cloth merchants. The four wheat grains were likely brought into the area through trade or by travelers.

1 Introduction

During the 1st and 2nd millennia AD, the Middle Niger Delta (MND) lay within the boundaries and influence of at least three major West African empires: Ghana, Mali, and Songhay (figure 1). Trade within these empires relied to some extent on the movement of goods along the Niger River system, and on support of agricultural crops from this fertile region. This paper examines the role that cotton and wheat may have played in these three empires, in the context of their recovery from the archaeological site of Dia.

Briefly, the Soninké kingdom of Ghana, also known as the kingdom of Wagadu, was first noted in Arabic chronicles by al-Fazari writing in the early 9th century, though the origins of the kingdom likely date some century's earlier (Levtzion, 1973, 1985; Lewicki, 1974; Munson, 1980; Fage, 1995). The whereabouts of the Ghana capital are debated, but al-Bakri wrote in 1067-68 that it was located in the towns of Koumbi Saleh (Mauritania), an urban center comprised of an African town housing the king and court, and another town 10 km distant occupied by Muslim traders (Levtzion, 1985; Shaw, 1985; Fage, 1995). One of the principal roles of this kingdom was the controlled exchange of salt brought by Tuareg caravans from the Sahara, and

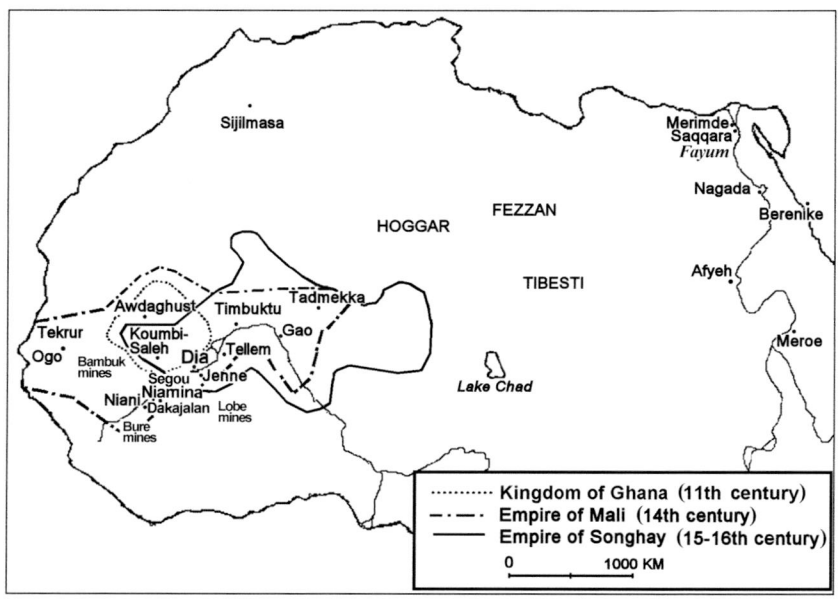

Fig. 1. The extent of the kingdom of Ghana, and the empires of Mali and Songhay, and locations of some of the sites discussed in the text.

gold mined at Bambuk on the Senegal and Faleme rivers, at Bure on the upper Niger, and possibly at Lobi south of Jenné (McIntosh, 1981; Fage, 1995). Trans-Saharan caravan routes ranging from Sijilmasa to Awdaghust, and into the MND trading cities of Jenné-jeno and eventually Timbuktu, carried gold, salt, copper, kola, and North African goods, and was sustained by crop products from the Sudan. Towards the end of the 11th century, deteriorating climate and social unrest, and possibly warfare between Ghana and newly Islamized Saharan tribes known as the Almoravids, led to the decline of Ghana (Levtzion, 1985; Fage, 1995). In the 13th century, the legendary hunter Sundiata regained the lands of Ghana from the intermediate state of Sosso, and became founder of the empire of Mali (Levtzion, 1973; 1985).

The empire of Mali was considered by scholars of Timbuktu as a natural continuation of the kingdom of Ghana (Fage, 1995: p. 74). However, instead of being ruled by the northern Soninké, as was the case of Ghana, Mali was formed by the Malinké of the upper Niger, who over the course of the century established what has been called a 'peripatetic' capital (Conrad, 1994; Levtzion & Hopkins, 2000). Debate concerning the earliest capital focuses on the cities of Niani and Dakajalan, and on an area of the Niger between Bamako and Segou, perhaps near the town of Niamina (Hunwick, 1973; Conrad, 1994). At its height, the Mali Empire controlled trade networks in gold and salt that stretched beyond those controlled by Ghana, from Tadmekka in the southern Sahara, to Awdaghust and the Atlantic, south to forests of Guinea and east to what is now modern Nigeria (Fage, 1995: p. 74). Most importantly, Mali now exercised authority over the southern Saharan trade centers of Timbuktu and Jenné, which allowed the domination of salt exchange between these two cities, as well as command over the gold trade between Jenné and the Akan forest to the south (Levtzion, 1985). Like Ghana, the infrastructure of Mali, and later Songhay, depended on the agricultural products of its population, which it gained through taxation (Levtzion, 1973; Lovejoy, 1985). These products consisted mainly of pearl millet (*Pennisetum glaucum*), sorghum (*Sorghum bicolor*), African rice (*Oryza glaberrima*), and fonio (*Digitaria exilis*), but also of cloth from textile industries centered in places such as Timbuktu and Jenné (Levtzion, 1973).

Early in the 15th century, the farmers and fisherfolk of the independent state of Songhay at Gao, began expanding further upstream into the MND, warring with Mali, but also with the Tuaregs, Fulbe, and Mossi (Levtzion, 1973; Lovejoy, 1985). Between 1450 and 1473, the Songhay took command of Timbuktu and Jenné, but struggles continued between Mali, Songhay, and the Fulbe over the Niger waterways of the Massina (Macina) and the Sahelian hinterlands until the end of the 15th century (Levtzion, 1973; Davidson, 1998). Near the end of the 16th century, the empire of Songhay fell and trans-Saharan trade routes shifted elsewhere, diminishing commerce in the cities of Jenné, Timbuktu, and Gao (Fage, 1995; Davidson, 1998).

2 The study site

Compared to Jenné and Timbuktu, the town of Dia likely played a small, albeit important role in the commercial enterprises of the above-mentioned West African empires. According to oral tradition, Dia, also known as Ja, Dya, Diakha, and Zagha, was one of the first towns founded by the Soninké in their expansion and development of the Kingdom of Ghana (Wagadu). Their arrival at Dia may have occurred sometime around the 9th century, but more certainly between the 9th and 13th centuries (Salvaing, 1983). From Dia, two specialist groups emerged, the Diakhanké (Jakhanke), originating around 1200, and the Dyula (Dioula, Jula, Wangara), who may have originated as a sect or division of Diakhanké (Curtin, 1971, 1975; Levtzion, 1973; Willis, 1979; Perinbam, 1980). The role of the Diakhanké is debated, with Sanneh (1989) claiming a mainly clerical vocation, while Curtin (1971) emphasizes their function as long-distance traders. Indeed, it seems possible that the Diakhanké existed in both capacities, though the scholars may have garnered greater esteem and historical mention (Sanneh 1989). The Dyula were, more clearly, specialist traders credited with founding Jenné, and eventually dispersing south and southeast towards the forests and gold fields of the Black Volta (Curtin 1971; Levtzion 1973; Fage 1995). The commerce of both the Diakhanké, as described by Curtin (1971), and the Dyula focused on cotton textiles, gold, and kola (Brooks, 1993). Their diaspora from Dia reputedly led to the founding of Soninké colonies in Tichitt, Walata, Sandsanding, and Jenné, sometime in the 13th century (Curtin, 1971; Perinbam, 1980).

Dia also has special significance as a holy place and important center of religious devotion and teaching, certainly of Islam, but also of animism. During his travels through Dia around 1352-53, Ibn Battuta remarked, "Kabara and Zagha [Dia] have two sultans who owe obedience to the King of Mali. The people of Zagha are old in Islam. They are pious and interested in learning," (Levtzion & Spaulding, 2003: p. 70). Ibn Battuta further wrote, "Dia, an old Muslim town in Massina was autonomous under the rule of its Cadi, and the Sultan could not pursue into the town one who had sought its sanctuary," (Levtzion, 1985: p. 164).

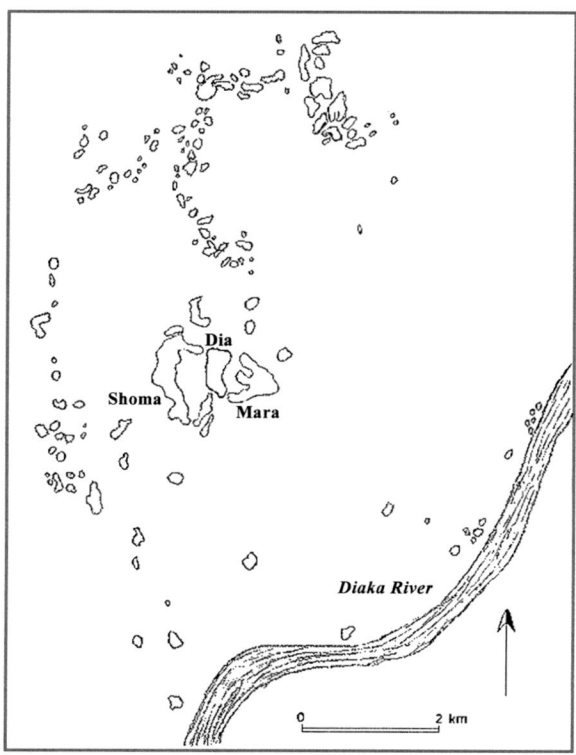

Fig. 2. Situation of the mounds of Shoma, Dia, and Mara, and the nearby satellite sites. Adapted from Haskell, et al., 1988.

Dia's reputation as a spiritual center had further developed by the early 17th century when it was known, as it is today, for its population of 'marabouts' (Muslim divines), and as the hub for obtaining religious charms and amulets (Gallais, 1967; Willis, 1985; Sakai, 1990;). However, despite the number of references to Dia as a center of Islamic learning, the relatively complete conversion to Islam by Dia's occupants probably did not occur until the 19th century, until which time, a mix of Islam and traditional beliefs likely prevailed (Arazi, 2002).

Interestingly, Curtin (1971: p. 230) describes Dia as "a natural centre for early commerce south of the Sahara, able to serve both as a river port and a desert port, like Gao and Timbuktu in later centuries." But, modern Dia, a town of roughly 8,000 fishers, farmers, pastoralists, ironsmiths, and potters, lacks a market. It is uncertain how long there has been no market in Dia, but early French colonial efforts to establish one there met with resistance. This resistance allegedly was connected with sorcery in an attempt to keep out strangers (Sakai, 1990: p. 211) and protect the town from the profanations of commerce (Gallais, 1967: p. 555). The nearest market today is in Tenenkou, about 16 km north.

From 1998 to 2001, Dia was excavated by a team of Dutch, English, French, and Malian researchers, as part of a project initiated by the Rijksmuseum voor Volkenkunde, Leiden (Bedaux et al., 2001; Bedaux et al., 2005). The Université du Mali, the Institut des Sciences Humaines, and the Musée National du Mali in Bamako, the Mission Culturelle in Djenné, the Universities of Paris I and VI, the C.N.R.S., University College London, and Leiden University are associated with the project. Haskell, et al. (1988), in a program studying early urbanism in the MND, also carried out test excavations at the site in 1986-87. Both projects also are involved in promoting the cultural heritage of the MND, and in halting pillaging and illicit trade in antiquities.

Archaeologically, Dia consists of three mound sites totaling 100 hectares in area: Shoma (49 ha), Dia (23 ha), and Mara (28 ha), and numerous satellite sites (figure 2). Excavation on Shoma and Mara indicates that Shoma was occupied first, from about 800 to 400 BC until around AD 1900, and that Mara was inhabited roughly AD 500 to 1900. It is unclear if occupation of the middle mound of Dia, which remains populated today, was contemporaneous with the other two mounds, as excavation and coring were not allowed. Though five horizons were distinguished and many different taxa examined, this paper focuses only on cotton and wheat found in Horizon 3 (AD 500 to 1000) and Horizon 4 (AD 1000 to 1600). These two crops, and perhaps African rice, mark the extent of likely plant products traded or exchanged in the medieval period at Dia.

Botanical remains at Dia were recovered by simple bucket flotation using 0.33 mm mesh screen. In total, about 6,063 charred items were collected from more than 3,500 liters of soil. Preservation was generally poor; the density of plant remains per liter of excavated sediment was roughly 1.7 charred items. About 86% of the recovered specimens belong to five taxa: cotton (*Gossypium* sp.) (44%), rice (*Oryza* spp.) (23%), doum palm (*Hyphaene thebaica*) (8%), West African plum (*Vitex* spp.) (6%), and a wild grass (cf. *Sacciolepis* sp.) (5%). The majority of plant remains were identified at the J.W. Goethe-University, Department of Archaeology and African Archaeobotany, Frankfurt, with assistance from Katharina Neumann and Stefanie Kahlheber. An overview of the archaeobotanical research at Dia is forthcoming (see Murray, 2005).

3 Cotton

Most cotton-fiber production in the world today derives from the New World species *Gossypium hirsutum* and *G. barbadense*. Old World cottons, *G. arboreum*, from

South Asia, and *G. herbaceum*, from Africa, Arabia, and India, have shorter seed hairs than the New World species and this has led to their replacement as fiber crops (Hutchinson *et al.*, 1947; Zohary & Hopf, 2000). The use of cotton has some antiquity; some of the earliest known cotton textiles and string come from Harappan sites dating mid- to late-3rd millennium BC (Fuller & Madella, 2001).

Archaeological recovery of cotton in Africa has occurred only occasionally. The earliest known cotton seeds and seed hairs come from the Nubian site of Afyeh, dating 2600 to 2400 BC (Chowdhury & Buth, 1971). These seeds, identified as *Gossypium arboreum* L. var. *soudanense* or *G. herbaceum* L. var. *africanum* (Watt) Volleson, were found in animal dung and appear to represent the use of cotton seeds as animal fodder, rather than the cultivation of cotton for fiber (Zohary & Hopf, 2001). Carbonized cotton textiles recovered from a cemetery in Meroë date 300 BC to AD 300, and written references to cotton occur from about AD 350 (Griffith & Crowfoot, 1934). It is argued that these textiles were possibly made from *G. arboreum* var. *soudanense*, but definitive determination was not possible.

Remarkably well preserved cotton textiles from the Tellem burial caves of the Bandiagara (Mali), date from the 11th to the 12th century AD (Bedaux, 1991). Although there is little stratigraphy in the Tellem caves, C^{14} dates on some of the material, and the fact that many of the caves are difficult to reach and are considered holy by the Dogon, suggests disturbance and mixing with later deposits may have been minimal. The cotton objects from these caves consist mainly of blankets, tunics, skirts, and head coifs and caps. The different cotton textiles are basically identical in weaving technique, decoration and form, suggesting they were woven by a single group of people. Features found on the pieces however, bear similarities with weavings of various modern groups of central West Africa, further suggesting some level of regional homogeneity (Bolland, 1991). Based on the facial morphology of some of the skeletons, it is thought the Tellem may have come to the Bandiagara from further south, perhaps set in motion by the fall of the Ghana empire. Their occupation of the Bandiagara seems to date to the 11th to 15th centuries, but whether the Tellem migrated elsewhere or died out is unknown. It appears rather certain that the Dogon and Tellem are unrelated, though some Tellem individuals may have later assimilated into Dogon groups (Bedaux, 1991).

Cotton pollen, thirteen disques à cordeler, and five spindle whorls were recovered from 11th century levels at Ogo, in the Middle Senegal Valley (Chavane, 1985). Disques à cordeler are round and flat perforated ce-

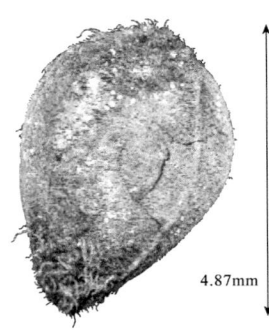

Fig. 3. Cotton seed recovered from Dia, probably Gossypium arboreum *or* G. herbaceum.

ramic disks, with a diameter between 8 and 16 cm, and thickness of about 2 cm (Chavane, 1985: p. 143). The large central perforations range between one and four holes, and the smaller lateral perforations number between four and six. There is generally some sort of décor incised on one side of the disk. Disques à cordeler are thought to be used in the manufacture of cordage, perhaps in a manner similar to spindle whorls. These disks are less common outside of the Middle Senegal Valley, though similar disks were found in the Malian Méma sites of Diabali and Kobadi, and at Niono, some of these predating those of the Middle Senegal (Chavane, 1985; R. McIntosh, pers. comm.). Spindle whorls from Ogo resemble the shape, size, and manufacture of those found in MND sites.

A single charred cotton seed and 22 fragments came from excavations at Gao (Mali), dating AD 1399-1445 (Fuller, 2000; Insoll, 2000). The seeds were recovered from an ashy deposit found in a ceramic vessel at the base of a refuse pit at the site of Gadei. These finds came from a single flotation sample collected from the site, thus it is likely that other cotton seeds were present in this and other contexts.

Cotton was the most commonly recovered taxon at Dia (figure 3). Approximately 560 whole cotton seeds and over 1000 seed fragments were found on both Shoma and Mara, in at least 27 Horizon 4 (AD 1000 to1600) pits and middens, as well as in contexts identified as kitchens, courtyards, floor surfaces, ashy areas, and hearths. Paul Fryxell (University of Texas-Austin) verified several seeds to genus, but was unable to determine species level based on the charred nature of the seeds. The seeds compare well with those found by Fuller (2000) at Gao. Two seeds yielded AMS dates of AD 1283-1397 and AD 1403-1465 (655±38 BP, AA50175; 480±33 BP, AA50174) indicating that they belong to the Old World species, *Gossypium arboreum* or *G. herbaceum*.

Today, cotton is grown mainly in the southern part of Mali, in the region south of Bamako. Although this

Fig. 4. Spindle whorls from Horizon 4 (AD 1000-1600), Dia.

author did not observe cotton growing at Dia, Gallais (1967: p. 553) wrote of cotton growing on the mounds of Mara and Shoma in the late 1950's, thus it seems possible that it was grown at Dia in the past. Monteil (1927) also indicates that cotton can be produced with irrigation in the MND, whereas in the south of Mali it is dry land cultivated. It is also possible that raw cotton was brought to Dia as an item of trade or commerce.

The abundance of cotton seeds and spindle whorls at Dia suggests manufacture of cotton yarn or string, and possibly cloth. The spindle whorls (figure 4) are the common black terracotta type found in many other regions of West Africa (Bedaux et al., 1978; McIntosh & McIntosh, 1980; Chavane, 1985; McIntosh, 1995). They are present in a wide variety of shapes and sizes, and most are elaborately decorated with incised lines, triangles, circles, and dots. Some of the spindle whorls stratigraphically date to late Horizon 3, but the majority comes from Horizon 4.

Cotton seeds and spindle whorls may serve as proxy evidence for string or yarn production, but there is no archaeological evidence for cotton cloth manufacture, as weaving materials such as looms and other related equipment were not recovered. Some support for textile production throughout the Sahel and savanna comes from Arabic texts. Islamic chronicles state that in the Sahel, thriving markets existed for cotton cloth and that it was a principal commodity of trade between towns to the north and south, and west to the Atlantic (Curtin, 1975; Brooks, 1993). High quality cotton weavings made by the Malinké for trade in salt during the Mali Empire were traded west to the Senegambia (Curtin, 1975). These apparently became so valuable that European traders could more profitably trade in cloth than in gold (Curtin, 1975).

Evidence for cotton weaving at Dia also comes from Arabic chronicles. The Diakhanké, as mentioned previously, reportedly began a great deal of the cloth trade and they claim Dia as their place of origin (Curtin, 1975). As a place of 'origin,' Dia, or the region of Dia, may simply have been the locality in which the Diakhanké developed as a clerical group, as Sanneh (1972) asserts that the Diakhanké were primarily scholars, and not traders. He suggests they came to Dia from the west, and that they probably had some familiarity with the states of Tekrur and Silla (Senegal). This is where al-Bakri wrote that cotton cloth production was underway by the 11th century (Levtzion & Hopkins, 2000: pp. 77-78), and if the Diakhanké were merchant-clerics, as suggested by Curtin (1975), they may have gained familiarity with textile production through their association with these coastal polities. On the other hand, the presence of Tellem cotton weavings dating to the 11th century presents the possibility that cotton production was a local development (Bedaux et al., 1978), one that might have been adopted by the Diakhanké.

The timing of the Diakhanké presence in Dia is difficult to discern, but Salvaing (1983) suggests sometime between the 9th and 13th centuries. Around AD 1350, Ibn Battuta wrote that large groups of Soninké came to Diakha (the region of Dia) after the fall of Mali, signifying that possibly there was more than a single diaspora of Diakhanké into or out of Dia (Sanneh, 1972). These latter dates coincide with the occurrence of cotton seeds and spindle whorls from Shoma and Mara, which might mean that it was later Diakhanké merchants that brought cloth production to the region or further developed a Tellem technology. It is certainly also possible that cotton textile production at Dia was not associated with Diakhanké, or furthermore, that the cotton seeds recovered from this site do not represent the manufacture of cloth. Despite the incomplete chronology and historical incongruities, the archaeological recovery of cotton seeds and spindle whorls at Dia provides an additional resource for understanding textile trade and the movement of people in the MND.

4 Wheat

As is well known, the first domesticated wheats (*Triticum* sp.) were recovered from sites in the Fertile Crescent dating roughly 10,000 years ago (Smith, 1995; Zohary & Hopf, 2000). The first evidence of wheat in West Africa, presumably introduced from the Near East via North Africa or Egypt, comes from Arabic chroniclers, who observed it in towns and cities along the southern Sahara and in the Sudan beginning late 1st millennium AD. Research into archaeological recovery of wheat in West Africa has revealed no finds, perhaps due to the negligible time dedicated to archaeobotanical recovery in this part of the world, but also because wheat was probably not a widely used or cultivated crop.

The earliest recovery of wheat in Africa comes from Egypt. Emmer and barley, and possibly freethreshing wheat were recovered from numerous sites in northern and southern Egypt: the Fayum, Merimde, Nagada, Kom el-Hisn, and Saqqara, dating to the 3rd millennium BC and older (for reviews see Vartavan & Amorós, 1997; Zohary & Hopf, 2000). Later sites in Egypt include the Roman port site of Berenike, where barley, durum wheat, and emmer from early to mid-1st millennium AD contexts, provide evidence of foodstuffs consumed by local inhabitants (Cappers, 1999). Careful consideration of chaff and weed seeds suggests that the cereals recovered from Berenike were imported from elsewhere, perhaps somewhere along the Nile Valley.

The introduction of wheat came much later to the rest of Africa. Barley, durum wheat, and emmer were in use in North Africa only by the early 1st millennium AD, as evidenced by grains and chaff recovered from Romano-Libyan farm sites (Van der Veen et al., 1996). Emmer, barley, and bread wheat were also found in samples from Aksum, Ethiopia, dating to a similar period, the mid 1st millennium BC for the emmer and barley grains and chaff, and 6th-7th centuries AD for the bread wheat (Boardman, 1999). The early movement of wheat into the Sudan is not well known. Oases and highlands in the Sahara may have provided early bridges for the movement of wheat cultivation west and south, and time spent in these more arid locales may have produced varieties better adapted to the conditions in which they are now grown (Zeven, 1980). According to Lewicki (1974), medieval Arabic sources state that wheat was cultivated by AD 976 in Awdaghust, but consumed mainly by nobility. Al-Bakri (Levtzion and Hopkins, 2000, p. 68; McDougall, 1985) recorded, "Wheat is grown there [Awdaghust] by digging with hoes (*fa's*), and it is watered with buckets (*dalw*)." Bovill (1995: p. 70) states that Awdaghust, roughly a fifteen-day march from Koumbi Saleh, was well-watered and could support a wide range of crops, including wheat, "a luxury," he says, "that was reserved for the rich".

The presence of wheat in the capital town of Mali early in the 14th century was documented by Ibn Battuta, who recorded that it was consumed there, but that it was rare (Lewicki, 1974). Similarly, Al-Omari wrote of wheat and its scarcity in ancient Mali, and ad-Dimashqi, writing in the 14th century, noted wheat (*qamh*) growing in Kaukau (Gao) (Lewicki, 1974, pp. 22-23). Lewicki (1974) concludes that though wheat was known in West Africa during medieval times, it was little cultivated. He suggests that it was an expensive food, perhaps even a luxury brought from North Africa.

Today, wheat is cultivated in scattered areas of the Sahara, though usually in small quantities. Zeven (1974; Dalziel, 1948) says that durum wheat is cultivated in North Africa, and bread wheat can be found growing in Saharan oases, the Sudan, in Ethiopia, and in North Africa. Munson (1980) makes the case that if it is possible to grow wheat under these conditions in modern times, it is possible that it was grown in the past under similar rainfall regimes, and that perhaps it was even more widespread during periods of higher precipitation.

As mentioned previously, archaeological recovery of wheat in West Africa until recently has been rare or non-existent. At Dia, four wheat grains were recov-

Fig. 5. Ventral and dorsal views of two wheat grains recovered from Dia, probably Triticum aestivum *(bread wheat) or* T. turgidum conv. durum *(durum wheat).*

ered in midden and pit contexts on Shoma and Mara (figure 5). One grain yielded a date AD of 779-1157 (1070±73 BP, AA50176). Based on morphology alone, it is often difficult to distinguish naked wheat grains to species, therefore the grains were submitted for identification to C.C. Bakels and W. Kuijper (University of Leiden), who verified the grains as *Triticum aestivum* L. or *T. turgidum* ssp. *durum* (Desf.) MacKey.

Due to the low number of grains recovered, it is probable that wheat was not grown at Dia. The grains were probably present as items of trade or as a gift, or perhaps as food brought by travelers for their own consumption. Certainly, Dia during the 8th to 12th centuries was within the interaction sphere of the kingdom of Ghana, and because of its position along the Niger River, it probably acted as a trade center, however small. Its reputation as one of the hubs of Islamic learning at this time may also have attracted students and learned scholars, or those seeking spiritual guidance.

5 Summary

The MND played an important role in the development of trade and commerce in West Africa during the rise and fall of three major empires. Although small compared to the larger and better-known cities of Jenné, Timbuktu, and Gao, the town of Dia, according to oral tradition and Arabic texts, appears to have functioned as a minor port town, as a major hub of Islamic learning, and possibly as a center of textile manufacturing. Archaeobotanical recoveries of cotton and spindle whorls at Dia are interpreted as evidence for production of cotton yarn and possibly cloth. The wheat finds are more difficult to interpret, but seem most practically explained as items of trade or as food brought by travelers.

6 Acknowledgements

Many thanks to the National Science Foundation (No. 0089599), The Wenner-Gren Foundation (No. 6783), Sigma Xi, the University of Wisconsin-Madison Anthropology Department, the Vilas Travel Grant, and the Rijksmuseum voor Volkenkunde for support of this research. Special thanks to K. Neumann, S. Kahlheber, C.C. Bakels, W. Kuijper, P. Fryxell, the members of Projet Dia and the people of the town of Dia.

7 References

Arazi, N., 2002. Islam and Alternative Religious Practices during the Second Millennium AD in the Inland Niger Delta of Mali. *Paper presented at the Mande Studies Association (MANSA)*, Leiden University, June, 2002.

Bedaux, R.M.A., 1991. The Tellem Research Project: the archaeological context. In: R. Bolland (ed.), *Tellem Textiles: Archaeological Finds from Burial Caves in Mali's Bandiagara Cliff*. Leiden, Mededelingen van het Rijksmuseum voor Volkenkunde 27, pp. 14-36.

Bedaux, R.M.A., Constandse-Westermann, T.S., Hacquebord, L., Lange, A.G. & J.D. van der Waals, 1978. Recherches Archaeologiques dans le Delta Interieur du Niger (Mali). *Palaeohistoria* 20, pp. 91-220.

Bedaux, R., MacDonald, K., Person, A., Polet, J., Sanogo, K., Schmidt A. & S. Sidibé, 2001. The Dia archaeological project: rescuing cultural heritage in the Inland Niger Delta (Mali). *Antiquity* 75, pp. 837-848.

Bedaux, R.M.A., Polet, J., Sanogo, K. & A. Schmidt, 2005. *Recherches archéologiques à Dia dans le Delta intérieur du Niger, Mali: bilan des saisons de fouilles 1998-2003*. Leiden Mededelingen van het Rijksmuseum voor Volkenkunde Series 33.

Boardman, S., 1999. The Agricultural Foundation of the Aksumite Empire, Ethiopia. In: M. van der Veen (ed.), *The Exploitation of Plant Resources in Ancient Africa*. New York, Kluwer Academic/Plenum Publishers, pp. 137-147.

Bolland, R., 1991. The Tellem Textiles. In: R. Bolland (ed.), *Tellem Textiles: Archaeological Finds from Burial Caves in Mali's Bandiagara Cliff*. Leiden, Mededelingen van het Rijksmuseum voor Volkenkunde, pp. 52-77.

Bovill, E.W., 1995. *The Golden Trade of the Moors: West African Kingdoms in the Fourteenth Century*. Princeton, Marcus Wiener Publishers.

Brooks, G.E., 1993. *Landlords and Strangers: Ecology, Society, and Trade in Western Africa, 1000-1630*. Boulder, Westview Press.

Cappers, R., 1999. Trade and Subsistence at the Roman Port of Berenike, Red Sea Coast, Egypt. In: M. van der Veen (ed.), *The Exploitation of Plant Resources in Ancient Africa*. Kluwer Academic/Plenum Publishers, New York, pp. 185-197.

Chavane, B.A., 1985. *Villages de l'Ancien Tekrour: Recherches archéologiques dans la moyenne vallée du fleuve Sénégal*. Paris, Éditions Karthala.

Chowdhury, K.A. & G.M. Buth, 1971. Cotton seeds from the Neolithic in Egyptian Nubia and the origins of Old World cotton. *Biological Journal of the Linnean Society* 3, pp. 303-312.

Conrad, D.C., 1994. A Town Called Dakajalan: The Sundiata Tradition and the question of Ancient Mali's Capital. *Journal of African History* 35/3, pp. 355-377.

Curtin, P.D., 1971. Pre-colonial trading networks and traders: the Diakhanké. In: C. Meillassoux (ed.), *The Development of Indigenous Trade and Markets in West Africa*. London, Oxford University Press, pp. 228-239.

Curtin, P.D., 1975. *Economic Change in Precolonial Africa: Senegambia in the Era of the Slave Trade*. Madison, The University of Wisconsin Press.

Dalziel, J.M., 1948. **The useful plants of West Tropical Africa.** London, Crown Agents for the Colonies.

Davidson, B., 1998. *West Africa before the Colonial Era: A history to 1850*. London, Longman.

Fage, J.D., 1995. *A History of Africa*, 3rd Edition. London, Routledge.

Fuller, D.Q., 2000. The Botanical Remains. In: T. Insoll (ed.), *Urbanism, Archaeology and Trade: Further Observations on the Gao Region (Mali). The 1996 Fieldseason results*. Cambridge, BAR International Series 829, pp. 28-35.

Fuller, D.Q. & M. Madella, 2001. Issues in Harappan Archaeobotany: Retrospect and Prospect. In: S. Settar & R. Korisettar (eds.), *Indian Archaeology in Retrospect: Vol. 2, Archaeology of the Harappan Civilization*. New Delhi, Manohar and Indian Council of Historical Research, pp. 317-390.

Gallais, J., 1967. *Le delta intérieur du Niger: Étude de géographie régionale, Tome II*. Dakar, Mémoires de L'Institut Fondamental D'Afrique Noire (IFAN).

Griffith, F.LL. & G.M. Crowfoot, 1934. On the early use of cotton in the Nile Valley. *Journal of Egyptian Archaeology* 20, pp. 5-12.

Haskell, H.W., McIntosh, R.J. & S.K. McIntosh, 1988. *Archaeological Reconnaissance in the Region of Dia, Mali*. Final Report to the National Geographic Society, Washington D.C.

Hunwick, J.O., 1973. The Mid-Fourteenth Century Capital of Mali. *Journal of African History* 14, 2, pp. 195-206.

Hutchinson, J.B., Silow, R.A. & S.G. Stephens, 1947. *The Evolution of Gossypium and the Differentiation of the Cultivated Cottons*. London, Oxford University Press.

Insoll, T., 2000. *Urbanism, Archaeology and Trade: Further Observations on the Gao Region (Mali). The 1996 Fieldseason Results*. Cambridge, BAR International Series 829.

Levtzion, N., 1973. *Ancient Ghana and Mali*. London, Methuen & Co, Ltd.

Levtzion, N., 1985. The Early States of the Western Sudan to 1500. In: J.F.A. Ajayi & M. Crowder (eds.) *History of West Africa, volume 1*. Longman, Harlow and Ibadan, pp. 129-166.

Levtzion, N. & J.F.P. Hopkins, 2000. *Corpus of Early Arabic Sources for West African History*. Princeton, Markus Wiener Publishers.

Levtzion, N. & J. Spaulding, 2003. *Medieval West Africa: Views from Arab Scholars and Merchants*. Princeton, Marcus Wiener Publishers.

Lewicki, T., 1974. *West African Food in the Middle Ages According to Arabic Sources*. Cambridge, Cambridge University Press.

Lovejoy, P.E., 1985. The internal trade of West Africa before 1800. In: J.F.A. Ajayi & M. Crowder (eds.), *History of West Africa, volume 1*. Harlow and Ibadan, Longman, pp. 648-690.

McDougall, E.A., 1985. The View from Awdaghust: War, Trade and Social Change in the Southwestern Sahara, from the Eighth to the Fifteenth Century. *Journal of African History* 26, pp. 1-31.

McIntosh, S.K., 1981. A Reconsideration of Wangara/Palolus, Island of Gold. *Journal of African History* 22, pp. 145-158.

McIntosh, S.K., 1995. *Excavations at Jenné-jeno, Hambarketolo, and Kaniana (Inland Niger Delta, Mali), the 1981 season*. Berkeley, University of California Press.

McIntosh, S.K. & R.J. McIntosh, 1980. *Prehistoric Investigations at Jenné, Mali*. Cambridge, BAR International Series 89, Cambridge Monographs in African Archaeology 2.

Monteil, C., 1927. *Le Coton chez les Noirs*. Paris, Librairie Émile Larose.

Munson, P.J., 1980. Archaeological Data on the Origins of Cultivation in the Southwestern Sahara and Their Implications for West Africa. In: B.K. Swartz, Jr. & R.E. Dumett (eds.), *West African Culture Dynamics: Archaeological and Historical Perspectives*. The Hague, Mouton Publishers, pp. 101-121.

Murray, S., n.d. La Flore, 2005. In: R.M.A. Bedaux, J. Polet, K. Sanogo & A. Schmidt (éds.), *Recherches archéologiques à Dia dans le Delta intérieur du Niger, Mali: bilan des saisons de fouilles 1998-2003*. Leiden, Mededelingen van het Rijksmuseum voor Volkenkunde.

Perinbam, B.M., 1980. The Julas in Western Sudanese History: Long-Distance Traders and Developers of Resources. In: B.K. Swartz, Jr. & R.E. Dumett (eds.), *West African Culture Dynamics: Archaeological and Historical Perspectives*. The Hague, Mouton Publishers, pp. 455-475.

Sakai, S., 1990. Traditions orales à Ja: Histoire et Idéologie dans une ancienne cité Islamique. In: J. Kawada (ed.), *Boucle Du Niger – approches multidisciplinaires*. Tokyo, Institut de Recherches sur les Langues et Cultures d'Asie et d'Afrique, pp. 211-258.

Salvaing, B., 1983. A propos de Dia et ses lettrés au XIX siècle. *Annales de l'Université d'Abidjan* (Serie I) 11, 119-135.

Sanneh, L., 1972. The Diakhanke and the Ummah Al-Muhammadiyah: A Preliminary Study of the Clerical and Educational Role of the Diakhanké. *Conference on Manding Studies, School of Oriental and African Studies*. London, pp. 1-22.

Sanneh, L., 1989. *The Jakhanke Muslim Clerics: A Religious and Historical Study of Islam in Senegambia*. Lanham, MD, University Press of America.

Shaw, T., 1985. The prehistory of West Africa. In, J.F.A. Ajayi & M. Crowder (eds.) *History of West Africa, volume 1*. Harlow and Ibadan, Longman, pp. 48-86.

Smith, B.D., 1995. *The Emergence of Agriculture*. New York, Scientific American Library.

Van der Veen, M., Grant, A. & G. Barker, 1996. Romano-Libyan Agriculture: Crops and Animals. In: G. Barker, D. Gilbertson, B. Jones & D. Mattingly (eds.), *Farming the Desert: The UNESCO Libyan Valleys Archaeological Survey, Volume 1: Synthesis*. Tripoli, UNESCO Publishing, Department of Antiquities, pp. 227-263.

Vartavan, C. de & V.A. Amorós, 1997. *Codex of ancient Egyptian plant remains*. London, Triade Exploration.

Willis, J.R., 1979. Introduction: Reflections on the diffusion of Islam in West Africa. In: J.R. Willis (ed.), *Studies in West African Islamic History, Vol. 1, The Cultivators of Islam*. Franc Cass, pp. 1-34.

Willis, J.R., 1985. The Western Sudan from the Moroccan invasion (1591) to the death of Al-Mukhtar Al-Kunti (1811). In: J.F.A. Ajayi & M. Crowder (eds.), *History of West Africa, volume 1*. Harlow and Ibadan, Longman, pp. 531-576.

Zeven, A.C., 1974. Indigenous Bread Wheat Varieties from Northern Nigeria. *Acta Botanica Neerlandica* 23, pp. 137-144.

Zeven, A.C., 1980. The Spread of Bread Wheat over the Old World Since the Neolithicum as Indicated by its Genotype for Hybrid Necrosis. *Journal D'Agriculture Traditionnelle et de Botanique Appliquée* XXVII, 1, pp. 19-53.

Zohary, D. & M. Hopf, 2000. *Domestication of Plants in the Old World: The origin and spread of cultivated plants in West Asia, Europe and the Nile Valley*. Oxford, Oxford University Press.

Identifying African rice domestication in the Middle Niger Delta (Mali)

S.S. MURRAY

Anthropology Department, University of Wisconsin, Madison, United States of America

Archaeobotanical research at the site of Dia, in the Middle Niger Delta (Mali), recovered a high incidence of African rice grains without *paleas* or *lemmas*, present from the earliest levels (800-400 BC), indicating probable cultivation of this now staple food. As it is difficult to discern naked rice grains to wild or domestic species (*Oryza barthii, O. longistaminata,* or *O. glaberrima*), a morpho-metrical study was conducted. Ratios of the dimensions (length, width, thickness) of whole, ancient grains were compared to those of modern African rice species, and although there is overlap in some measurements, several trends seem clear. In addition, all complete rice grains ranging up to the latest levels at Dia (c. 17th century AD) were measured, providing an interesting perspective on the evolution of African rice over more than 2000 years.

1 Introduction

Until recently, when its cultivation was largely replaced by the Asian species (*Oryza sativa* L.), African rice (*O. glaberrima* Steud.) was an important staple food for people across much of West Africa. This is true to the extent that it is said, "to rice-eating people no meal is a meal without rice" (Harlan, 1993: p. 58). Certainly historically, African rice farming was spread across broad, but patchy regions of West Africa, from the Atlantic coast of sahelian Mauritania to Lake Chad, south to the Bight of Benin, and west along the coast to Senegal (Pearson *et al.*, 1981; Portères, 1976).

Dependence on this important cereal, as well as on the annual and supposed perennial wild ancestors (*O. barthii* A. Chev. and *O. longistaminata* A. Chev. & Roehr.), appears to have great antiquity, and although archaeological data on the place and time of domestication is virtually non-existent, research suggests that African rice may have had several centres of primary genetic transformation. This paper presents a study of ancient African rice grains recovered from one possible centre, the Middle Niger Delta (MND), where today rice is still largely cultivated by traditional methods.

2 African rice domestication

Based on traditional cropping techniques, morphologic and genetic studies of modern rice varieties, and archaeobotanical research, the MND has long been considered a primary centre of African rice diversification and domestication (Andah, 1993; Clark, 1976; Harlan, 1995; Harris, 1976; McIntosh & McIntosh, 1988; Ogbe & Williams, 1978; Portères, 1962, 1970, 1976; Shaw, 1976). In this region today, cultivation is based mainly on traditional floodwater farming techniques that appear to have some antiquity (Andah, 1993; Harris, 1976). For pearl millet (*Pennisetum glaucum*) and sorghum (*Sorghum bicolor*), the conventional method of cultivation is known as *décrue*, a practice that involves planting of crops in the humid soils of the receding flood (Harlan & Pasquereau, 1969). African rice however, is referred to as a crop of the flood, or *crue*. It is planted in moist rain-fed soils before the rise of floodplain waters, typically with no irrigation or water-control mechanisms. As the inland delta becomes inundated and fields are submerged, the rice plants slowly mature, with some "floating" varieties steadily elongating to meet the height of the water (Harlan and Pasquereau, 1969). While the annual wild ancestor, *Oryza barthii,* also occurs alongside domestic rice fields, its natural habitat appears to be the extensive networks of channels, ponds, swamps, and waterholes that form seasonally during the floods, but which become desiccated as the dry season commences (Harlan and Pasquereau, 1969; Harris, 1976).

The MND is also considered an original centre of rice diversification based on morphologic and genetic analysis. According to Portères (1970: p. 46; 1976), modern domestic varieties in the MND possess genetically dominant traits of large, thick rigid and erect panicles, spikelets loosely attached to pedicels, anthocyanic (red-brown) pigmentation, and varieties that float with rising flood waters. Recessive characters, such as adherence of spikelets to stalks, absence of pigmentation, and inability to 'float', were similar to traits observed in the Sénégambia, suggesting that rice more likely derived from the MND than from the latter region. In an attempt to date rice diversification in the primary (MND) and secondary (Sénégambia) centres, Portères assumed that if rice was present at secondary centre sites dating 1500-800 BC, then it must have been domesticated in the primary centre prior to 1500 BC to have sufficient time to spread along the Sénégal and Gambia Rivers. Subsequent radiocarbon dating however, revealed that the Sénégambian sites date to the first millennium AD, invalidating Portères' chronology for African rice domestication (Hill, 1978; McIntosh & McIntosh, 1979: p. 161; Posnansky & McIntosh, 1976).

Thus far, the earliest archaeological evidence of domesticated African rice, based on well-preserved, diagnostic *paleas* and *lemmas*, occurs at Jenné-jeno dating about 250 BC (McIntosh, 1995). One other region however, has yielded early dates for African rice use, though at present, cautious identification assumes a wild state (*Oryza barthii* or *O. longistaminata*). Gajiganna in North-East Nigeria has a long sequence of rice impressions beginning 1800-1400 BC and continuing until 1000-800 BC (Klee & Zach, 1999; Klee et al., 2004), while excavation at nearby Kursakata, dating 1000-400 BC, has revealed the remains of various wild grasses, including rice (Klee & Zach, 1999; Neumann et al., 1996). Etymological analysis of the words for rice in West Africa and study of its varietal types and forms, suggests that Asian rice (*O. sativa*) was probably introduced into West Africa by the Portuguese sometime in the 16th century AD (Portères, 1970).

3 The study site

The site of Dia is located in the upper MND, a few kilometers from the banks of the Diaka River, a major tributary of the Niger (figure 1). Approximately 70 km south-west of Dia is the archaeological site of Jenne-jeno, to the north is Timbuktu, and to the west is the Méma, a now extinct branch of the Middle Niger floodplain with some of the earliest evidence of late stone age sites in the area. Dia consists of three ancient tell sites, totaling about 100 hectares in size (figure 2). The modern town of Dia is settled on the middle mound, and the sites of Dia-Shoma and Dia-Mara are located to the west and east, respectively. Floodplain soils surround the three sites, and in the fields, domesticated African and Asian rices are the main crops. Wild African rice species grow in marshes adjacent to the

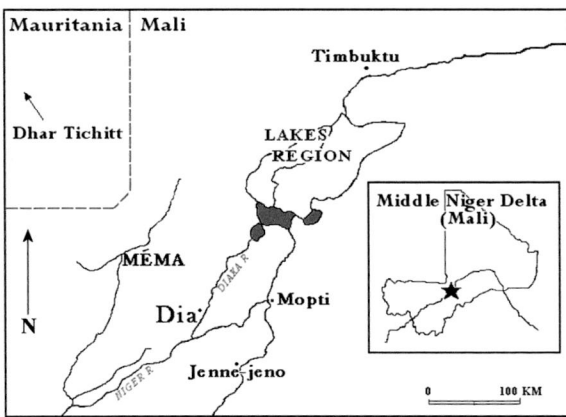

Figure 1: Location of Dia and the Middle Niger Delta.

cultivated fields. The local vegetation consists mainly of leguminous shrubs and trees, and grasses such as *Cenchrus biflorus*.

Modern Dia is considered a traditional and religious Islamic town, yet strong in animism (Gallais, 1967b: p. 553). It numbers about 8000 inhabitants of diverse ethnic background, without electricity, running water, or a market. Two groups at Dia are greatly dependent on the annual floods: the Bozo fisher-folk, reputed to be the earliest inhabitants of the area, and the Marka (or Nono) rice farmers, who claim an ancient relationship with the Bozo, and who are regarded as the earliest rice cultivators of the MND (Gallais, 1967a; McIntosh, 1998; Sundström, 1972). According to oral tradition and archaeological survey, Dia is one of the oldest towns in the MND, and populations emigrating from Dia are thought to have founded Jenné (Bedaux et al., 2001; Levtzion, 1973). Dia-Shoma's earliest levels contain quartz microliths, stone and bone beads, bone points, grinding stones, and pottery known as Faita that is similar to that found in late stone age sites of the Mema. Associated with these finds are terracotta cattle figurines, bone from cattle, ovicaprines, fish (*Clarias, Tilapia*, and *Lates nilotica*), and small to large wild animals (waterbuck, Roan antelope, topi), and some iron slag (Bedaux et al., 2001; MacDonald, pers. comm.).

From 1998-2002, a team of Dutch, English, French, and Malian researchers undertook an archaeological excavation at Dia (Bedaux et al., 2001). Thirteen units, ranging in size from roughly 10-25 square meters, were excavated, 10 of these on Shoma and three on Mara (figure 2). The sequence on Shoma dates from 800-400 BC (where the radiocarbon curve flattens) to around AD 1900, and Mara dates from about AD 500-1900. Excavation on the middle mound of Dia was not possible; thus, the occupational sequence is unclear for this portion of the site. Five horizons were distinguished between Shoma and Mara: Horizon 1

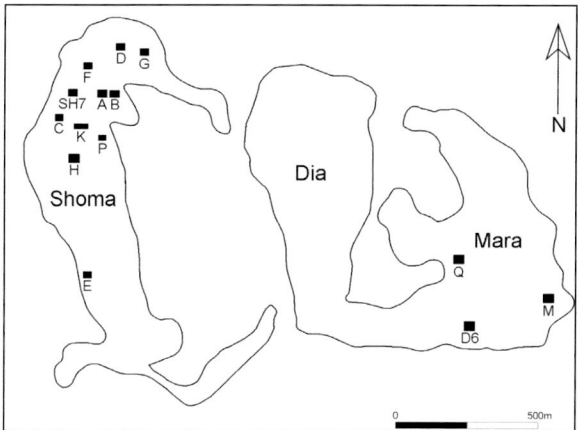

Figure 2: Situation of Dia mounds and excavated units.

(800-400 BC – AD 0) represents the earliest deposits from the site, when fisher-forager and agriculturalists likely first moved into the area. Horizon 2 (AD 0–500) and Horizon 3 (500–1000 AD) mark a period when Shoma and Mara were either seasonally inhabited or partially abandoned. During Horizon 4 (AD 1000–1600), it appears that the town of Dia was at its most extensive occupation on both Shoma and Mara. Most of the deposits of Horizon 5 (AD 1600–1900) were lost to heavy erosion, but historical records and oral traditions suggest that Shoma was abandoned for most of this horizon (Bedaux et al., 2001)

Over the course of the first three excavation seasons, more than 3500 liters of sediment were processed by simple bucket flotation, recovering roughly 6063 charred botanical items. A total of 1376 whole and fragmented rice grains were recovered from all levels on both Shoma and Mara, this number comprising about 23% of all botanical finds at Dia. Roughly 24% of the rice grains were recovered from the earliest horizon (Horizon 1) on Shoma (figure 3), and 51% were recovered from Horizon 4. AMS dates on three grains indicate that African rice was present from the earliest occupation levels dating 800-400 BC (2522±47 BP, AA44337; 2,485±52 BP, AA44338; 2,520±48 BP, AA44339). Dates on another 7 grains place rice in secure contexts through to the mid-1st millennium AD. Rice from Horizon 4 was dated through stratigraphic position and associated radiocarbon dates.

Determination of the ancient grains to wild or domestic species was difficult. Virtually all rice grains, particularly those from Horizon 1, were recovered naked – without the *paleas* and *lemmas* necessary for distinguishing African rice grains to wild or domestic species. Typically, spikelets of modern wild African rice (*Oryza barthii*) have hairs and awns, and the modern domesticate (*O. glaberrima*) is smooth and without awns, although wild characters such as brittle rachii

Figure 3: Ancient African rice grain from Horizon 1 (K128), Dia, Middle Niger Delta.

and hispid hulls also occasionally occur (Katayama & Sumi, 1995; Ogbe & Williams, 1978; Portères, 1976).

4 The metrical study

An examination of the literature on African rice morphology shows that grain dimension (length, width, thickness) in wild and domestic species overlaps extensively (Lu-Rong Bao, pers. comm.; Katayama, 1992; Katayama & Sumi, 1995; Ogbe & Williams, 1978). Figure 4, for example, shows the length and thickness of the uncharred modern domesticate *Oryza glaberrima* and the wild ancestor *O. barthii*. Interestingly, Katayama and Sumi (1995; see also Katayama, 1992, 1994) have found that ratios of these dimensions (L/W, L/T, W/T) show differences between Asian (*O. sativa*) and African species (*O. glaberrima* and *O. barthii*), and between species of *Series Glaberrima* (*O. glaberrima* and *O. barthii*) and *Series Sativa* (*O. sativa* and the wild African species *O. longistaminata*). Although there is much variability, these ratios point to an evolutionary trend towards

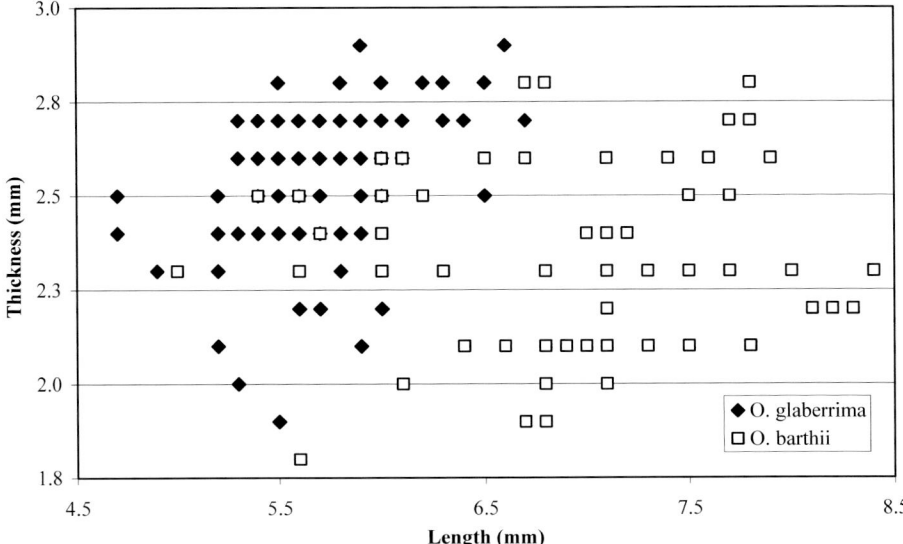

Figure 4: Length and thickness of uncharred modern domesticated and wild African rice.

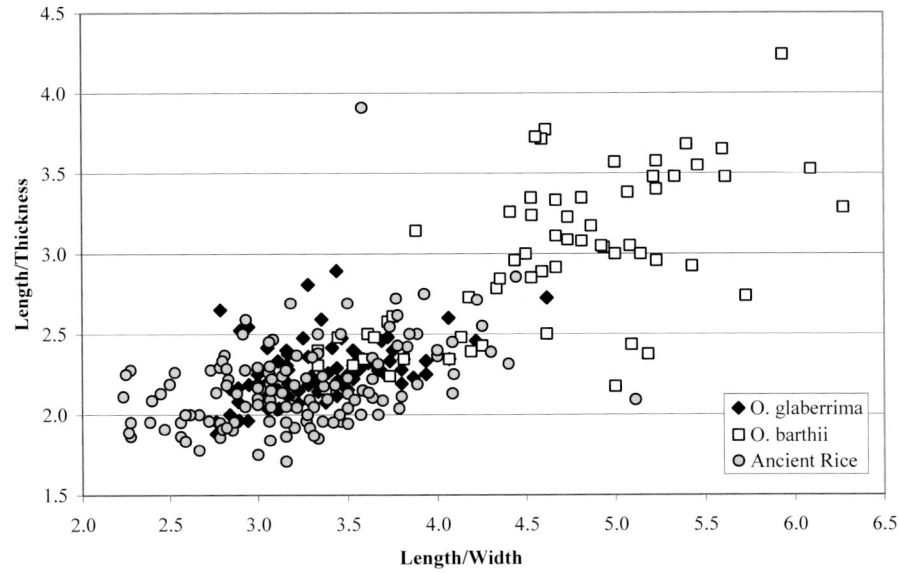

Figure 5: Ratios of uncharred modern domesticated and wild African rice, and charred ancient rice from Dia.

thicker grains in *O. glaberrima* and *O. barthii*, compared to *O. sativa* and *O. longistaminata*. Comparing ratios of *O. glaberrima* and *O. barthii*, these two species often can be distinguished by the manner in which volume is made larger: *O. glaberrima* adds to overall volume by increasing thickness and width, while *O. barthii* increases volume more through length.

In an attempt to identify the archaeological rice grains from Dia to wild or domestic species, the ancient grains were measured and compared to ratios of modern grains. Only unbroken and undistorted ancient grains were chosen for measurement (total n = 134; Horizon 1=57, Horizon 2 =28, Horizon 3=8, Horizon 4=41). These grains, as well as modern dehulled grains of *Oryza glaberrima* (n = 91) and *O. barthii* (n = 68), were measured with a binocular microscope at 10x magnification using an eyepiece micrometer. The modern grains were obtained from Niger, Nigeria, and Mali by the International Rice Research Institute (IRRI).

To mimic possible size changes in the ancient grains due to charring, the modern *Oryza glaberrima* and *O. barthii* grains were first measured, and then charred in a household oven at 120°C for 30 minutes. The temperature was gradually increased to 150°C for another 30 minutes, then 175°C for 15 minutes, 200°C for 15 minutes, and finally 230°C for 30-45 minutes. The modern grains were re-measured, and a change of 3-4% in overall dimension was found, with size typically decreasing with charring.

In prehistory, it is likely that charring of grains occurred in a wide variety of ways, at high and low tem-

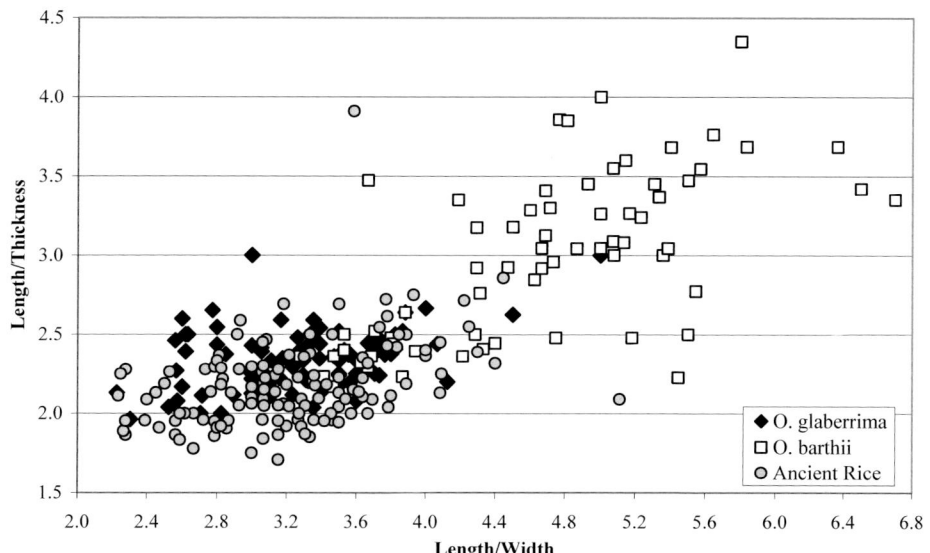

Figure 6: Ratios of charred modern domesticated and wild African rice, and charred ancient rice from Dia.

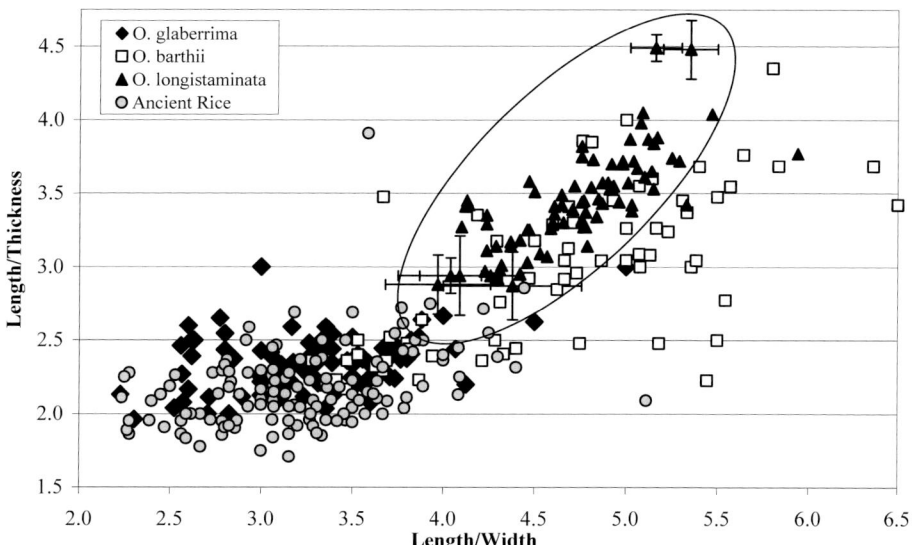

Figure 7: Ratios of charred modern (Oryza glaberrima and O. barthii), uncharred modern (O. longistaminata), and ancient rice.

peratures and rates of speed, and in different mediums, such as in sand or in ceramic pottery. Since the ancient grains measured for this study were undistorted grains, it seemed important to compare them to similar modern grains, in other words, not grains heated in such a way as to become misshapen through charring. In the charring experiments conducted for this study, minimal distortion seemed to occur more often when grains were charred slowly and temperature increased gradually.

Overall, when ratios of the ancient and modern rices were compared, the majority of ancient grains more closely resembled the modern domestic African rice than they resembled the wild grains, and this pattern was little altered whether the modern grains were uncharred or charred (figures 5-6). The ancient grains were also compared to the wild African rice species, *Oryza longistaminata*. As the United States Department of Agriculture (USDA) considers *O. longistaminata* a weedy pest, it was impossible to obtain shipments of this grain into the USA for charring and measurement. Instead, the ancient grains were compared to averages and standard deviations of 83 accessions (30 grains each) of uncharred *O. longistaminata* collected in Nigeria, Ivory Coast, and Sénégal by Katayama (1992).

Figure 7 shows that the ancient grains do not overlap with *O. longistaminata*, and would unlikely do so even with a 3-4% change in grain size due to charring. Based on the distribution of the ancient grains, it is also clear that both wild and domestic rice species are present in the ancient sample. This is not unexpected

Oryza glaberrima - charred n=91	Length (mm)	Width	Thickness	Volume	L/T	L/W	W/T
Mean	5.8	1.8	2.5	25.8	2.4	3.3	0.7
Median	5.8	1.7	2.5	25.5	2.4	3.3	0.7
Standard Deviation	0.5	0.2	0.2	4.8	0.2	0.5	0.1
Minimum	4.6	1.2	1.8	14.4	2.0	2.2	0.5
Maximum	6.8	2.5	3.0	41.6	3.0	5.0	1.0
Oryza barthii - charred n=68	Length	Width	Thickness	Volume	L/T	L/W	W/T
Mean	6.7	1.5	2.3	22.3	3.0	4.7	0.6
Median	6.7	1.5	2.3	23.2	3.0	4.7	0.6
Standard Deviation	0.8	0.2	0.3	5.4	0.5	0.8	0.1
Minimum	4.9	0.9	1.7	9.7	2.2	3.3	0.4
Maximum	8.7	1.8	2.8	33.6	4.4	6.7	0.9
Ancient grains - charred n=134	Length	Width	Thickness	Volume	L/T	L/W	W/T
Mean	4.4	1.4	2.0	12.7	2.2	3.2	0.7
Median	4.4	1.4	2.1	12.8	2.1	3.2	0.7
Standard Deviation	0.5	0.3	0.3	4.5	0.3	0.5	0.1
Minimum	3.3	0.9	1.1	4.0	1.7	2.2	0.4
Maximum	5.6	2.0	2.7	23.3	3.9	5.1	1.1

Table 1: Means, medians, and standard deviations of modern and ancient African rice.

as it is common today to find wild rice growing along the margins of the domestic fields, where it is harvested along with *Oryza glaberrima*, sold at market, and consumed.

5 Discussion and conclusions

Although the metrical study suggests the majority of archaeological rice from Dia is domesticated, two puzzling details remain. First, the ancient grains are consistently smaller than the modern grains used in this study, most notably in length (table 1). Second, statistical analyses of the ancient grains by horizon showed no significant difference in length, width, or thickness from the earliest to latest levels (table 2). In other words, there was no significant change in size or shape of African rice recovered from Dia over the course of more than 2000 years.

Table 1 and figure 8 give dimensions of the modern wild and domestic species, and the ancient rice grains. The most prominent difference between the species is the dimension of length; *Oryza barthii* is long and thin, whereas *O. glaberrima* is shorter and thicker. How did it occur that the domestic grain became shorter than the wild grain? As mentioned previously, *O. glaberrima* and *O. barthii* show an evolutionary trend towards thicker and wider grains compared to other rice species, but this trend is especially marked in *O. glaberrima*. Is it possible that in cultivation of some wild populations,

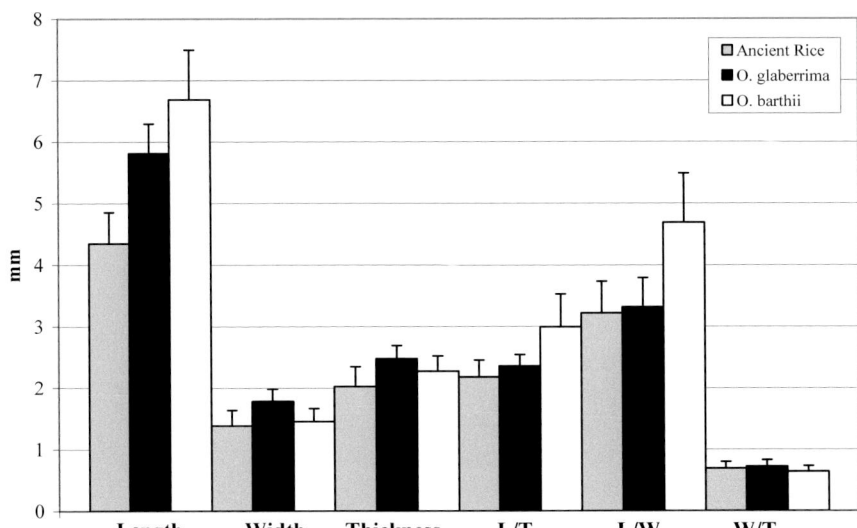

Figure 8: Average dimensions of modern and ancient African rice.

Horizon 1 n=57	Length	Width	Thickness	Volume	L/T	L/W	W/T
Mean	4.4	1.4	2.1	13.1	2.1	3.3	0.7
Median	4.5	1.4	2.2	13.5	2.1	3.3	0.6
Standard Deviation	0.5	0.2	0.3	4.5	0.2	0.5	0.1
Minimum	3.3	0.9	1.3	4.3	1.7	2.2	0.4
Maximum	5.6	1.8	2.7	22.1	2.9	5.1	1.0
Horizon 2 n=28	Length	Width	Thickness	Volume	L/T	L/W	W/T
Mean	4.2	1.4	2.0	11.7	2.1	3.1	0.7
Median	4.3	1.4	2.0	11.6	2.1	3.1	0.7
Standard Deviation	0.5	0.2	0.2	4.0	0.2	0.4	0.1
Minimum	3.3	0.9	1.4	5.0	1.9	2.3	0.6
Maximum	4.9	2.0	2.3	22.5	2.7	4.1	0.9
Horizon 3 n=8	Length	Width	Thickness	Volume	L/T	L/W	W/T
Mean	4.3	1.4	1.9	11.2	2.3	3.2	0.7
Median	4.0	1.4	1.8	10.2	2.3	3.0	0.8
Standard Deviation	0.8	0.2	0.2	3.8	0.2	0.7	0.1
Minimum	3.4	0.9	1.5	4.9	1.9	2.3	0.6
Maximum	5.5	1.6	2.2	15.7	2.8	4.0	0.8
Horizon 4 n=41	Length	Width	Thickness	Volume	L/T	L/W	W/T
Mean	4.4	1.4	2.0	12.9	2.2	3.2	0.7
Median	4.4	1.4	2.0	12.9	2.2	3.1	0.7
Standard Deviation	0.4	0.3	0.4	5.0	0.4	0.5	0.1
Minimum	3.4	0.9	1.1	4.0	1.8	2.3	0.5
Maximum	5.2	2.0	2.7	23.3	3.9	4.4	1.1

Table 2: Means, medians, and standard deviations of the ancient grains by horizon.

human selection emphasized width and thickness as a means of producing a more efficient increase in volume, and while length also increased, it did so at a much slower rate; thus *O. glaberrima* evolved? Katayama and Sumi (1995: pp. 812–813) offer a similar argument in stating that the flatter grain shape "had been held in the process of evolution from *O. breviligulata* [= *O. barthii*] to *O. glaberrima*," and that while volume is greater in the domesticate, the difference in volume between the species is small and "stresses the youth of domestication". However, Katayama and Sumi (1995) also imply that *O. glaberrima* is decreasing in size, presumably length, which suggests that length is diminishing, rather than slowly increasing. This is of interest because it might help explain why the ancient grains from Dia are small compared to the modern species. The ancient grains possess a shape comparable to modern *O. glaberrima*, but they average 75-80% of the overall size of the modern domesticate, and remain so until at least the 16th century AD. This trait was also observed in a sample of 54 grains of archaeological African rice from Toguéré Galia, a site roughly 100 km south-east of Dia, provisionally dated 11-12th century AD (Bedaux et al., 1978). Although more research needs to be done, the average length of 4.7 mm and thickness of 2.2 mm places the Toguéré Galia grains within the range of the ancient grains from Dia. Similarly, Bedaux et al. (1978: pp. 172–173) observed that the Toguéré Galia rice averaged about 75% of the length of modern domesticated rice. Is increased length in modern African rice species a phenomenon of the last few centuries, or does regional variability better explain the small dimensions of the ancient grains? Perhaps the evolution of African rice is more complicated, with genetic mixing between the domestic and wild species creating new races, intermediate crops, and weedy relatives adapted to a wide variety of conditions (Andah, 1993: p. 249).

During discussion of this paper at the 4th IWAA, a wide range of suggestions were offered to explain the differences between the modern and ancient African rice. Roger Blench commented that in ethnobotanical studies, increased size is often not a characteristic actively sought by humans in their manipulation of plants. And, René Cappers offered that perhaps the small ancient grains represent remains from processing, such as 'trailings' from winnowing or sieving. Alternatively, Stefanie Kahlheber mentioned in her paper (this volume) that only slight increases in size were seen through time in archaeological pearl millet from Oursi, and this was echoed in conversations with Dorian Fuller and Mark Nesbitt regarding their own research.

Overall, the lack of change through time supports the idea that rice at Dia was fully domesticated from the earliest occupation, as some change in size or shape should be evident if it had undergone in situ domestication, or if early rice at Dia was wild and domesticated

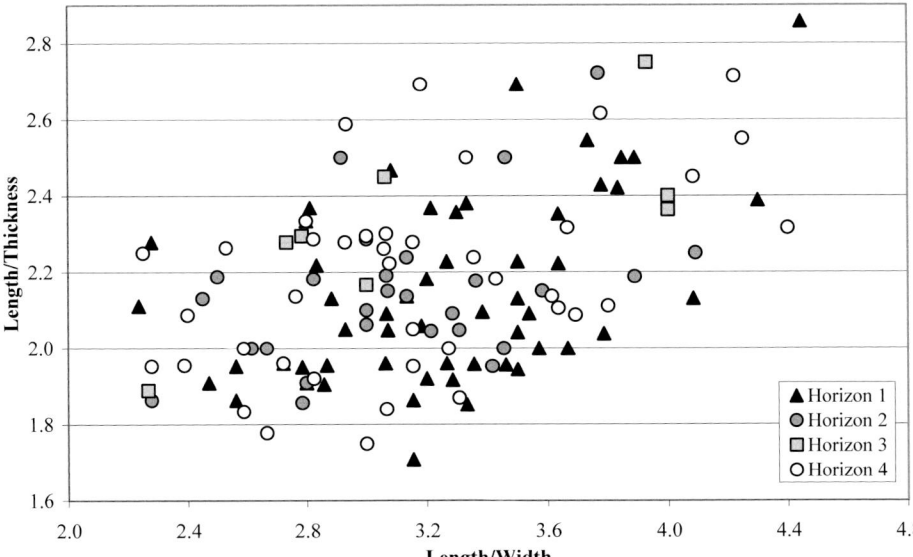

Figure 9: Ratios of the ancient rice grains through time by horizon.

rice was introduced at some point during occupation (table 2 and figure 9). As a result, it appears unlikely that African rice underwent genetic transformation at Dia; more likely, domestication occurred elsewhere, possibly in the MND, and then spread to Dia around 800-400 BC. If domestication occurred in the MND, it may have been the result of long years of cultivation by local fisher-foragers in the region, perhaps the proto-Bozo or Nono. Or, perhaps domesticated rice was introduced from elsewhere into the upper MND, such as from the Méma to the west, though as of yet only fonio (*Digitaria* sp.) has been recovered there (MacDonald, pers. comm.), or the Lakes Region to the north, as proposed by Gallais (1967a: p. 102).

As this study uses only the parameter of grain dimension to distinguish the ancient grains from Dia as domesticated, the results should be regarded cautiously. Further research is necessary, and may incorporate the measurement of additional varieties of wild and domestic African rice, as well as individual grain dimensions of *Oryza longistaminata*. Certainly, the study of archaeological African rice samples recovered in the future will provide further understanding to this research.

6 Acknowledgements

I would like to thank the National Science Foundation (Grant No. 0089599), The Wenner-Gren Foundation (Grant No. 6783), Sigma Xi Grant-in-Aid of Research, the Rijksmuseum voor Volkenkunde (Leiden), the University of Wisconsin-Madison Vilas Travel Grant, and the University of Wisconsin-Madison Anthropology Department for funding of this PhD dissertation research. I am especially grateful to Katharina Neumann and Stefanie Kahlheber (Frankfurt), C.C. Bakels and Wim Kuijper (Leiden), James Burton and David Meiggs (Madison), The International Rice Research Institute (IRRI), the members of Project Dia (particularly Annette Schmidt), and the towns-people of Dia.

7 References

Andah, B.W., 1993. Identifying early farming traditions of West Africa. In: T. Shaw, P. Sinclair, B.
Andah & A. Okpoko (eds), *The Archaeology of Africa: Food, metals and towns*. London, Routledge, pp. 240–254.

Bedaux, R.M.A., Constandse-Westermann, T.S., Hacquebord, L., Lange, A.G. & J.D. Van der Waals, 1978. Recherches Archéologiques dans le Delta Intérieur du Niger (Mali). *Palaeohistoria* 20, pp. 91–220.

Bedaux, R., MacDonald, K., Person, A., Polet, J., Sanogo, K., Schmidt, A. & S. Sidibé, 2001. The Dia archaeological project: rescuing cultural heritage in the Inland Niger Delta (Mali). *Antiquity* 75, pp. 837–848.

Clark, J.D., 1976. Prehistoric Populations and Pressures Favoring Plant Domestication in Africa. In: J.R. Harlan, J.M.J. de Wet & A.B.L. Stemler (eds), *Origins of African Plant Domestication*. The Hague, Mouton Publishers, pp. 67–105.

Gallais, J., 1967a. Le Delta intérieur du Niger; *Étude de Géographie Régionale* (Mémoires de IFAN 79) 2 tomes. Dakar, IFAN.

Gallais, J., 1967b. Le Delta intérieur du Niger. *Étude de Géographie Régionale* (Mémoires de IFAN 79) 2 tomes. Dakar, IFAN.

Harlan, J.R., 1993. The Tropical African Cereals. In: T. Shaw, P. Sinclair, B. Andah & A. Okpoko (eds), *The Archaeology of Africa: Food, metals and towns.* London, Routledge, pp. 53–60.

Harlan, J.R., 1995. *The Living Fields: Our Agricultural Heritage.* Cambridge, Cambridge University Press.

Harlan, J.R. & J. Pasquereau, 1969. Décrue Agriculture in Mali. *Economic Botany* 23, pp. 70–74.

Harris, D.R., 1976. Traditional Systems of Plant Food Production and the Origins of Agriculture in West Africa. In: J.R. Harlan, J.M.J. de Wet & A.B.L. Stemler (eds), *Origins of African Plant Domestication.* The Hague, Mouton Publishers, pp. 311–356.

Hill, M.H., 1978. Dating of Sénégambian Megaliths: A Correction. *Current Anthropology* 19, pp. 604–605.

Katayama, T.C., 1992. Grain Morphology of Wild Rice in African Countries (II). *Memoirs of the Faculty of Agriculture, Kagoshima University* 28, pp. 15–45.

Katayama, T.C., 1994. Grain Morphology of Wild Rice in African Countries (IV). *Memoirs of the Faculty of Agriculture, Kagoshima University* 30, pp. 1–30.

Katayama, T.C. & A. Sumi, 1995. Studies on Agronomic Traits of African Rice (*Oryza glaberrima* Steud.) III. Some grain morphological aspects of domestication and decrement. *Japanese Journal of Crop Sciences* 64, pp. 807–814.

Klee, M. & B. Zach, 1999. The Exploitation of Wild and Domesticated Food Plants at Settlement Mounds in North-East Nigeria (1800 Cal BC to Today). In: M. van der Veen (ed.), *The Exploitation of Plant Resources in Ancient Africa, Proceedings of the 2nd International Workshop on Archaeobotany in Northern Africa, held June 23-25, 1997.* New York, Kluwer Academic/Plenum Publishers, pp. 81–88.

Klee, M., Zach, B. & H-P. Stika, 2004. Four thousand years of plant exploitation in the Lake Chad Basin (Nigeria), part III: plant impressions in potsherds from the Final Stone Age Gajiganna Culture. *Vegetation History and Archaeobotany* 13, pp. 131–142.

Levtzion, N., 1973. *Ancient Ghana and Mali.* London, Methuen & Co. Ltd.

McIntosh, R.J., 1998. *The Peoples of the Middle Niger: The Island of Gold.* Oxford, Blackwell Publishers.

McIntosh, S.K., 1995. *Excavations at Jenne-jeno, Hambarketolo, and Kaniana (Inland Niger Delta, Mali), the 1981 Season.* Berkeley, University of California Press.

McIntosh, S.K. & R.J. McIntosh, 1979. Initial perspectives on prehistoric subsistence in the Inland Niger Delta (Mali). *World Archaeology* II, pp. 227–243.

McIntosh, S.K. & R.J. McIntosh, 1988. From Stone to Metal: New Perspectives on the Later Prehistory of West Africa. *Journal of World Prehistory* 2, pp. 89–133.

Neumann, K., Ballouche, A. & M. Klee, 1996. The emergence of plant food production in the West African Sahel: new evidence from northeast Nigeria and northern Burkina Faso. In: G. Pwiti & R. Soper (eds), *Aspects of African Archaeology: Papers from the Tenth Congress of the PanAfrican Association for Prehistory and Related Studies.* Harare, University of Zimbabwe Press, pp. 441–444.

Ogbe, F.M.D. & J.T. Williams, 1978. Evolution in Indigenous West African Rice. *Economic Botany* 32, pp. 59–64.

Pearson, S.R., Stryker, J.D. & C.P. Humphreys, 1981. *Rice in West Africa: Policy and Economics.* Stanford, California, Stanford University Press.

Portères, R., 1962. Berceaux Agricoles Primaires Sur le Continent Africain. *Journal of African History* 3, pp. 195–210.

Portères, R., 1970. Primary Cradles of Agriculture in the African Continent. In: J.D. Fage & R.A. Oliver (eds), *Papers in African Prehistory.* Cambridge, Cambridge University Press, pp. 43–58.

Portères, R., 1976. African Cereals: *Eleusine*, Fonio, Black Fonio, Teff, *Brachiaria*, paspalum, *Pennisetum*, and African Rice. In: J.R. Harlan, J.M.J. De Wet & A.B.L. Stemler (eds), *Origins of African Plant Domestication.* The Hague, Mouton Publishers, pp. 409–452.

Posnansky, M. & R.J. McIntosh, 1976. New Radiocarbon dates from North and West Africa. *Journal of African History* 17, pp. 161–165.

Sundström, L., 1972. *Ecology and Symbiosis: Niger Water Folk.* Uppsala, Studia Ethnographica Upsaliensia.

Crop production on the Senegal River in the early First Millennium AD: preliminary archaeobotanical results from Cubalel

M.A. Murray, D.Q. Fuller & C. Cappeza
Institute of Archaeology, University College London, United Kingdom

This paper reports the archaeobotanical results from flotation samples collected at Cubalel on the Senegal River. These samples are dominated by the grains and chaff of domesticated pearl millet (*Pennisetum glaucum*), from the earliest levels onwards, as well as other economic taxa, such as fonio (*Digitaria* cf. *exilis*), *Zizyphus* and *Celtis* fruit stones and numerous wild/weed species. The large quantities of *Pennisetum* chaff, including involucres, involucre bases, peduncles and bristles, provides the basis for a consideration of pearl millet crop processing activities and the use of processing by-products as animal fodder and fuel.

1 Introduction

The Iron Age site of Cubalel is located on the Senegal River, which is the natural border with Mauritania in northern Senegal. Cubalel was excavated by Susan and Rod McIntosh of Rice University, Texas. The site comprises a series of eight tumuli (or tells) situated along the river, covering an area of over two hectares. Excavations of the tumuli have uncovered a six metre sequence of stratigraphy spanning 800-900 years, dating from AD 0-100 to AD 900. The area of Cubalel has some of the most productive agricultural land in the Middle Senegal Valley floodplain. This is the first archaeobotanical data reported for this region.

2 Sampling and sorting

In total, 289 archaeobotanical samples were recovered from six of the eight Cubalel tumuli by Mary Anne Murray in 1991. 23 samples were collected from C1, eight from C2, 128 from C3A (the most thoroughly excavated tumulus), 68 from C3B, 51 from C6, seven from C8 and four from a separate trench. Nearly every context was sampled and these included hearths, pit fills, occupation floors and ash deposits. The botanical remains were recovered by bucket flotation using 1 mm and 250 mµ mesh sieves. Due to the use of bucket flotation, sample sizes were generally smaller than those taken for machine flotation. They ranged from between about 4 and 12 litres each, yet multiple samples were often taken from the same contexts. The plant samples were analysed under a low power (10x to 64x) binocular Wild MC3 microscope. All items, such as the seeds and chaff of cereals and other food plants, wild/weed species, wood charcoal, other plant parts and animal dung were extracted from each sample. Identifications of plant taxa were made on the basis of morphological characteristics and the comparison of the ancient specimens with modern comparative reference material. Several specimens were photographed using a scanning electron microscope (SEM) (figures 1-6). The samples were quite rich in organic remains as the average volume of material to be sorted for each sample was about 50 ml. Of the 80 samples analysed thus far, many consisted primarily of wood charcoal, while fewer were rich in seeds, chaff and other items. Table 1 lists the taxa found at Cubalel to date.

3 Discussion

The crop assemblage from Cubalel provides evidence for the importance of modern West African savannah cultivars during the Iron Age. These include pearl millet (*Pennisetum glaucum*) (figures 1-4), which had a single domestication in the far western Sahel, perhaps in Mauritania (Tostain, 1992), by the second millennium BC, and fonio (*Digitaria* cf. *exilis*) (figure 5), which has several possible centres of origin across the West African savannah belt (Harlan, 1971). What is of interest, however, is that the likely ecological contexts of origin of these two cereals are different, and thus Cubalel attests to the bringing together of these cereals into a sin-

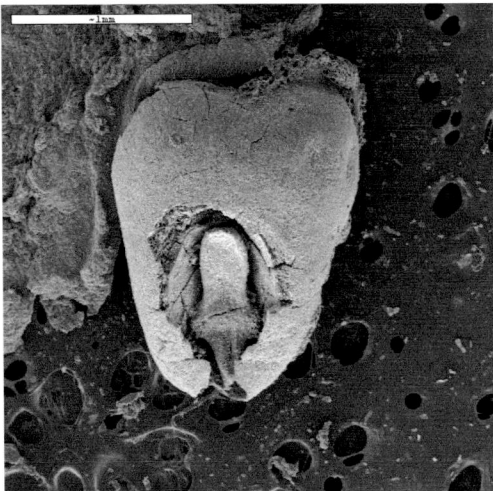 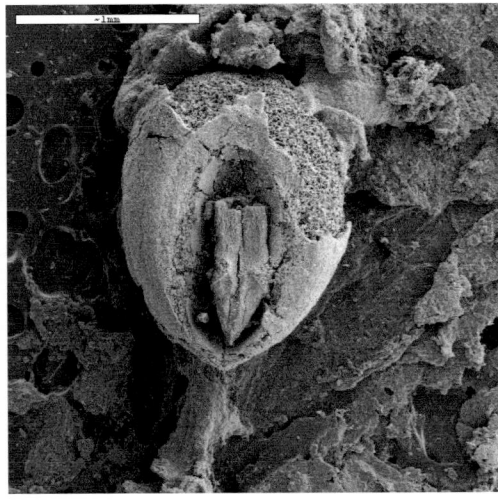

Figure 1: Examples of larger, plumper Pennisetum glaucum *grains from Cubalel.*

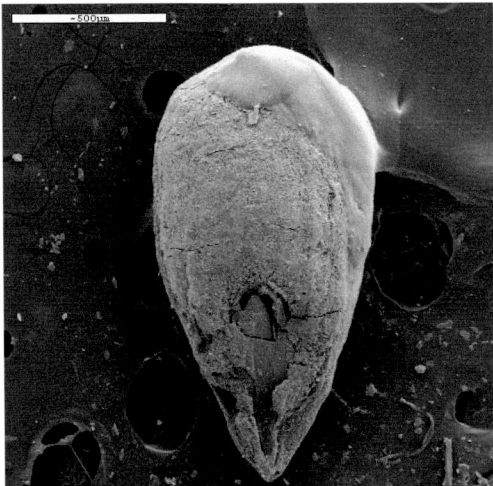 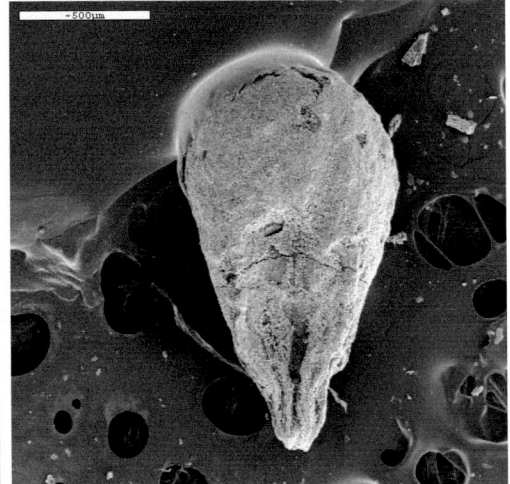

Figure 2: Examples of smaller, thinner Pennisetum glaucum *grains from Cubalel.*

 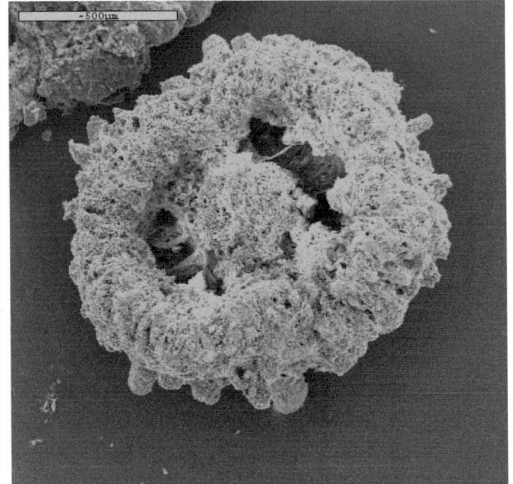

Figure 3: Examples *of* Pennisetum glaucum *chaff from Cubalel: left, domestic* P. glaucum *involucre, with preserved right spikelet; right, two-grained involucre apex, viewed from above.*

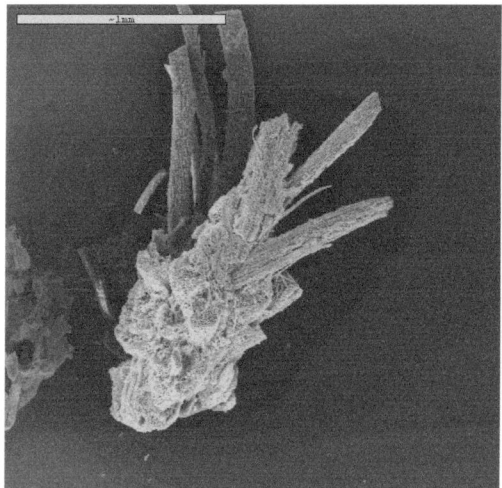 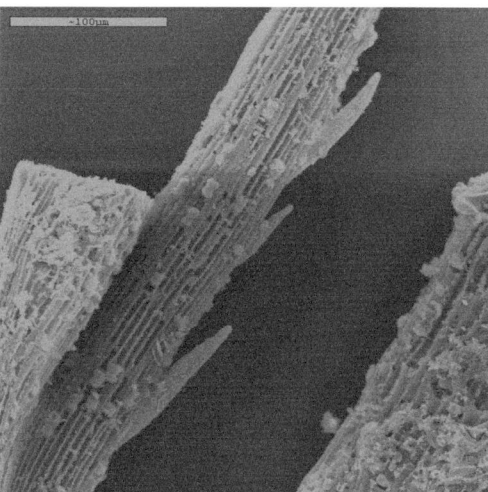

Figure 4: Pennisetum glaucum *involucre apex, with bristles, viewed from the side (left), with close-up of bristle showing unicellular, serrate hairs (right).*

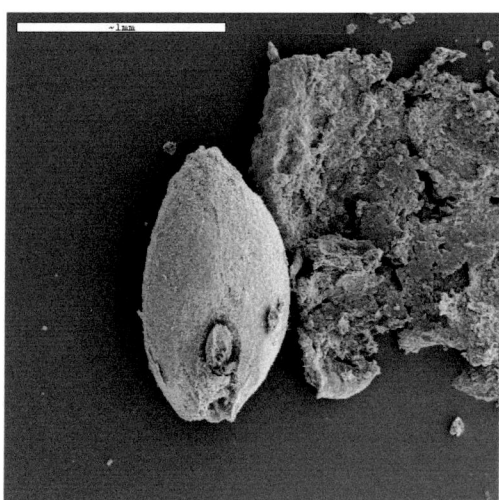 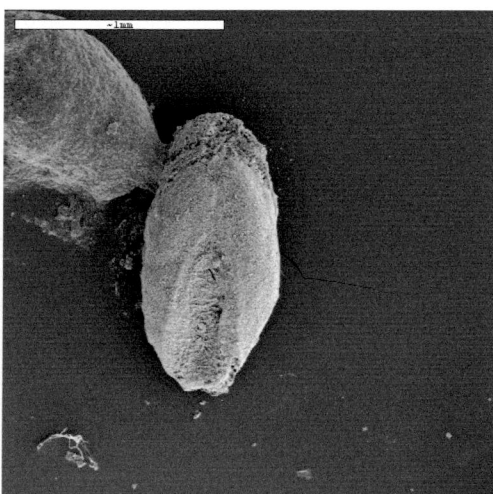

Figure 5: Examples of Panicum *sp. grains, cf.* P. laetum.

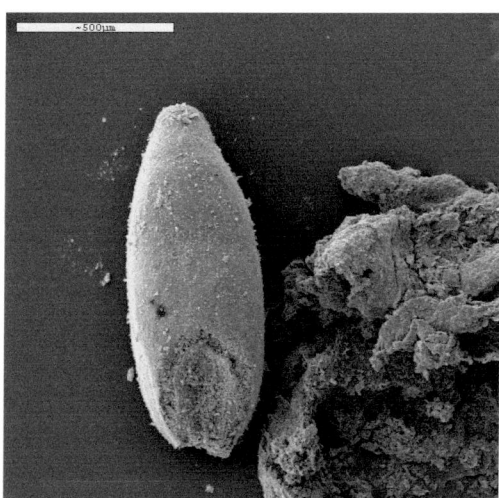 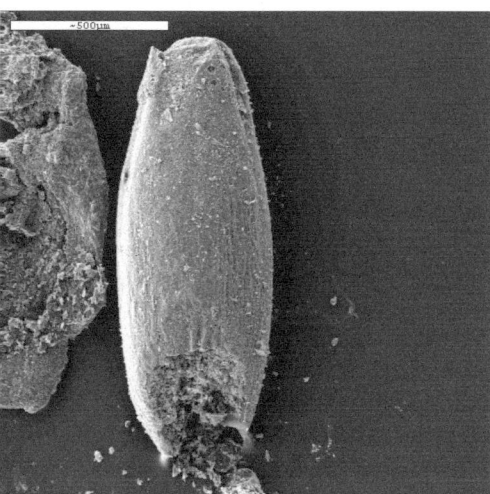

Figure 6: Examples of Digitaria *sp. grains, cf.* D. exilis.

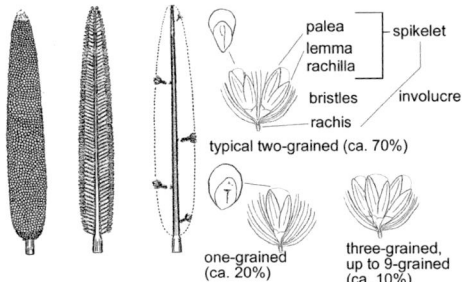

Fig. 7. A diagram of a Pennisetum spike and involcres with parts labelled. Examples of one, two and three-grained involucres shown with the approximate proportions of these involucre types recorded in modern cultivated Pearl Millet (based on Godbole 1925).

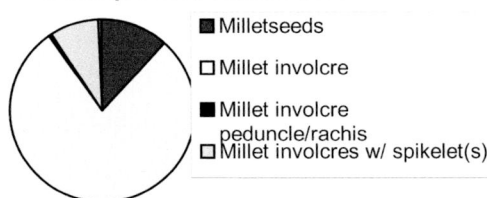

Figure 8. Chart of proportions of millet crop items in sample 176, Tumulus C3A, which is rich in threshing by-products from involucres (total number of items: 17,773).

gle economic system by the later Iron Age (early First Millennium AD). Elsewhere these taxa are reported to co-occur in Phase IV of Jenne-Jeno in Mali some time after AD 800 (McIntosh, 1981). The other grain encountered in quantity, which may have been of economic significance is *Panicum* cf. *laetum* (figure 6), traditionally an important wild food resource in Senegal.

The absence of crops typical of *decrue* (or flood recession) agriculture, such as sorghum or rice, is surprising considering the location of the site on *fonde* land. *Fondes* are high uninundated levees above the floodplains along the bank of the Senegal River where only limited rain fed cultivation millet is practiced, while more extensive agriculture consists of the *decrue* cultivation of sorghum and rice on the floodplain. Further analysis of the samples may give more evidence for these two forms of agriculture.

Indeed, more evidence is required to discuss differences between the six tumuli sampled, as well as temporal differences throughout the 800-900 year stratigraphic sequence.

4 Pearl millet processing and expected grain to chaff ratios

Because pearl millet ears are dense, they can readily be harvested to avoid weeds (see Reddy, 1994 for an ethnographic account of harvesting and processing pearl millet). Domestication has selected for stalked involucres (*i.e.* with a lengthened rachis), as shown in figure 3. These involucres are broken during threshing, which also generally releases most of the grains. Each involucre contains one or more spikelets (Godbole, 1925), with two spikelet involucres being by far the most common, then one grained spikelets are the next most common while three or more grained involucres are rare (and up to nine grains have been reported from individual involucres) (figure 7). From this the expected ratio of grains to involucres in the threshing product should be close to 2:1. As has been shown with wheat chaff, we would expect involucres to be differentially destroyed relative to grains during charring, quite likely on a substantial scale and thus charred ratios of the threshing product may be on the order of 100:1 to 200:1, and thus any assemblages with high proportions of preserved involucres may be considered chaff-rich. After winnowing, however, the by-product should consist almost entirely of involucres. Therefore archaeological assemblages with involucre counts greater than (or even equal to) grain counts are most likely to represent the charred waste of winnowing by-products, while very substantially grain-rich assemblages, or those without involucres are more likely to represent either the winnowing or threshing product. In samples with low grain counts, it may not be feasible to infer processing stage.

5 Crop processing waste at cubalel

Among the crop rich samples at Cubalel, are samples that include hundreds of pearl millet grains and thousands of pearl millet chaff, mainly involucre apex portions with the stubs of bristle bases. The common proportions of such material can be illustrated by the examples of sample 176 (figure 8) and sample 137 (figure 9), two pit fills from tumulus C3A. By the reasoning outlined above this material clearly indicates that such samples represent the by-products of winnowing. These chaff items were also recovered within charred dung matrix, which suggests that winnowing by-products were used as fodder, and that animal dung was used as fuel.

6 Taxa descriptions

The following section offers brief descriptions of the taxa present at Cubalel (table 1), as well as some of the more uses of these taxa in West Africa.

Figure 9. Chart of proportions of millet crop items in sample in samples 137, Tumulus C3A, which is rich in threshing by-products from involucres (total number of items: 5,524).

Cereals

The remains of *Pennisetum* are the most common components in the assemblage, including seeds, involucres, involucre bases, involucre stalks, and the bristles that surround the involucre bases (figures 1-4, 7). Much of the material appears to be cultivated pearl millet, *Pennisetum glaucum*. The *Pennisetum* seeds that fall outside the size range of cultivated pearl millet (Brunken, 1977) may be the result of shrinkage due to charring. Some of these seeds may also belong to the subspecies *stenostachyum*, which originated as a hybridisation between cultivated pearl millet and the wild subspecies *monodii*. The morphologically intermediate *Pennisetum stenostachyum* is a common mimetic weed in pearl millet crops (Brunken *et al.*, 1977). The smallest *Pennisetum* seeds may possibly be the wild subspecies *monodii*. Other remains of *Pennisetum*, too, indicate the presence of *Pennisetum glaucum*, as stalked involucres are a characteristic of the cultivated subspecies and these elements are ubiquitous throughout the assemblage.

More than 50% of the samples contain *Digitaria* spp., many of which most closely resemble fonio (*Digitaria exilis*), one of the two cultivated *Digitaria* species used as cereals in West Africa (figure 5). As yet none appear to be black fonio (*Digitaria iburua*), the other cultivated *Digitaria*. Others specimens are labelled as *Digitaria* type. In this region, the genus *Digitaria* contains generally good fodder plants (Dalziel, 1937) and some, as with other wild grasses in the Sahel, are collected for food (Boré, 1983).

The remains of other cereal include a possible fragment of rice (*Oryza* sp.) from the earliest phase of occupation in tumulus C3A. The fragment is small and seems too narrow to be domesticated rice. Rice had already been domesticated by the time of the first phase of occupation at Cubalel. As yet, there is no evidence that rice was a major crop (or weed) at the settlement as there no other rice anywhere in the Cubalel sequence, even in the later phases.

Table 1: Taxa present at Cubalel.

CEREALS and OTHER GRASSES
Pennisetum glaucum seeds
Pennisetum glaucum involucre bases
Pennisetum glaucum involucre base stems
Pennisetum glaucum involucres
cf. *Pennisetum glaucum* bristles, many in dung fragments
cf. *Oryza* sp.
Digitaria cf. *exilis*
cf. *Panicum* sp.
cf. *Setaria* sp.
cf. *Cenchrus* sp.
Dactyloctenium sp.
cf. *Bothriochloa* sp.
cf. *Echinochloa* sp.
cf. *Chloris* sp.
cf. *Brachiaria* sp.
cf. *Paspalum* sp.
Graminae indeterminate
FRUIT
Zizyphus sp. seeds (charred)
Zizyphus sp. seeds (mineralised)
Zizyphus sp. kernels (charred)
Zizyphus/Celtis (mineralised)
cf. *Zizyphus* thorns
cf. *Celtis* (charred)
cf. *Celtis* (mineralised)
OTHER TAXA
Leguminosae
Solanaceae
Cyperaceae
Caryphyllales
Fistymbristis sp. (*hispidula*?) (uncharred?)
Seeds indeterminate
OTHER ITEMS
cf. Root/tuber fragments
cf. Nut shell fragments
cf. Leaf fragments
Dung fragments

Other Graminae

Other Graminae are present in more than 80% of the samples studied to date although many of these remain challenges for further identification. The grasses may have arrived on site as collected foods, weeds of crops or as animal fodder preserved in charred animal dung.

Panicum spp. - *Panicum* is a genus with over forty species in West Africa, including those used as food, such as *P. laetum*, *P. subalbidum*, and *P. turgidum* (Boré, 1983; Cissé, 1991). The Cubalel specimens appear to be *P. laetum* (figure 6) and may have been used as a wild cereal.

Setaria spp. – There are eleven West African *Setaria* species (Hutchinson & Dalziel, 1936). The three more common to the area include *Setaria pallidefusca,* which is highly drought resistant and makes good animal pasture and fodder (Dalziel, 1937). The edible seeds are also collected for food in the Inland Niger Delta (Boré, 1983; Cissé, 1991). *Setaria verticillata* and *S. sphacelata* also provide animal fodder (Dalziel, 1937).

Echinochloa sp. – Of the five species present in West Africa (Hutchinson & Dalziel, 1927-1936), the most economically important is *E. stagnina,* which is a swamp grass whose grains are often prepared like rice (Boré, 1983; Cissé, 1991). It is also a rich fodder, either green or as hay, and the stems are used for thatch (Dalziel, 1937, p. 527). *E. colona* is also appreciated as a minor cereal and in some areas it is tended in a type of protocultivation (Boré, 1983, p. 30). Dalziel (1937, p. 527) claims that it had also been cultivated in Egypt as a cereal. *E pyramidalis* is another species used for food.

Brachiaria sp. – Several specimens are present in West Africa. *B. ramosa* has been recovered in abundance at Jenne-Jeno (McIntosh, 1995) and together with *B. lata* is listed by Bore (1983) as one of the wild grasses still used for food in the Inland Niger Delta.

cf. *Paspalum* sp. – These seeds were found in fourteen samples. Three species are listed by Dalziel (1937) as useful West African plants - *P.conjugatum, P. scrobiculatum* and *P. vaginatum.* We have only had access to comparative specimens the first two species, neither of which closely corresponds with the archaeological specimens. The third species is a grass of coastal areas. Boré (1983), however, mentions *Paspalum orbiculare,* a grass of damp areas that can be found in millet fallows and in rice fields as one of the plants collected for food in the Inland Niger Delta.

cf. *Bothriochloa* sp. – These seeds were found in seventeen samples. The genus *Bothriochloa* are grasses of no economic importance and may be present as a weed of crops.

Other grasses include cf. *Dactyloctenium* sp., cf. *Cenchrus* sp., and cf. *Chloris pilosa.*

Leguminosae
Remains from the Leguminosae family have been found in more than 40% of the samples. Most of them are seed fragments from the subfamilies Mimosoideae and Caesalpinoideae. Genera, such as *Acacia, Prosopis,* and many other trees typical of the Sahel and open dry savannah belong to these subfamilies. Many small legumes are also commonly found in the samples and have yet to be identified.

Fruits
Fruit remains are a common component in the samples. These primarily consist of the fragmented seeds and kernals of *Zizyphus.*

Zizyphus sp. – This is the most frequent taxon in this category (present in nearly 80% of samples), as indicated by the presence of both charred and mineralized seeds and kernels, as well as the charred thorns of the plant. Three species are present in West Africa: *mauritiana, mucronata* and *spina Christi.* All of which have edible fruits although *mucronata* is reported to have less value, both in terms of nutritional benefit and taste, compared to the other two species (Dalziel, 1937; Von Maydell, 1986). Although it is very difficult to distinguish between the seeds of Z. *mauritiana* and Z. *mucronata.* It may be more likely that the remains present at Cubalel are Z. *mauritiana,* the species most commonly used for food. The fruits of this species, a small tree, are edible fresh, dried or ground into a flour that makes a long lasting bread, cakes and beverages (Cissé, 1991; Dalziel, 1937; Von Maydell, 1986). *Celtis* sp. is also present in the assemblage.

A considerable variety of species of fruits and seeds of woody plants has been recognized in the plant remains at the site of Saouga in Burkina Faso, showing the importance of this type of resource at the end of the first millennium A.D. (Neumann *et al.*, 1998). A re-examination of the fruit and nut fragments from Cubalel may reveal several more useful taxa.

Other remains which are frequently present in the assemblage have been identified only to the family level, such as Solanaceae or Cyperaceae or to the order level, such as Caryophyllales (which includes families, such as Aizoaceae and Portulacaceae). These and other taxa will be identified in greater detail at a later date.

The presence of root/tuber remains is of interest as it may represent an important, though as yet unrecognised, food source. The analysis of this material with scanning electron microscopy (SEM) may help to identify it further.

7 Conclusions

The main cereal at Cubalel was pearl millet (*Pennisetum glaucum*). The large numbers of seeds and the various components of pearl millet chaff indicate the importance of the cereal, as well as the stages of its processing. The remains of millet chaff are also found in animal dung from the site, indicating the importance of the species as an animal fodder as well. The absence of crops typical of *decrue* (or flood recession) agriculture, such as sorghum or rice, as opposed to the normally rain fed millet needs to be explored further.

Several other grasses present may have been important foods, such as fonio (*Digitaria* cf. *exilis*) and further analysis will provide a clearer picture of these taxa. It is also clear that various fruits, particularly *Zizyphus,* were another major component of the diet at Cubalel.

The use of animal dung as fuel is clearly shown throughout the sequence, as is the use of wood as fuel, as wood charcoal is also ubiquitous throughout. Indeed, many of the samples contain only wood charcoal.

After a full analysis of the plant remains from Cubalel, it will be possible to answer questions concerning the type of agriculture practiced at the settlement, the use of pearl millet by-products as fodder, fuel, etc., differences in fuel use (wood vs. animal dung), the role of wild foods, differences in the presence of taxa through time, such as the co-occurance of pearl millet and fonio in the sequence, but also those between the eight Iron Age tumuli located along the Senegal River at this important Middle Senegal Valley settlement.

8 References

Boré, Y., 1983. Recensement des graminées sauvages alimentaires (cereales mineures utilisées en 5è, 6 è et 7e regions). Unpublished PhD from Ecole Normale Superieure.

Brunken, J., 1977. A systematic study of *Pennisetum* sect. *Pennisetum* (Graminae). *American Journal of Botany* 64, pp. 161-76.

Brunken, J., Wet, J.M.J. de & J. R. Harlan, 1977. The Morphology and Domestication of Pearl Millet. *Economic Botany* 31, pp. 163-174.

Cissé, A., 1991. *Inventaire et etude biologique des plantes sauvages alimentaires consommés dans le sahel malien en periode de dissette.* Bamako, Ministere de l'Education, UNESCO.

Dalziel, J. M., 1937. *The Useful Plants of West Tropical Africa.* London, The Crown Agencies for the Colonies.

Godbole, S. V., 1925. *Pennisetum typhoideum.* Studies on the Bajri Crop. I. The Morphology of *Pennisetum typhoideum. Memoirs of the Department of Agriculture in India (Agricultural Research Institute, Pusa)* 14 (8), pp. 247-268.

Harlan, J., 1971. Agricultural Origins: Centers and Noncenters. *Science* 174, pp. 468-474.

Hutchinson, J. & J.M. Dalziel, 1927-36. *Flora of West Tropical Africa.* 2 vols. London, The Crown Agencies for the Colonies.

Maydell, H.-J. von, 1986. *Trees and Shrubs of the Sahel.* GTZ, Verlag Josef Margraf.

McIntosh, S., 1981. *Excavations at Jenne-Jeno, Hambarketolo, and Kaniana (Inland Niger Delta, Mali), The 1981 season.* Publications in Anthropology 20, Berkeley, University of California.

Neumann, K., Kahlheber, S. & D. Uebel, 1998. Remains of woody plants from Saouga, a medieval West African village. *Vegetation History and Archaeobotany* 7, pp. 57-77.

Reddy, S. N., 1994. Plant Usage and Subsistence Modeling: An Ethnoarchaeological Approach to the Late Harappan of Northwest India. Ph.D dissertation. University of Wisconsin. Ann Arbor Michigan: University Microfilms.

Tostain, S., 1992. Enzyme div ersity in pearl millet (*Pennisetum glaucum* L.) 3. Wild millet. *Theoretical and Applied Genetics* 83, pp. 733-742.

Early domesticated pearl millet in Dhar Nema (Mauritania): evidence of crop processing waste as ceramic temper

D.Q. FULLER, K. MACDONALD

Institute of Archaeology, University College London, London, United Kingdom

R. VERNET

Université de Nouakchott, Nouakchott, Mauritania

New evidence relating to the early cultivation of pearl millet (*Pennisetum glaucum*) in Southeast Mauritania is reported based on material recovered during archaeological fieldwork in 2000 in Dhar Nema. Evidence of Tichitt tradition ceramics from the sites of Djiganyai and Oued Bou Khzama are predominantly chaff-tempered, and examination of the impressions through casting and SEM indicates tempering with the winnowing by-products of fully domesticated pearl millet. At the latter site chaff-tempered pottery and tuyères are associated with early iron-smelting by the mid-first millennium BC. At the earlier site Djiganyai, direct AMS dates on the organic fraction of sherds put the earliest millet chaff-tempered pottery back to ca. 1750-1600 cal. BC, providing a *terminus ante quem* for domestication and even earlier cultivation of this species. The use of winnowing by-products implies large-scale processing of this crop in the vicinity of ceramic production suggesting local cultivation, perhaps in wadis that flow into the nearby Amourj Palaeolake. Earlier untempered wares may imply the lack of local cultivation, the absence of the domesticated form that requires threshing, or different tempering choices.

1 Introduction

A two year programme of research of the Dhar Néma region of Southwest Mauritania (figure 1) based on surface survey and limited test-pitting has produced important new evidence about the Tichitt Tradition (2000-200 BC) relating to early agriculture, iron metallurgy and social complexity (MacDonald *et al.*, 2003). Fieldwork by MacDonald, Vernet and colleagues included survey on foot of two 5 km radius areas on the Néma escarpment and two areas of the floodplain, around ancient palaeolakes that were surveyed by vehicle. 25 sites were located, of which three were test excavated, including Djiganyai and Bou Khzama. From both of these sites large surface collections of pottery, as well as excavated material was examined for plant impressions, and the present paper focuses on the discussion of the contribution of these remains to the history of agriculture within this region and the domestication of Pearl Millet (*Pennisetum glaucum*). The chronological framework suggested by pottery stylistic analysis has been tested through direct AMS dates that also secure the antiquity of the identified plant impressions. The current evidence is part of a growing body of evidence indicating the widespread cultivation of domesticated Pearl millet in West Africa by the early to mid second millennium BC.

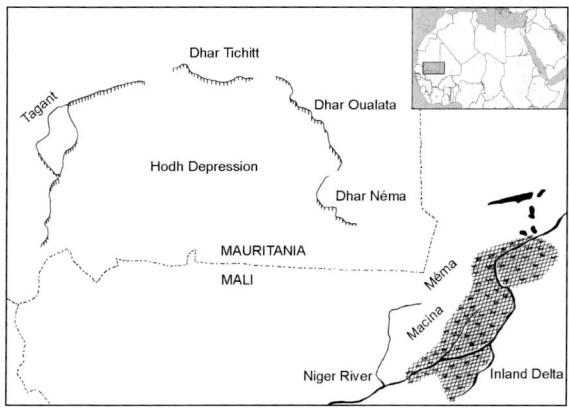

Figure 1: Map showing the location of the Dhar Nema sites.

Figure 2: Example of sherd with impressions from Oued Bou Khzama.

2 An Early Tichitt site: Djiganyai (Dj)

This site of 3 ha was located by a wadi leading to the Palaeolake Amourj. Test excavations indicated 1 m of intact deposit, including three horizons of distinct material culture. This sequence begins with a Pre-Tichitt Ceramic Late Stone Age Occupation, followed by Classic Tichitt and Late Tichitt Ceramic phases, and thus represents the first large site with stratified Tichitt tradition materials. The lowest occupation (Pre-Tichitt Phase) includes pottery with little or no recognizable vegetable temper, despite a high organic content (directly dated back to ca. 1950-1750 BC), while subsequent Classic and Late Tichitt material is often heavily tempered with vegetable material, which has been examined in the current archaeobotanical study. Test excavations also provided bone evidence for domestic fauna (cattle, sheep/goat, dog), hunted game (gazelles, warthog, giraffe), and aquatic resources (*Clarias* catfish, *Tilapia* carp, gastropods and bivalves). Due to the exploratory nature of the excavation intensive sampling for plant macroremains was not possible, although hackberry tree fruit pits were also recovered (*Celtis* cf. *integrifolia*). Despite examination, the sand-tempered pottery of the lowest level was not found to contain plant impressions. But this absence of evidence need not imply an absence of agriculture, as chaff-tempering practices may only have developed in the next phase. Plant impressions are profusely preserved from the second (Classic Tichitt) and later phases.

3 A Late Tichitt site: Oued Bou Khzama (OBK)

This site of ca. 5 ha is located on the flanks of the Néma escarpment, and features two small enclosures, agricultural terraces, and schist grain bin stands. It has evidence for sixteen iron furnace mounds, and excavations of one slag mound revealed remains of an oval low shaft furnace, with tuyères. The Late Tichitt ceramic material from the site and the tuyères proved to be tempered with domesticated pearl millet. The AMS dates confirm ceramic typological dates of 800-200 BC, and suggest the beginnings of iron metallurgy in the first half of the First Millennium BC.

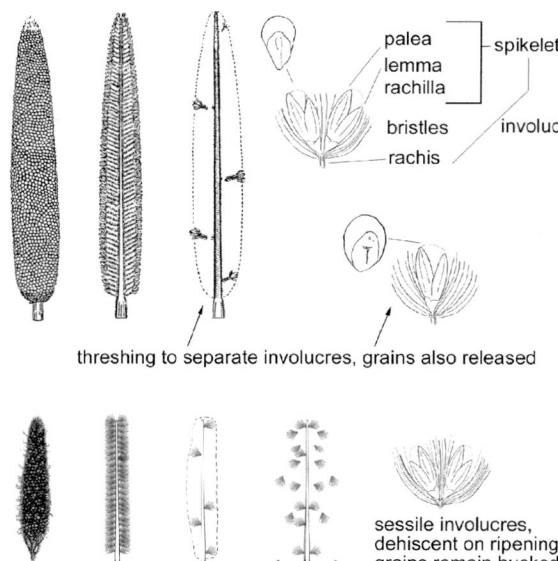

Figure 3: Diagram comparing domesticated (top) and wild (bottom) forms of pearl millet. Wild millet is shown schematically to indicate natural shedding of sessile involucres. This diagram does not represent the order in which involucres mature and are shed.

4 Laboratory methods

The sherds were brushed to remove sand and examined under a binocular microscope to identify examples with promising impressions (*e.g.* figure 2). These impressions were then cast using a dental casting agent of Polyvinylsiloxane. This material is easy to use, does not require a clay mold rim, is quick drying and easy to remove. Multiple casts were taken from each impression site, as early casts tended to remove embedded sand and lack anatomical details. It was found that this process also tended to remove small parts of the sherd surface with each generation of casting. Casts were then sorted under a binocular microscope and the best examples were trimmed and mounted on a SEM stub, sputter coated and imaged with a SEM.

5 Radiocarbon dates

Selected sherds, with identifiable pear millet impressions and also of interest in terms of stylistic information relating to conventional ceramic dating, were selected for direct dating of the organic fraction using AMS-dating. The results so far are given in the following list, with calibration ranges of the 1-sigma intercepts (Stuiver *et al.*, 1998):
1. GX-29358-AMS. Djiganyai, Context 6, Sherd with indeterminate organic temper:
   3550+/- 40bp; 1970-1780 cal BC
2. GX-29359-AMS. Djiganyai, Sf-S2, Sherd with millet temper:
   3370+/- 40bp; 1740-1620 cal BC
3. GX-28140-AMS. Djiganyai, Sf-S9, Sherd with millet temper:
   3260+/- 40bp; 1620-1510 cal BC
4. GX-28137-AMS. Bou Khzama, Sf-S1, Sherd with millet temper:
   2340+/- 40bp; 760-400 cal BC

6 Domestication traits, threshing and tempering

The domestication trait *par excellence* in seed crops is the loss of natural seed dispersal mechanisms, selected for by regimes of harvesting (by methods such as sickles or uprooting) and sowing from stored seed. In pearl millet this selective pressure created a stalk (lengthened rachis) at the base of the involucres (Brunken *et al.*, 1977), whereas in wild *Pennisetum* the sessile involucres are shed on ripening (figure 3). Adapted to animal (and wind) dispersal these spiny involucres tend to retain their seeds, whereas the less spiny domesticated forms readily lose their seeds (*i.e.* they are free-threshing). Grain plumpness, however, is a problematic character for identifying domestication due the existence of involucres with 1, 2 or more grains on a single ear (Godbole, 1925). Recent gene linkage studies, show that all of these domestication traits are controlled by a single set of linked gene loci (Poncet *et al.*, 2000).

Most impressions in the examined pottery consist of macerated plant material, including a large number of bristle fragments, which match those of *Pennisetum* as well as some larger fragments of spikelets (figures 4 and 6) and involucres (figures 5 and 7), including rachis stalks, comparable to domesticated pearl millet. All of this material is consistent with the use of the by-products of threshing and winnowing as ceramic temper. Characteristic bristles with uniseriate hairs are also visible (figure 8).

The presence of domestication traits such as involucre raches therefore attest to fully domesticated Pearl Millet in Dhar Nema by the Classic Tichitt period. This provides a *terminus post quem*, with the beginnings of cultivation necessarily earlier.

The absence of other plant species is to be expected as weeds are rarely incorporated in the crop-processing by-products of pearl millet on account of its dense spikes which are readily harvested alone (Reddy, 1994, 1997). Threshing is most likely to have occurred in the vicinity of cultivation to minimize transport of bulky harvested plants, while the ready availability of win-

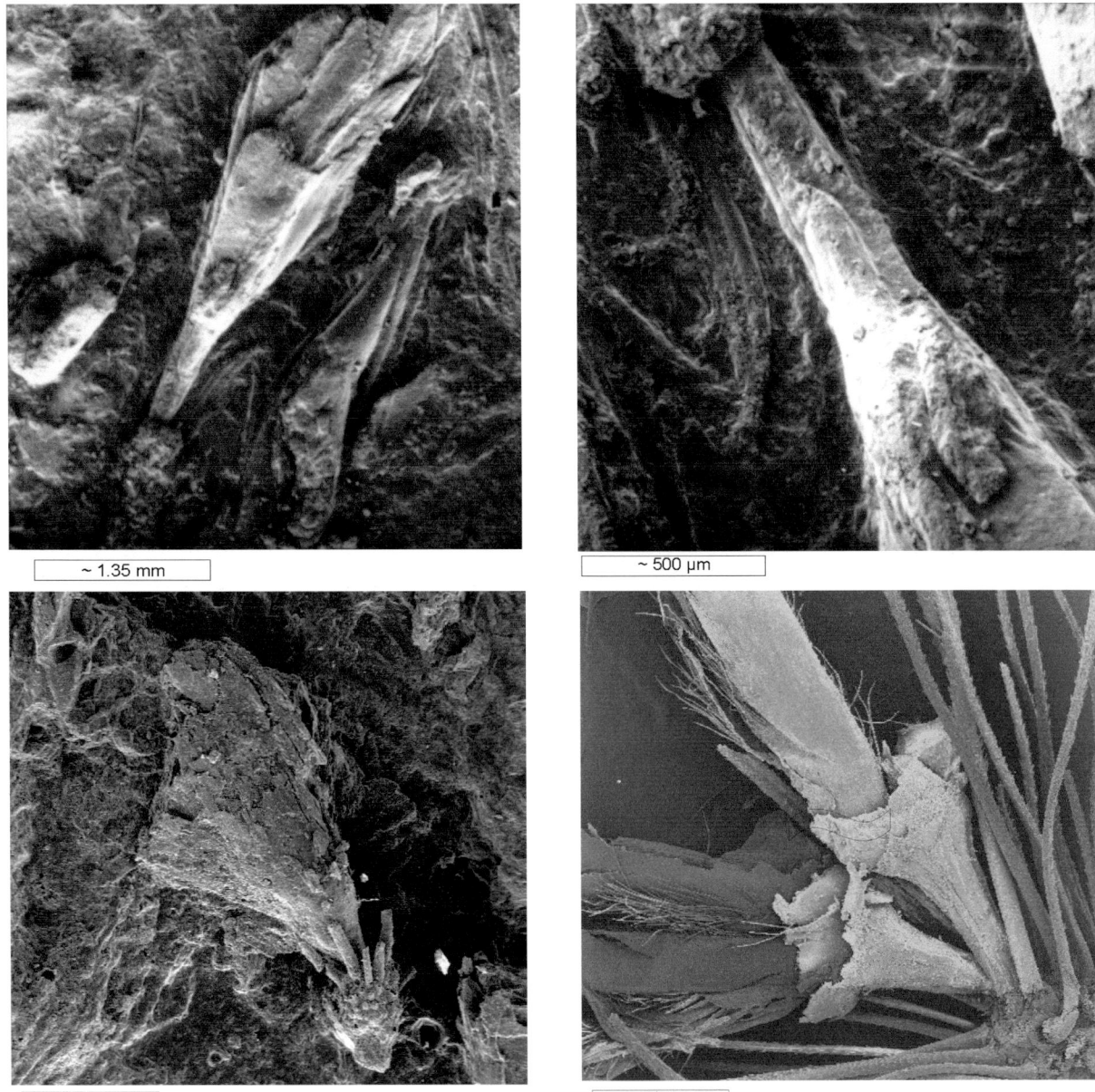

Figure 4 (page 74, first row): Impression of two partial spikelets, with bristles (Left), with Close-up of spikelet base, attached to involucre, with some bristles visible (right), from Djiganyai (Dj-00 context 2), directly dated specimen, 3260 bp.

Figure 5 (page 74, second row): Impression of a spikelet and involucre apex, showing rows of radiating bristle bases sherd from Djiganyai (Dj-00 context 6), directly dated specimen 3370 bp (Left); Modern reference material of Pennisetum glaucum (right), two spikelets in a single involucre arising above bristles. Note fine hairs on the margins of the lemma and palea.

Figure 6 (page 75, first row): Impression of paired pearl millet spikelets (mainly lemmas), separated from involucre (left), from OBK(1). Arrow indicates location of hairs shown in close-up (right). Direct AMS date: 2430 bp.

Figure 7 (page 75, second row)): Impression of involucre apex, with bristles, surrounding spikelets (left), from excavated tuyère, OBK, submitted for AMS date; and impression of involucre apex, with bristle bases, and fragment spilet chaff (right), from same.

Figure 8 (page 75, third row): Impression of bristle fragment, showing serrate hairs, from OBK(1), directly dated specimen, 2430 bp (left); Close-up of bristle in modern material, showing the numerous unicellular serrate trichomes on the bristle surface (right).

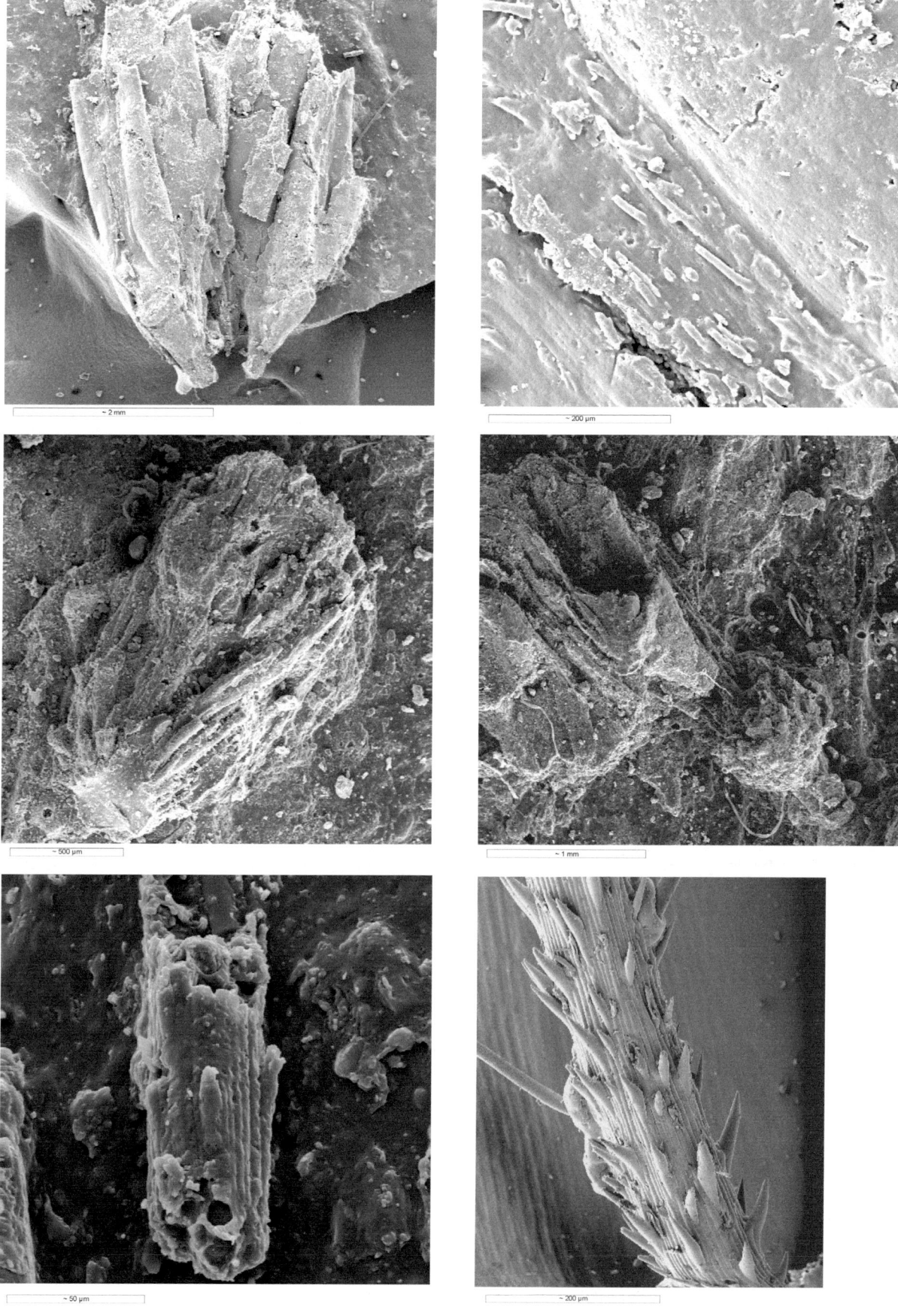

nowing waste is likely to have been near to pottery production sites. This implies cultivation and pottery making in the same locales, such as in the vicinity of Djiganyai. The millet impressions in the ceramics from Bou Khzama represents the same range of threshing and winnowing by-products as that described for Djiganyai.

Recent isozyme surveys of wild populations and domesticated varieties have identified only a limited number of modern wild populations that are close to the domesticated crop (Tostain, 1992, 1998), notably two foci, one in the far west of Africa (Mauritania) and the other in the region west of Lake Chad. For the eastern domestication, the *terminus ante quem* is at present provided by finds in India from the first half of the second millennium BC, which remains earlier than the limited evidence in Africa (Fuller, 2003). This implies that the present evidence, like earlier reports (Amblard & Pernes, 1989; Neumann, 2003) thus provides a minimum domestication age of ca. 1700 BC for the western of these pearl millet centres of origin.

7 References

Amblard, S. & J. Pernes, 1989. The identification of cultivated pearl millet (Pennisetum) amongst plant impressions on pottery from Oued Chebbi (Dhar Oualata, Mauritania). *The African Archaeological Review* 7, pp. 117-126.

Brunken, J., De Wet, J.M.J. & J.R. Harlan, 1977. The Morphology and Domestication of Pearl Millet. *Economic Botany* 31, pp. 163-174.

Fuller, D.Q., 2003. African crops in prehistoric South Asia: a critical review. In: K. Neumann, A. Butler & S. Kahlheber (ed.), *Food, Fuel and Fields. Progress in Africa Archaeobotany* (= Africa Praehistorica 15). Heinrich-Barth-Institut, Köln, pp. 239-271.

Godbole, S.V., 1925. *Pennisetum typhoideum*. Studies on the Bajri Crop. I. The Morphology of *Pennisetum typhoideum*. *Memoirs of the Department of Agriculture in India* (Agricultural Research Institute, Pusa) 14(8), pp. 247-268.

MacDonald, K., Vernet, R., Fuller, D. & J. Woodhouse, 2003. New Light on the Tichitt Tradition: A preliminary report on survey and excavation at Dhar Nema. In: P. Mitchell, A. Haour & J. Hobart (ed.), *Researching Africa's Past. New Contributions from British Archaeologists* (= Oxford University School of Archaeology Monograph No. 57). Oxford University School of Archaeology, Oxford, pp. 73-80.

Neumann, K., 2003. The late emergence of agriculture in Sub-Saharan Africa: archaeobotanical evidence and ecological considerations. In: K. Neumann, A. Butler & S. Kahlheber (ed.), *Food, Fuel and Fields. Progress in African Archaeobotany* (= Africa Praehistorica 15). Heinrich-Barth-Institut, Köln, pp. 71-92.

Poncet, V., Lamy, F., Devos, K.M., Gale, M.D., Sarr, A. & T. Robert, 2000. Genetic control of domestication traits in pearl millet (Pennisetum glaucum L., Poaceae). *Theoretical and Applied Genetics* 100, pp. 147-159.

Reddy, S.N., 1994. Plant Usage and Subsistence Modeling: An Ethnoarchaeological Approach to the Late Harappan of Northwest India. Ph.D. thesis, University of Wisconsin. Ann Arbor Michagan: University Microfilms.

Reddy, S.N., 1997. If the threshing floor could talk: integration of agriculture and pastoralism during the Late Harappan in Gujarat, India. *Journal of Anthropological Archaeology* 16, pp. 162-187.

Stuiver, M., Reimer, P.J., Bard, E., Beck, J.W., Burr, G.S., Hughen, K.A., Kromer, B., McCormac, G., Van der Plicht, J. & M. Spurk, 1998. INTCAL98 Radiocarbon Age Calibration, 24,000-0 cal BP. *Radiocarbon* 40(3), pp. 1041-1083.

Tostain, S. 1992. Enzyme diversity in pearl millet (Pennisetum glaucum L.) 3. Wild millet. *Theoretical and Applied Genetics* 83, pp. 733-742

Tostain, S. 1998. Le mil, une longue histoire: hypotheses sur sa domestication et ses migrations. In: M. Chastanet ed., *Plantes et paysages d'Afrique. Une histoire a explorer*. Editions Karthala and Centre de Recherches Africaines, Paris, pp. 461-490.

Figs and their importance in the prehistoric diet in Gran Canaria Island (Canary Isles)

J. Morales
School of Archaeology & Ancient History, University of Leicester, United Kingdom

T. Delgado
El Museo Canario, Las Palmas de Gran Canaria (Spain)

1 Introduction

The archipelago of the Canary Islands is located in the Atlantic Ocean opposite the coasts of Africa, at 28° north latitude and only 100 km from the Sahara (figure 1). The earliest human occupation started with the arrival of people from Northern Africa around 500 BC. This period of occupation ended with the Spanish conquest in the 15th century AD (Morales Mateos, 2003b). The earliest immigrants brought with them several domesticated plants (wheat, barley, lentils, field bean) and animals (sheep, goat, pig, dog). Whether the fig, a fruit of Mediterranean origin (Zohary & Hopf, 2000), was also introduced to the Canary Islands at this time or much later is a matter of some debate. Early European ethnohistorians (Abreu, 1977) considered that Majorcan missionaries introduced the fruit in the 14th century, but the first European travellers and explorers, who came to the islands in the 14th and 15th centuries AD already mention the exploitation of this fruit. These texts emphasise the importance of this fruit in the diet of the pre-Hispanic people in Gran Canaria. In fact, Spanish conquerors were conscious of this and cut down fig trees as a strategy to oppress the population (Morales Padrón, 1993).

The purpose of the present paper is to clarify the

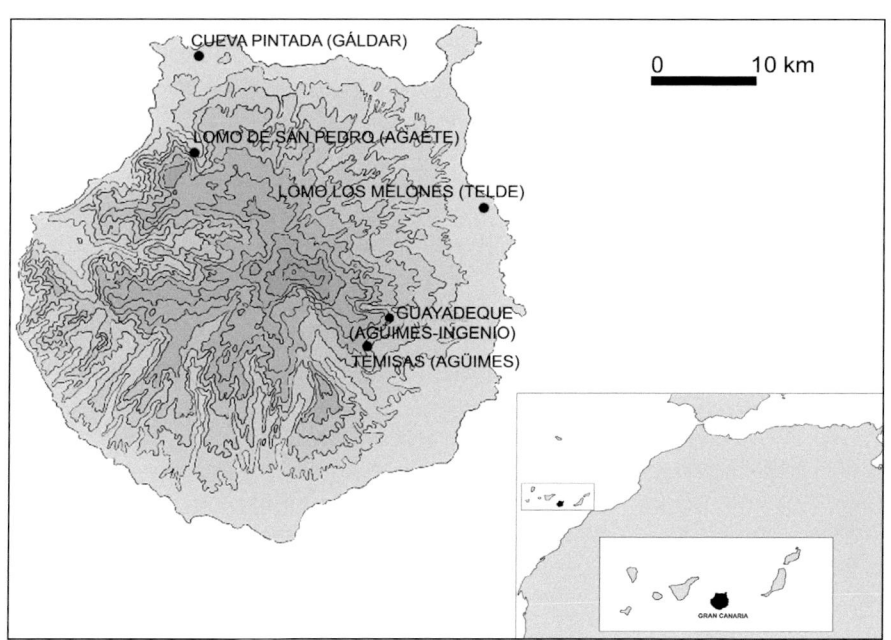

Figure 1: The Canary Islands and the archaeological sites mentioned in the texts.

Table 1: Fig varieties and sexual characteristics.

	fig varieties	flower morphology	presence of wasp	*profichi* (April to June)
wild plants	Caprifig	male flower with five stamens; female flower with short style	male and female wasp (wasps' eggs can develop only in short style flowers)	production of pollen in the male flower but no fecundation, fig not edible
cultivated plants	Smyrna	male flower not presented; female flower with long style	only female wasp, from a caprifig syconia (wasp can not lay their eggs in long style flowers)	fecundation by pollen from a caprifig, edible fig but it can not be dried
	San Pedro	male flower not presented; female flower with long style	only female wasp, from a caprifig syconia (in the mammoni crop)	not fecundation (parthenocarpic), edible fig but it can not be dried
	common type	male flower not presented; female flower with long style	no wasps involved	not fecundation (parthenocarpic), edible fig but it can not be dried

importance of this crop for the pre-Hispanic population on the island. The research presents the archaeobotanical evidence for figs at different sites in Gran Canaria, from both domestic structures and human skeletons.

2 Ethnohistorical evidence

The ethnohistorical texts referring to figs are quite helpful. They were written between the 14th and 17th centuries by European explorers and the first colonists, who collected information about pre-Hispanic societies. The first writings from Italian explorers in 1341 AD indicate the presence of figs in pre-Hispanic houses from Gran Canaria (Boccacio, 1998). Other authors report on a daily diet composed of goat products, barley and dried figs (Morales Padrón, 1993).

Figs have long been exploited for their nutritional value, both in their fresh and dried state. In prehistoric times most figs were eaten dried. According to some texts the figs were first dried in the sun on rush mats. They were then stored in cylindrical receptacles made of plant fibre called *carianas*. Finally, they were pressed into loaves or threaded on strings, which allowed the fruits to be kept for long periods (Abreu, 1977).

At the end of the 16th century, Abreu documented the different names given to figs by the aboriginal descendants. They were called *arehormaze* when they were fresh; *tehaunenen* when they were already dried and had their sweet flavour (Abreu, 1977). This linguistic complexity is believed to reflect the important socio-economic status of the fruit (Lorenzo Santos, 1993).

After the European conquest, the cultivation of the fig tree held a very important role in the economy of the Canary Islands; it served as food as well as fodder and fuel and was also for medicinal purposes. Varieties of fig from the mainland were introduced in all the islands and became a daily component of the diet. In famine times, as it is documented in the beginning of the 18th century AD, figs became a staple food (Viera, 1982).

Nowadays, 35 varieties have been classified on the Canaries taking into account the colour and the form of the fruit (Perdomo, 2004). Cultivation takes place in different ecological environments that range from the sea-coast to the high mountains. Fig trees are cultivated on the borders of roads and fields, or confined to enclosures surrounded by stones. These circular structures protect the tree from the wind and livestock. In arid areas, they are cultivated in dry river-beds to exploit the periodical rains. However, cultivation is decreasing due to the changes in the way of life of the Canarian population. The import and presence of foreign food, the abandonment of agricultural tasks, the changing tastes and less apppreciation for traditional food have provoked a reduction in its exploitation and consumption.

3 Botany of the fig

The reproductive biology of the fig is complex. In a strict botanical sense fig "fruits" are actually inflorescences called syconia. They are hollow, fleshy structures composed of peduncular tissue, lined on the inside with hundreds of minute flowers. At one end is a small opening (ostiole) lined with dense, overlapping scales. Each tiny flower consists of a five-parted calyx and an ovary with a short style. Following pollination and fertilization the ovaries develop into minute one-seeded drupelets with an endocarp surrounding the seed (Condit, 1955; Galil, 1977). This is the fig remain

mamme (November to April)	*mammoni* (June to November)	seed characteristic
not fecundation (pollen is not viable), small fig not edible	fecundation by pollen from profichi (mammoni does not produce pollen), small and hardly edible fig	fertile seed (embryo and endosperm developed)
not developed	fecundation by pollen from a caprifig, edible fig (main crop) and it can be dried	fertile seed (embryo and endosperm developed)
not developed	fecundation by pollen from a carprifig, edible fig (main crop) and it can be dried	infertile seed profichi (hollow seed), fertile seed mammoni (embryo and endosperm developed)
not developed	not fecundation (parthenocarpic), edible fig (main crop) and it can be dried	infertile seed (hollow seed)

that archaeobotanists normally retrieve in archaeological digs (Zohary & Hopf, 2000).

Most species of figs in the world are monoecious (the genus *Ficus* has more than 600 species) and have male and female flowers within a single syconia. In *Ficus carica*, the male flowers and short-style female flowers are produced in syconia on "male" caprifig trees, while the long-style female flowers are produced in syconia on separate cultivated "female" trees (Smyrna, San Pedro and Common type). In fact, they are the two sexual forms of the same species, of which the female morph is known as the true fig and the male as caprifig.

Caprifigs of *Ficus carica* produce three crops of syconia per year: the profichi that ripen in early summer; the mammoni that ripen in the fall; and the mamme that over-winter on the tree and ripen in spring. Fig pollen is transferred from male flowers (only the profichi crop produces pollen) to female flower pistils by an insect called a fig wasp (*Blastophaga*). These wasps live, lay their eggs and feed within the fig syconia. Only the winged female wasps emerge and leave the fig, through the ostiole, (wingless males cannot leave the syconia). The female wasps fly to new syconia to oviposit eggs in some of the pistils, one egg per ovary. However, the profichi caprifig has many male flowers near the ostiole, and the pollen from these flowers are carried by the female wasps to the next syconium, in this way pollinating the ovaries. Although caprifig syconia incubate and perpetuate these tiny fig wasp, seeds may also develop in the short-style flowers. This is especially true of mammoni syconia in which pollination results in some ovaries (without wasps) developing seeds with viable embryos.

In the case of fruit setting for Smyrna figs, these trees produce only two crops of syconia annually, a breba crop (profichi) that ripens in early summer and a second (mammoni) crop that ripens in autumn. For fruit development to occur, the Smyrna fig needs to be pollinated with pollen from the caprifig. Because male flowers are not produced by Smyrna trees, branches with profichi figs of the caprifig tree are collected by the farmer and hung within the fig tree canopy in the late afternoon. The next morning the fig wasps emerge from the profichi figs and then transfer pollen to the young Smyrna pistils. Enough fertilization takes place to promote Smyrna fruit development. This process is called "caprification".

San Pedro fig types produce two crops of figs each year. The profichi crop is parthenocarpic (it proceeds directly without pollination and fertilization), but the second crop, like the Smyrna fig, requires caprification.

The common-type fig is nowadays the most extended cultivar of *Ficus carica* in the world. This cultivar of *Ficus carica* does not require caprification. Unlike the Smyrna and San Pedro types, the syconia remain on the branches and ripen without wasp pollination. The drupelets inside develop parthenocarpically, and the endocarps are generally hollow and without fertile seeds. Occasionally, female trees may produce apomictic seeds without pollination and fertilization: the seed embryo develops from an unfertilized egg (or another cell within the embryo sac), or from cellular tissue surrounding the embryo sac. Apomictic seeds enable the propagation of choice edible fig cultivars without the transmission of a mosaic virus disease through cuttings (Condit, 1955; Galil, 1977).

This complex reproduction biology allows us to recognize some botanical features of figs that can be preserved in archaeological material, which we can use to distinguish between different varieties of fig (table 1). We consider that it can be possible to distinguish between the seeds (endocarp[1]) of wild and cultivated fig varieties, at least between the common type fig and the rest. As we have seen, common type fig seeds are not pollinated and thus are hollow; while caprifig, Smyrna and San Pedro (only the mammoni crop) types require pollination for fruit development. The seeds of these types possess an embryo and endosperm, which can be easily seen through a binocular microscope. It is necessary to bear in mind that there are exceptions to this, such as unpollinated Smyrna seeds and San Pedro profichi; however, we regard it a useful criterion to use when samples contain a significant number of seeds.

4 Aims

Until recently the only archaeological evidence of fig consumption in Gran Canaria were dried fruits (figure 2) recovered from various collective silos in Acusa and Guayadeque (Jimenez, 1952). However, their exact context is not known, since they were collected randomly in the early part of the 20th century. In Tenerife island, the only archaeobotanical remains of fig consist of charcoal, recovered from the archaeological sites of Las Palomas and Don Gaspar in levels dating to around 560 AD (Machado et al. 1997). This general absence of data is due to the lack of proper recovery techniques that take into account the nature and size of the remains. Fig seeds measure c. 1.5 mm, which makes them difficult to detect without sieving and microscopic analysis.

With the aim of investigating the importance of figs in the prehistoric Gran Canaria diet, two complementary sources of information have been used: archaeobotanical materials recovered from pre-Hispanic domestic structures and human remains. The material comes from five different sites (figure 1).

5 The archaeobotanical evidence

5.1 General description of the sites

For this study, the structures from two archaeological sites have been sampled. A sampling strategy has been employed to obtain a representative assemblage of plant macro-remains. This recovery method has provided a large amount of carpological evidence that can be related to the pre-Hispanic time in Gran Canaria.

The site Lomo los Melones is located at the east coast of Gran Canaria. The climate is characterized by extreme dryness and by strong and constant North-East winds for most of the year. The excavated structure is domestic. It is a cross-shaped house built of stone. It has two occupation levels, the second one dating to

Figure 2: Dried figs from Guayadeque

cal. AD 1290-1430 (Mireles et al. in preparation).

The Archaeological Park Cueva Pintada is located in the north-west of Gran Canaria, one hundred metres above sea level. It is situated on a volcanic slope oriented to the South-West, protected from the dominant North-East winds. The climate is characterized by low levels of rainfall (200-300 mm per year). The settlement occupies 4500 square metres. It is composed of seven artificial caves and at least sixty features, most of which are individual or collective domestic structures. In the present study two domestic structures have been investigated. These structures have not been dated yet, however, it is estimated that structure (1) dates between the 6th and 13th century AD, while structure 2 dates from the 13th to 16th century AD (Saenz, I., personal comments).

5.2 Material and Methods

To recover the plant remains from both sites water flotation was used. The float was collected using a 0.1 mm sieve and, when dry, was sieved through 2, 1, 0.5 and 0.25 mm meshes, prior to sorting the plant material. The resulting residues were then examined under a binocular microscope. All identifiable and quantifiable plant taxa were extracted.

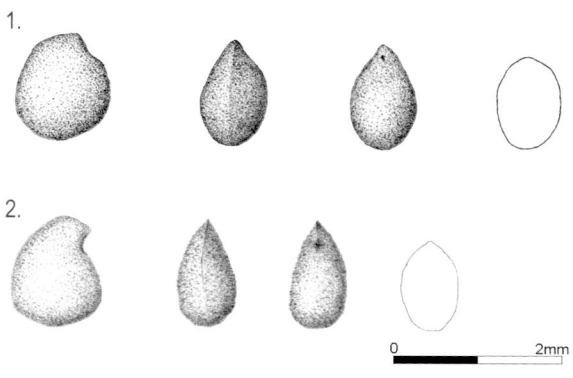
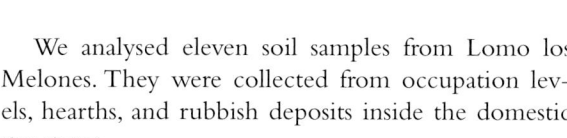

Figure 3: Ficus carica *seeds from Lomo Los Melones (1) and Cueva Pintada (2).*

Figure 4: Hollow charred Ficus carica *seed recovered in Cueva Pintada.*

We analysed eleven soil samples from Lomo los Melones. They were collected from occupation levels, hearths, and rubbish deposits inside the domestic structures.

A total of 101 samples have been analysed from the Archaeological Park Cueva Pintada. In structure (1) 245 litres of sediment was collected from one occupation level. We decided to take samples of the whole floor due to its archaeobotanical richness. In structure (2) 302 litres was collected from three different occupation levels.

In order to identify the fig seeds we have used both reference material from the island and also literature on archaeobotanical remains. All the recovered seeds in the archaeological sites are charred, while all the seeds retrieved from the human skeletons are desiccated.

Fig seeds are laterally compressed, ovate in outline and pointed at the apex (figure 3). They have a small circular hilum below the apex, a very characteristic feature of *Ficus carica*. The seeds display a fairly large variation in size and shape: the greatest dimension of 100 seeds from Archaeological Park Cueva Pintada varies from 0.7 to 1.6 mm. In addition to the seeds, one fruit peduncle, possibly from a fig fruit, was found in structure (1) from Cueva Pintada. The identification of the peduncle is based on its striated surface, size and also on the common occurrence of many fig seeds in the sample.

In the Canary Islands it must be noted that the existence of the caprifig is not mentioned in the ethnohistorical texts. Although in some cases several ethnohistorical authors allude to wild fig trees, we can interpret this reference as a way of marking the difference with the fig variety that grew on the mainland. We have studied fig remains from Gran Canaria island, and we have not recovered any pollinated seed. All the endocarps retrieved from archaeological sites and human skeletons are hollow, which means that they were not pollinated (figure 4). In the case of the dry figs collected in the silos, they lack male flowers and short style female flower??. *Blastophaga* insect remains were not found either. Thus, we conclude that during prehistory the pre-Hispanic people cultivated the Common type fig.

5.3 Results

In order to consider the results obtained from this study, we have to bear in mind the taphonomic factors affecting sample composition.

Firstly, geographical conditions may affect the preservation of the remains. Lomo los Melones site is located on the sea line and it is under the influence of sea winds. In different archaeological excavations in the Canaries we have discovered that in sites close to the sea (less than 10 metres) preservation of charred remains is very low. We have not yet identified the causes of this fact, but we think sea salts can be one of the factors. This statement may explain the scarce volume of charred material in this site, where each float had a volume of less than 5 ml. However, in Cueva Pintada site most of the floats had a volume of 25-30 ml.

Secondly, the difference between the amount and volume of the sample collected in both sites leads to differences in results. In Lomo los Melones we studied 11 samples, whereas in the Archaeological Park Cueva Pintada we studied 101.

Thirdly, the processing and use of the fig influences the outcome. Figs are eaten fresh or dried and fire is not generally used for the processing of the fruit[2], but fig seeds were recovered carbonized in both sites. We have to consider different possibilities to explain the carbonization of the seeds:

- Accidental charring during the handling of the fruits;
- Cleaning activities: unsuitable figs may have been thrown into the fire;

Table 2: Presence of Ficus carica in Lomo los Melones

Occupation level	No. samples	No. seeds	Sediment volume (litres)	No seeds/sample	No seeds/1 litre of sediment
I	3	7	14.5	2.3	0.5
II	8	129	51.5	16.1	2.5
Total	11	136	66	12.4	2.06

Table 3: Presence of Ficus carica in structure 1 from Cueva Pintada

Occupation level	No. samples	No. seeds	Sediment volume (litres)	No. seeds/sample	No. seeds/1 litre of sediment
II	49	903	245	18	3.6

Table 4: Presence of Ficus carica in structure 2 from Cueva Pintada

Occupation level	No. samples	No. seeds	Sediment volume (litres)	No. seeds/sample	No. seeds/1 litre of sediment
II	39	178	245	4.5	0.7
III	8	15	55	1.9	0.3
IV	1	2	2	2	1
Total	48	195	302	4.1	0.6

- Using dung or faeces as fuel: most of the seeds of the figs pass through the intestinal canal (in humans and domestic animals) largely undamaged;
- Some processing of the fig, such as drying in the oven or cooking to prepare jam or another kind of dish.

We will have not listed all the possible taphonomic factors and because of these complex processes our analysis may not be fully accurate. Despite these problems, it is clear that the fig seed results yield significant evidence of the use of this fruit during the occupation of both sites.

In Lomo los Melones, we found seeds in six of the samples. Regardless of the fact that the samples were relatively small (table 2), the large number of scattered fig seeds indicates the importance of their presence in the domestic structure. There are also higher frequencies of fig seeds in comparison to other crops. Moreover, fig charcoal has been identified at this site, confirming the importance of this tree in the economy of the people (Mireles et al. in preparation).

In the archaeological Park Cueva Pintada, we must point out the recovery of fig seeds in 95 of the 101 samples collected. It means fig seeds are present in almost all the samples. However it has to be underlined that seed density is higher in structure (1) than in structure (2) (tables 3 and 4). The large amount of fig seeds found in both sites emphasises the importance of this crop in relation to others. Although the number of seeds is rather low, considering that one fig can contain up to 2000 seeds, the constant presence of those seeds in the different domestic structures indicates the relevance of the fruit in the diet of the inhabitants.

6 The bioanthropological evidence

The study of the oral health of a large prehistoric skeletal series from Gran Canaria (a minimum number of 585 individuals) has shown the presence of fig seeds in four teeth belonging to four individuals. Fig seeds were preserved by desiccation and they were inside the pulpar cavity, exposed because of dental caries in three of the four and because of severe wear and tear in the fourth. This part of the study concentrates on the results of dental anthropology which was possible due to the excellent preservation of the teeth. These results allowed us to prove the direct consumption of figs.

6.1 Site descriptions

The skeletal remains analysed in this paper belong to the archaeological collection of El Museo Canario. They were retrieved randomly at the beginning of the 20th century, without any systematic method and the exact archaeological context of the material was not recorded. The sample consists of four crania from three different sites (figure 1). Two of them were obtained from the Guayadeque Valley (Agüimes-Ingenio), where there are several artificial (man-made) and natural caves, used for accommodation, grain storage and collective burial, all of which indicates a dense population living in this area in prehistoric times.

Figure 5: Ficus carica *seed inside pulpar cavity. Human skeleton from Guayadeque.*

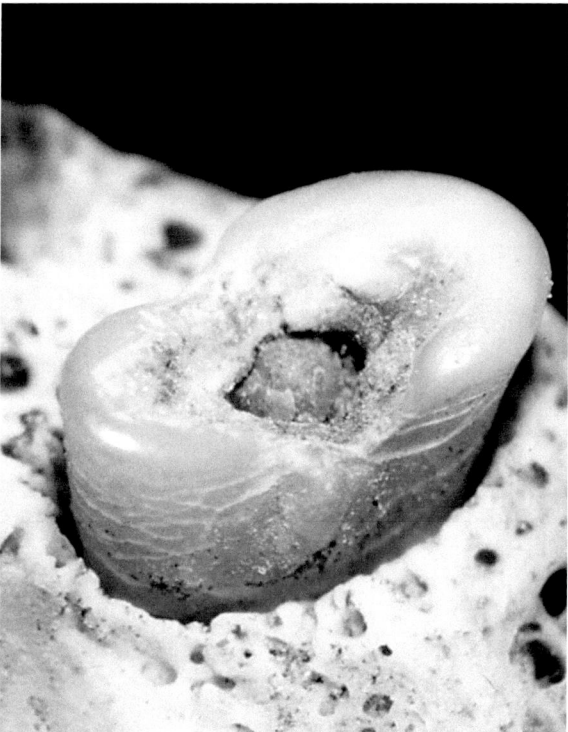

Figure 6: Ficus carica *seed inside pulpar cavity. Human skeleton from Temisas.*

The third cranium was recovered from Temisas (Agüimes), a large settlement of caves, used by ancient Canarians as domestic dwellings and food storing areas. The skull was found together with some skeletal remains in a natural cave close to those domestic areas.

Finally, the fourth cranium was recovered from the site known as Lomo San Pedro (Agaete), a set of burial caves. The excavation of one of these caves shows a collective burial space, where five individuals were deposited, most of them wrapped in plant fibre.

None of the skulls are dated directly, but for Guayadeque Valley site there are some radiocarbon dates, cal AD 520-770 and cal AD 780-1020, obtained from mummy wrappings and from wood (Martín, 2000), and 920 ± 70 BP (1030 d. C.) and 750 ± 50 BP (1180 d. C.), obtained from anthracological materials recovered from in a domestic area (Morales *et al.* 2001).

6.2 Materials and methods
According to the morphological criteria of the skull and jaws (Ubelaker, 1989), it could be established that the individual from Lomo de San Pedro and one from Guayadeque were males and the one from Temisas female. The sex of the second individual from Guayadeque could not be determined.

The degree of the ectocranial suture closure (Meindl & Lovejoy, 1985) and the dental wear and tear (Brothwell, 1972) indicate that the males from Lomo de San Pedro and Guayadeque and the female from Temisas were aged between twenty-five and forty-five, and the second individual from Guayadeque was in the age group seventeen to twenty-five.

The maxillas and mandibles were first examined with the naked eye under natural light conditions and finally with a binocular microscope. Before removing the seeds from the teeth, an internal identification of the pulpar cavity was carried out in order to avoid any loss of information during the removal process. Isolated analysis of one seed was not possible since its removal would most probably have caused its destruction (figure 5).

6.3 The results
A more indirect line of evidence is that of the occurrence of dental caries. This decay is caused by the local demineralisation of dental hard tissues brought about by the organic acids formed when bacteria metabolise the carbohydrates from the diet (Hillson,

1996). Therefore, diet plays an important role in the presence and frequency of dental caries. A diet rich in carbohydrate foods (cereals, fruits like figs, etc.) leads to significant rates of this pathological condition. Oral pathological studies have shown a high rate of caries among the aboriginal communities of Gran Canaria (Delgado, 2004). Remarkably, this community exhibits a caries frequency of 17.3 per cent (1164 carious teeth/6730 observed teeth). If we consider the origin of this dental decay, these results indicate the major role of carbohydrates in their diet.

The population was largely reliant on cereals, as is demonstrated by the studies of dental pathologies, especially of dental caries, but also by data obtained from other bioanthropological analysis, for example trace elements in bones (Delgado, 2004; Velasco, 1999). This shows the major role that agricultural products played in the diet of these human groups and therefore the great importance of agriculture in the subsistence economy of the pre-Hispanic population of Gran Canaria. But other plant resources from agricultural production like figs, could also be responsible for the state of dental health found in this population, characterized by a high rate of dental caries. Some analyses carried out on human groups that had a large intake of fermentable carbohydrates such as dates have shown high frequencies of dental caries (Littleton & Frohlich, 1989; Nelson et al., 1999).

Apart from cereals and fruits such as dates, figs can also contribute to high levels of dental caries. Two characteristics explain the significant cariogenic role of figs. Firstly, its chemical composition, rich in simple or low-molecular-weight carbohydrates enables rapid metabolisation by oral bacteria. Secondly, the sticky fig seeds adhere longer to teeth than other food. Thus, the natural oral cleaning mechanism is less efficient. These characteristics, in addition to the small size of the seeds, explain why they remained inside the dental pulpar cavity (figure 6). The high incidence of dental caries and the finding of carpological evidence of figs inside pulpar cavity of some teeth, suggest the important role of this fruit in the diet and so in the economy of this human group. Considering this and the fact that this fruit was consumed by people of both sex and different age ranges, we can assume that the fig was a staple food in the diet of the pre-Hispanic people of Gran Canaria.

7 Conclusion

It is clear from our evidence that the fig was introduced into the Canary Islands during the pre-Hispanic period. The earliest evidence for fig dates from the 6th-13th centuries in Gran Canaria (and AD 560 in Tenerife). The recovery of significant amounts of fig seeds in two archaeological sites, and their ubiquity in the samples confirms the importance of this fruit for the pre-Hispanic population, at least by the late pre-Hispanic period. Moreover, the presence of fig seeds inside the pulpar cavity of human teeth offered further evidence of figs as human food in pre-Hispanic times. Further, indirect evidence comes from the high occurrence of caries in the human teeth analysed. The status of figs in the diet is important considering that they were not only eaten fresh, but also dried. Drying the fruits increases the cariogenic properties of figs since simple sugars increase and their consistency becomes more sticky. Our findings emphasise the importance of using efficient recovery methods for plant materials from archaeological deposits and the thorough study of dental pathologies. The results demonstrate the need for a multidisciplinary approach to the study of the past, since information obtained from different sources permit a greater and better knowledge of ancient human lifestyles.

8 Notes

1. For ease of reading we use the term 'seeds' when referring to what botanically are 'endocarps'.
2. In the Canaries few years ago, during wet autumns, figs were dried in ovens. This fact makes figs not so sweet, but they can be preserved for a longer time than dried under the sun (Lorenzo, 1993).

9 References

Abreu, J., 1977. *Historia de la conquista de las siete islas de Canaria* [1634] Goya. Santa Cruz de Tenerife.

Boccacio, G., 1998. De Canarias y de las otras islas nuevamente halladas en el océano allende España [1341] In: J. A. Delgado Luis (ed.), *Colección A través del Tiempo, 16*. Excmo. Ayuntamiento Puerto de la Cruz, Excmo. Ayuntamiento Villa de Orotava. La Laguna.

Brothwell, D.R., 1972. *Digging up bones*. Trustees of the British Museum, London.

Condit, I.J., 1955. Fig Varieties: A Monograph. *Hilgardia* 11, pp. 323-538.

Delgado, T., 2004. *Economía, salud, nutrición y dieta de la población prehistórica de Gran Canaria. La aportación de la antropología dental*. Unpublished Ph.D. Universidad de Las Palmas de Gran Canaria.

Galil, J. 1977. Fig Biology. *Endeavour* 1, pp. 52-56.

Hillson, S., 1996. *Dental Anthropology*. Cambridge University Press, Cambridge.

Jiménez, S., 1952. El trigo uno de los alimentos de los grancanarios prehispánicos. *Revista de Historia* 18 (98-99), pp. 205-213.

Littleton, J. & B. Frohlich, 1989. An analysis of dental pathology and diet on historic Bahrain. *Paléorient* 15/2, pp. 59-84.

Lorenzo Santos, N., 1993. Proceso de secado y prensado de higos en Canarias. *Tenique* 1, pp. 105-121.

Machado, M.C., Arco Aguilar, M. del C., Vernet, J.L. & J.M. Ourcival, 1997. Man and vegetation in northern Tenerife (Canary Islands, Spain), during the pre-Hispanic period based on charcoal analyses. *Vegetation History and Archaeobotany* 6, pp. 187-195.

Martín, E., 2000. Dataciones absolutas para los yacimientos de Risco Chimirique (Tejeda) y playa de Aguadulce (Telde). *Vegueta* 5, pp. 29-46.

Meindl, R.S. & C.O. Lovejoy, 1985. Ectocranial suture closure: a revised method for the determination of skeletal age at death based on the lateral-anterior sutures. *American Journal of Physical Anthropology* 68, pp. 57-66.

Mireles, F., Machado, M.C. & J. Morales Mateos, (in preparation). Evidencias arqueobotánicas del yacimiento de Lomo Los Melones (Telde, Gran Canaria). *Tabona*, XI.

Morales Mateos, J., Alberto, V. & J. Velasco, 2001. Evidencias carpológicas de la actividad agrícola en la prehistoria de Gran Canaria: cebada, trigo y lentejas. Excavaciones en la antigua ermita de San Antón. *Tabona* 10, pp. 195-211.

Morales Mateos, J., 2003. Islands, plants and ancient human societies: a review of archaeobotanical works on the prehistory of the Canary Isles. In: Neumann, K., Butler A. & S. Kahlheber (eds.), *Food, fuel and fields. Progress in African archaeobotany*. Heinrich-Barth-Institut, Köln, pp. 139-148.

Morales Padrón, F., 1993. *Canarias. Crónicas de su conquista*. Cabildo Insular de Gran Canaria. Las Palmas de Gran Canaria.

Nelson, G.C., Lukacs, J.R. & P. Yule, 1999. Dates, caries, and early tooth loss during the Iron Age of Oman. *American Journal of Physical Anthropology* 108 (3), pp. 333-343.

Perdomo, A., 2004. La polifacética higuera: buena fruta, buena sombra…y mejor "pasto" para el ganado. *El Pajar. Cuaderno de Etnografía Canaria* 18, pp. 61-65

Ubelaker, D.H., 1989. *Human skeletal remains*. Manuals of archaeology 2. Smithsonian Institution, Taraxacum-Washington.

Velasco, J., 1999. *Canarios. Economía y Dieta de una sociedad prehistórica*. Cabildo Insular de Gran Canaria. Las Palmas de Gran Canaria.

Viera, J., 1982. *Diccionario de Historia natural de las Islas Canarias* [1810]. Mancomunidad de Cabildos, Madrid.

Zohary, D. & M. Hopf, 2000. *Domestication of Plants in the Old World. The origin and spread of cultivated plants in West Asia, Europe and the Nile Valley*. UCL Press London and Smithsonian Institution Press, Washington, D. C. Oxford.

The impact of hunter/gatherers on the vegetation in the Central Sahara during the Early Holocene

A.M. Mercuri
Laboratorio di Palinologia e Paleobotanica, Dipartimento del Museo di Paleobiologia e dell'Orto Botanico, Università di Modena e Reggio Emili, Italya

E.A.A. Garcea
Laboratorio di Archeologia, Dipartimento di Filologia e Storia, Università di Cassino, Italy

The Uan Tabu rockshelter is located in the Wadi Teshuinat, the largest wadi of the central Tadrart Acacus mountains in south-western Libya. This paper reports carpological and pollen data from the Early Holocene sequence of Uan Tabu. Data were integrated and compared with that of other sites in the area in order to draw a general picture on plant exploitation and landscape in the Early Holocene. Three main 'plant assemblage zones', corresponding to different concentrations and floristic lists of seed/fruit and pollen, were distinguished showing synchronous botanical changes which are correlated with the archaeological record. Since the beginning of the Holocene, the environment of the Tadrart Acacus has witnessed many changes brought about firstly by a changing climate and later by the influx of settlers to the region. In the cultural sequence, from the Early to the Late Acacus phases, behavioural changes towards increased plant harvesting and processing were observed. During the Late Acacus (ca. 8300-7490 BC; 8800-8600 uncal. years bp) plants were accumulated not only for food, but also for other uses: fodder, bedding, colouring, building, crafting, medical, and possibly votive purposes. It is therefore clear that plants must have played a crucial role in the cultural evolution of these human groups as already accepted elsewhere.

1 Introduction

Uan Tabu is one of the most important archaeological sites in the Tadrart Acacus mountain range, in the Libyan Central Sahara. The rockshelter was discovered by Fabrizio Mori (1965, 1998), who identified rock paintings from the Round Head and Pastoral phases. It is included in the concession area of the Joint Italo-Libyan Mission for Prehistoric Research in the Sahara and the Interuniversity Centre for Research on the Ancient Sahara and Arid Zones. This project was formerly directed by F. Mori and presently directed by M. Liverani, both from the University of Rome "La Sapienza", Italy.

The first sounding was dug during the 1960s (Mori, 1965) and a new trench was opened between 1990 and 1993 aimed at reconstructing the cultural sequence of the anthropogenic deposit and relating it to the rock paintings and to the past environmental context in general. The archaeological deposit covers a period of about 60,000 years, from the late Upper Pleistocene to the Early and Middle Holocene. The new multidisciplinary study was aimed at reconstructing the environmental and archaeological framework of the area in the late Quaternary (Garcea, 2001).

The Early Holocene chronology, based on archaeological studies and radiocarbon dates, established that the site was occupied by hunter/gatherers belonging to two cultural phases (table 1): a) Early Acacus, or "Epipalaeolithic", dating from 9550-9100 BC to 8300-7650 BC (Unit III, approx. lower 90 cm); b) Late Acacus, or "Mesolithic", dating from 8300-7650 BC to 7940-7490 BC (Units II and I, approx. upper 110 cm).

Archaeobotanical remains, including pollen, seeds/fruits and charcoal (analysed by Mercuri and Trevisan Grandi, 2001; Mercuri, 2001; and Neumann and Uebel, 2001, respectively) are concentrated in the Early Holocene layers, which were deposited during the occupation of Early and Late Acacus hunter-

Table 1: Archaeobotanical samples collected from the Holocene profile of the trench excavated at Uan Tabu in 1992. For radiocarbon dates cf. Bartolomei and Rizzo (2001).

Sedimentological units		Number of radiocarbon dates	Chronology	
			14C dates (uncal. Years bp)	Years BC (cal. By Oxcal 3.1 at 95.4% of confidence)
UNIT I	Organic unit	5	From 8850±100 bp to 8600±90 bp	From 8250-7650 BC to 7940-7490 BC
UNIT II	Organic unit	4	From 8870±100 bp to 8730±70 bp	From 8300-7650 BC to 8200-7550 BC
UNIT III	Colluviated unit	2	From 9810±75 bp to 8880±100 bp	From 9550-9100 BC to 8300-7650 BC
		11	From 9810±75 bp to 8600±90 bp	From 9550-9100 BC to 7940-7490 BC

gatherers (according to di Lernia and Garcea, 1997). These data provided important insight on the flora, vegetation, ecology, climate, human influence, and ethnobotany of the site. This information added new sets of evidence to previous archaeobotanical data from two other more or less contemporaneous sites in the Tadrart Acacus: (1) Uan Afuda, which was occupied between the late Upper Pleistocene and the Early Holocene (cf. analyses of pollen: Mercuri, 1999; seed/fruit and wood/charcoal: Castelletti *et al.*, 1999; fibres: Maspero, 1999; imprints: Magid, 1999) and (2) Ti-n-Torha/Two Caves, occupied during the Early and Middle Holocene (cf. analyses of pollen: Schultz, 1987; seed/fruit: Wasylikowa, 1992a, 1992b).

Altogether, data showed that the environment developed in the Early Holocene according to climatic and anthropic factors, and that plant cover transformations and human influence on the landscape have always been strictly interconnected. Thus, on the one hand, subsistence strategies adjusted to climatic and environmental changes, and on the other, the plant landscape was slowly and continuously being shaped by hunter/gatherers (Mercuri, 1999).

As mentioned above, carpological and pollen data from Uan Tabu were firstly presented in two separated papers within the monography which includes the multidisciplinary studies on the site (Garcea 2001). The aim of this paper is to integrate these archaeobotanical data and to compare them with those reported from other sites in the area so as to draw a general picture of the regional plant landscape in the Early Holocene. The main synchronous changes in the cultural sequence of Uan Tabu are also evidenced and correlated with the archaeological record in order to evaluate the impact of hunter/gatherers on the vegetation of the Tadrart Acacus.

2 Materials and methods

2.1 Site location

The Tadrart Acacus massif (800-1400 m a.s.l.) is the easternmost relief of the Central Saharan mountain ranges, situated in the Fezzan region in south-western Libya, adjacent to Tassili n'Ajjer in Algeria (figure 1). Uan Tabu (Lat. 24°51'35"N; Long. 10°31'42"E) is a rockshelter located on the left bank of the Wadi Teshuinat (915 m a.s.l.), which is the largest wadi of the central Tadrart Acacus. The rockshelter faces southwest and is about 50 m wide, 10 m high, and 4 m deep (Garcea, 2001) (figures 2 and 3).

2.1.1 Present vegetation

As for the present flora of Fezzan, Corti (1942) listed 230 species belonging to 165 genera and 50 families of Tracheophyta, among which 90 species live in the nearby region of Ghat, the lowland area with the richest flora in the region. On the other hand, the flora of the Tadrart Acacus massif is not well documented. Its plant cover is very discontinuous, mainly consisting of Saharo-montane vegetation and wadi vegetation which are characteristic of the Saharan Transitional zone (White, 1983).

Most of the Wadi Teshuinat is bare of vegetation, except for the centre of the valley where some vegetation is occasionally present (figure 2). It is an *Acacia* community type with affinities with the Saharo-montane vegetation. In particular, a few specimens of *Acacia tortilis* (Forsk.) Hayne ssp. *raddiana* (Savi) Brenan grow near the Uan Tabu rockshelter (figure 3). During February 1992, *Panicum turgidum* Forsk., *Pulicaria undulata* (L.) DC, *Pulicaria crispa* Schultz. and *Zilla spinosa* (L.) Prantl in particular were observed in front of the site. Furthermore, *Calotropis procera* Ait., *Cassia obovata* Collad., *Citrullus colocynthis* (L.) Schrader and *Schouwnia*

Depth (cm) - squares A-D	Cultural horizon	Number of archaeological layers	Number of pollen samples (PS)	Number of seed/fruit samples (MS)
84-10	Late Acacus	8 (from 8 to 1)	8 (PS from 8 to 1)	6 (MS from 6 to 1)
118-85		5 (from 13 to 9)	3 (PS from 11 to 9)	5 (MS from 11 to 7)
200-119	Early Acacus	7 (from 20 to 14)	5 (PS from 16 to 12)	7 (MS from 18 to 12)
200 cm		20	16	18

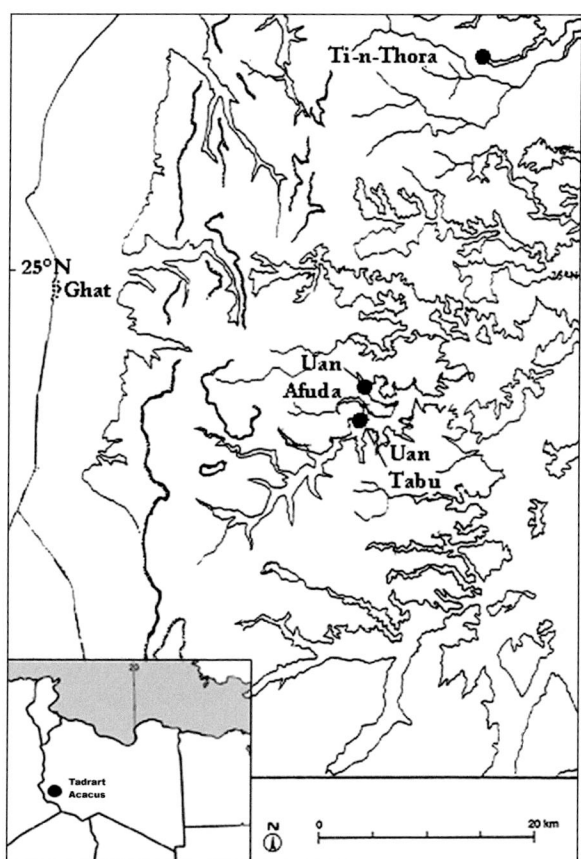

Figure 1: Location map of the Tadrart Acacus, and the sites mentioned in the text.

purpurea (Forsk.) Schweinf. were present farther from the site, along the wadi (Mercuri and Trevisan Grandi, 2001).

2.1.2 Sedimentological units and sampling

The Early Holocene archaeological deposit was ca. 2 m deep, showing three sedimentological units (Cremaschi, 1998; Cremaschi and Trombino, 2001): Unit III is an organic colluvial unit with a relatively high content of mineral coarse material, pedorelicts and organic micromass; Unit II is an organic unit with coarse material, ash, clay and organic material; Unit I is an organic unit with abundant ash and straw, and a very low content of coarse material.

Vertical sampling was carried out along this sequence. From the same cultural layer, samples were taken for pollen analysis and separate samples were collected for carpological analysis. Altogether, 16 pollen samples (from PS16 to PS1) were collected with a horizontal corer, and 18 carpological samples (from MS18 to MS1) were taken with a bucket (table 1). A total of eleven radiocarbon dates from nine layers referred were derived from these samples.

2.1.3 Palynological analysis

The pollen samples (PS) were treated according to the method elaborated by the Institute of Earth Sciences, Vrije Universiteit of Amsterdam (Lowe *et al.*, 1996). They were prepared with with Na-pyrophosphate, HCl 10%, sieved through a 7 μm nylon mesh, acetolysis,?? submitted to heavy liquid separation (Na-metatungstate hydrate), HF 40% and ethanol 98%, added to glycerol, dried in stove and mounted in glycerol jelly. *Lycopodium* tablets were added to establish the pollen concentration (pollen per gram = p/g). Microscopic analyses were carried out at 400 and 1000 magnifications. Pollen identification was based on keys, atlases (e.g.: Bonnefille and Riollet, 1980; Reille, 1992, 1995, 1998; El Ghazali, 1993; Ayyad and Moore, 1995), and reference slides. Information on habitus, distribution, and ecology are obtained from different floras (Corti, 1942; Ozenda, 1958; Ali and Jafri, 1976-77; Jafri and El-Gadi, 1977-78; Turril and Milne-Redhead, 1952).

Figure 2: The Wadi Teshuinat with some vegetation in the bottom of the valley and Acacia *trees near the rockshelter.*

Grass pollen types (Gramineae) were distinguished into six morphological types, based on reference slides, morphobiometrical studies on African wild species, and literature (e.g.: Bonnefille and Riollet, 1980; Bottema, 1992). Provisional names of the pollen types were chosen according to reference pollen and macroremains recorded in the same deposit. The maximum diameter, the porus and anulus diameter, and the sculpture of exine (scabrate or verrucate) were observed to make the following distinctions from the smallest grains: 1) maximum diameter <25 μm, scabrate or verrucate exine (Sporoboleae cf.); 2) 25-36 μm, verrucate exine; 25-39 μm, scabrate exine (*Urochloa/Dactyloctenium* cf.); 3) 36-75 μm, verrucate exine (*Panicum* s.l.); 4) 40 - 49 μm, scabrate exine (Andropogoneae cf./ Cerealia-type I); 5) 50 - 59 μm, scabrate exine (Cerealia-type II); 6. non-measurable, very crumpled big pollen (Gramineae undiff.). 'Cerealia type' (or 'Cereal type' *sensu* Faegri *et al.*, 1989: 285) is a morphological concept referring to pollen with a maximum diameter of more than 40 μm with scabrate exine (here into the categories 4) and 5) mentioned above). Other large-sized pollen has verrucate exine and is included into the *Panicum* s.l.; those with a diameter greater than 70 μm most probably belong to *Panicum*, *Setaria*, and *Echinochloa* species. Thus, the grass pollen grains with a maximum diameter less than 40 μm comprise Cerealia types and some *Panicum* s.l. As they include some pollen of wild cereals, they were considered as the best marker of the grasses used for food.

2.1.4 Macrofossil analysis

Each carpological sample consisted of 6 litres of soil that was dry sieved in the field through a 2-mm mesh. Although specimens smaller than 2 mm, if present, were lost, the sampling method was systematic and facilitated the transport of all material to our laboratory. Smaller seeds/fruits were also looked for in a few bulks of sediment (about 0.5 l each), but no new taxa were recorded. The carpological list of Uan Tabu is quite comparable to those from Uan Afuda and Ti-n-Thora (Wasylikowa 1992a, 1992b; Castelletti *et al.* 1999), suggesting that no major loss of plant material has occurred. A few groups of fruits visible with the naked eye were also collected.

In the laboratory, residues from sieving were examined with the naked eye and under a stereo-microscope Wild M10 (25-80 magnifications) in order to isolate and identify plant macrofossils. Seeds, fruits, stems, twigs, and wood were identified and hand-sorted from the macrofossil samples (MS). Water-flotation was not used because a large part of the carpological remains were destroyed or damaged when immersed in water. All specimens were numbered and stored. Fragments of macrofossils were counted as 1/3 or 1/2 of the intact specimen, depending on their size. Concentrations were reported as seeds/fruits (= sf) in 1 litre (expressed as sf/l).

Identification was based on fossil descriptions and drawings by Wasylikowa (1992a, 1992b), and other publications/keys on macrofossil records (for grasses, Hubbard, 1992; Barakat and Fahmy, 1999), descriptions of several floras (Flora of Libya, Flora of Tropical East Africa, Flora of Turkey: respectively, Sheriff and Siddiqi, 1988; Clayton and Renvoize, 1982; Clayton *et al.*, 1974; Verdcourt, 1991; Davis, 1985), and the refer-

Figure 3: The rockshelter at Wadi Teshuinat.

ence collection. Results are still preliminary and identifications are still in progress as many records have not been identified yet. The number of species, distribution and habitat are mainly obtained from the Flora of Tropical East Africa by Clayton and Renvoize (1982).

The terminology follows Wasylikowa (1992a, 1992b). The Gramineae include *Brachiaria* type A and type B, *Urochloa* type, *Urochloa/Brachiaria*, a type with intermediate features of the previous two, *Cenchrus*, *C.* cf. *ciliaris*, *C.* cf. *biflorus*, *Digitaria* type, *Echinochloa* sp., *Panicum* sp., *Pennisetum* type, *Setaria* type. The term 'type' following plant name means morphological resemblance to a taxon named but similar fruits may also be found in other genera not seen by the authors. When there is a closer resemblance to a particular taxon the term 'cf.' is used rather than 'type'.

2.1.5 Zones, sub-zones and the interpretation of data
In the whole sequence, zonation was established by the 'constrained incremental sum of squares cluster analysis' (CONISS; Tilia by Grimm, 1991-93) and by the visual examination of diagrams. Zones and sub-zones were named according to previous definitions (Mercuri and Trevisan Grandi, 2001; Mercuri, 2001). Early Holocene samples were attributed to: a) one Pollen Zone UTB2, which was sub-divided into four sub-zones *a*, *b1*, *b2*, and *b3*; b) three Carpological Zones UTB-mc 2, 3, and 4. Zones and sub-zones corresponded to different concentrations and floristic lists of samples.

In table 2, the main results from pollen and seeds/fruits analyses are compared zone by zone (or sub-zone by sub-zone). In addition, selected concentration data from the two types of analyses are presented in the diagram in figure 4. Curves are tentatively arranged along an age scale with the aim of showing the synchronous change in the amount of records passing from the Early to the Late Acacus phases. The age scale is expressed as uncal. years bp because the calibration gives a range of data (by Oxcal 3.1; see also table 1), and the medium point is not the most probable (Telford et al., 2004). Age scales are not commonly drawn to present the data from archaeological sites because it is well known that humans are the major agent of deposition, and on-site deposits are not continuous. Nevertheless we think that, in this context the age scale can be accepted due to the high number of radiocarbon dates (more than 50% of the analysed samples are radiocarbon dated).

The interpretation of data from the various zones was based on two approaches. Firstly, the plant landscape was inferred by the floristic list and the assemblage of taxa, which were tentatively compared with the present vegetation (e.g.: Corti, 1942; White, 1983). The level of accuracy of such a method is affected by the fact that past vegetation may not have been completely similar to the present one (Mercuri and Trevisan Grandi, 2001). Secondly, the exploitation, use, and knowledge of plants was mainly inferred by anthropogenic indicators of seeds/fruits and pollen of plants selected and accumulated in the shelters by humans for different purposes. The frequency of these plants and the ability to relate them to human activities makes them the most useful group for reconstructing the relationships between plants and humans in their dwelling sites. The use of such plants can be supported either by the high amount of these plants

Table 2: Synoptic table of pollen and macrofossil data from Early Acacus (Unit III) and Late Acacus (Units II and I) showing similarities and trends. Pollen sub-zones and seed/fruit (= macro★) zones are named according to Mercuri and Trevisan Grandi (2001) and Mercuri (2001). Charcoal data by Neumann and Uebel (2001) are included in the macrofossil zones.

Sedimento-logical Units	Cultural horizon	Pollen sample no. (PS) with ≥ 500 p/g	**Pollen sub-zones ,UTB2'** (pollen types= total number in the zone; percentages= mean of each sub-zone)	
UNIT I	Late Acacus	3 (PS 5, 2, 1)	b3	**Tamarix**, Ficus, **Cassia**, **Acacia**, Indigofera, Pistacia, etc.; shrubs considerably increase (20%; 17 pollen types) with **Artemisia**, Capparis, Maerua, Zygophyllum, etc.; herbs decrease but are more varied (71%; 47 pollen types); Typha 17%, Gramineae 29%; Compositae 16%; high pollen concentration and very high floristic richness
		2 (PS 8, 7)	b2	**Tamarix**, **Ficus**, Cupressus, and Quercus undiff.; shrubs halve but are more varied (4%; 9 pollen types) with Zygophyllum, **Artemisia**, Calligonum, Chenopodiaceae, Compositae, Moltkiopsis, Salvadora persica, etc.; herbs further increase (95%, 21 pollen types); abundance of **Typha** 69%; Gramineae 21%; Compositae 3%; very high pollen concentration and high floristic richness
UNIT II		2 (PS 11, 10)	b1	**Ficus**, Cupressus, Populus, and Celtis; **Zygophyllum** is the only shrub (8%); herbs increase (91%, 20 pollen types) with Gramineae 53% and Typha 28%; Compositae 3%; high pollen concentration and low floristic richness
UNIT III	Early Acacus	1 (PS 14)	a	Cupressus, Quercus ilex t.; shrubs include Olea and Capparis (6%); herbs prevail (88%, 10 pollen types) with **Gramineae** (32%) and **Typha** (24%); low pollen concentration and very low floristic richness
		8		

which were recorded in the shelters or in other cases by ethnobotanical data, but the inference cannot be completely confirmed.

Six categories were distinguished in the pollen sum of anthropogenic indicators (Mercuri, 1999): 1) Plants used for food/manufacturing/fuel, including *Panicum* s.l. pollen. Plants used as fuel are mainly recognised according to the ethnobotanical notes by Corti (1942); 2) Grazing/good pasture plants; 3) Medicinal plants; 4) Toxic/medicinal plants; 5) Plants used for colours/tannins; 6) Ruderal/nitrophilous/trampled area plants. The first five categories refer to species selected, accumulated and used for different purposes, whereas the last category refers to synanthropic species which naturally grow near human settlements. Medical remedies of plants are described according to present ethnographic data (Fiori, 1912; Corti, 1942; Uphof, 1959). The people who lived at Uan Tabu may or may not have been fully aware of these and other properties, which we might presently ignore. However, they certainly had knowledge and expertise of the numerous attributes and qualities of wild plants, as the great variety of plant species accumulated at the site indicate.

3 Results and discussion

Pollen and seeds/fruits provided interesting information on the environment and the knowledge on plant uses, i.e. the 'botanical culture', of the hunter-gatherers who lived at Uan Tabu. Analytical tables and descriptions have been reported elsewhere (Mercuri and Trevisan Grandi, 2001; Mercuri, 2001). This paper describes the key elements of pollen and carpological flora, as well as the trend of the records along the sequence, enabling us to examine the archaeobotanical data as a whole. (table 2). In the discussion, the data from Uan Tabu are compared with those from Uan Afuda and Ti-n-Thora (Mercuri, 1999; Castelletti et al., 1999; Schultz, 1987; Wasylikowa, 1992b).

3.1 Main Pollen Flora

Pollen was recorded in eleven samples from Unit III to Unit I, whereas five samples from hearths and ash-rich layers were sterile. Only eight samples showed a good amount of pollen, between 500 and 240,000 pollen/gram, the mean concentration being about 42,000 p/g. The data reported below refer to these eight samples.

A total of 3933 pollen grains were counted, and 87

Macro★ sample no. (MS) with twigs or seeds/fruits	Macrofossil zones ‚UTB – mc'	
5 (MS 6, 5, 3, 2, 1)	4	Abundant uncharred spikelets/ florets of **Brachiaria** and **Urochloa** types, and many more charred grains; charred/uncharred **Cenchrus**, **Setaria**, Echinochloa, Panicum, Digitaria, and few Pennisetum; uncharred Dactyloctenium, Andropogoneae, Sporobolus, and Tragus; mostly uncharred, and a few charred, Boraginaceae; charred seeds of Acacia tortilis subsp.raddiana, and fruits of Cyperaceae; uncharred and few charred Bifora, Tribulus terrestris, Citrullus colocynthis, and Liliaceae; uncharred Caryophyllaceae, Leguminosae, Torilis, Chenopodium murale, C.hybridum, Picris, and other Compositae; twigs of **Tamarix**; many stems of Gramineae; high concentration and high floristic diversity [charcoal: abundant Tamarix, with Calotropis procera, and Leptadenia pyrotechnica, and few Salvadora persica]
2 (MS 11, 8)	3	Few uncharred spikelets/florets of **Brachiaria** and **Urochloa** types, and more charred grains; few uncharred Setaria and one charred grain of Panicum; many uncharred and charred **Boraginaceae**; charred seeds of **Leguminosae**, and Cyperaceae; charred/uncharred Chenopodium and Citrullus colocynthis; many stems of Gramineae; very low concentration and higher floristic diversity [charcoal: abundant Tamarix, with few Calotropis procera]
4 (MS 18,17, 13,12)	2	Uncharred spikelets/florets of **Brachiaria** and **Urochloa** types, and few charred grains at the top of the zone; frequent detached lemnas and paleas; few uncharred Cenchrus, Echinochloa, and one charred grain of Panicum; few Boraginaceae, and one fruit of Compositae; few stems of Gramineae; low concentration and floristic diversity [charcoal: very abundant Tamarix, with few Calotropis procera and Chenopodiaceae]
11		

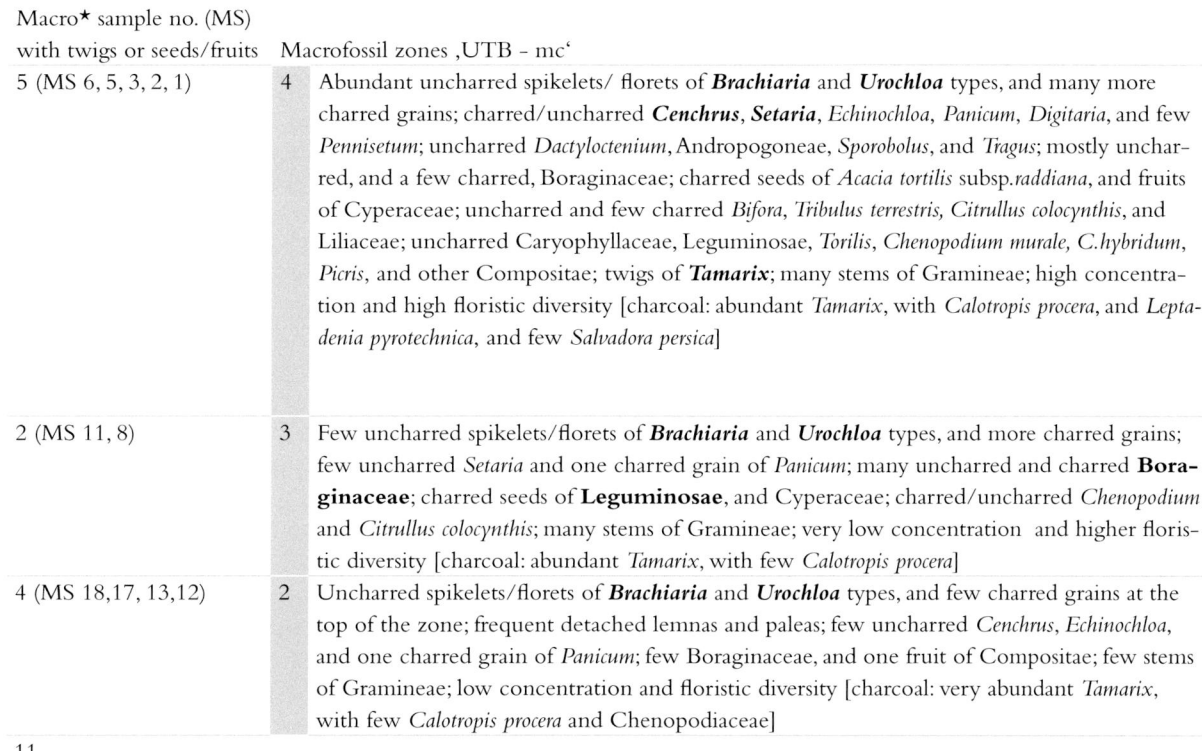

Figure 4: Age and concentration diagram showing selected archaeobotanical data (explanation in the text). Pollen concentration is expressed as pollen per gram (p/g); twigs and seed/fruit concentrations are expressed as number per litre (twigs/l or sf/l). Diagram drawn with Palaeo Data Plotter – Beta Test version 2.0 by Steve Juggins, 2002.

taxa belonging to 43 families of Spermatophyta were identified. A maximum of 52 pollen types were recorded in PS2/Layer 2-Unit I. *Typha*, including *T. latifolia* type which reached a maximum of 81% (193,400 p/g in PS7/Layer 6/6a-Unit I). Pollen clusters of a single or mixed types were occasionally recorded in Units II and I.

Trees (mean = 5%) were mainly *Tamarix, Ficus, Cupressus,* and *Cassia,* together with *Acacia, Indigofera* cf., *Quercus ilex* type and *Quercus* undiff., *Pistacia, Populus,*

Celtis, *Pinus*, and *Carpinus* cf. *orientalis*. Shrubs and lianas (mean = 11%) mainly included *Zygophyllum*, *Artemisia*, *Capparis*, and *Rhus*, together with *Cocculus*, *Aerva* cf. *persica*, *Calligonum*, *Moltkiopsis* cf. *ciliata*, and *Salvadora persica*. *Maerua* and *Olea* shrubs/trees were also present. Herbs (mean = 84%) were mainly represented by Gramineae (34%; six pollen types) and *Typha* (34%; three pollen types), together with *Hypericum*, *Echium plantagineum* type, *Parietaria* type, *Tribulus* cf. *terrestris*, Cyperaceae, Compositae, Cruciferae, Caryophyllaceae, Liliaceae, and Solanaceae.

Four main features characterised the pollen spectra in the deposit: 1) the state of preservation was variable: badly-preserved pollen was more frequent at the bottom, while well-preserved pollen was common in the upper layers; 2) Gramineae and Typhaceae prevailed among the herbs, together with Compositae and Cyperaceae; 3) concentrations were variable, depending on human action: frequent fires destroyed pollen, and plant accumulation concentrated it; 4) a quite rich floristic diversity was observed, mainly increasing from the bottom to the top of the sequence.

3.1.1 Prevalent types (vegetation and its uses)

These taxa were present in many spectra since they were the main elements of the regional plant landscape. They occasionally had abrupt and scattered high values in one or more spectra, suggesting plant accumulation and use in the archaeological site. Altogether, the leading taxa, i.e. types with > 5% in at least one sample (Accorsi *et al.*, 1999) were the following:

1) GRAMINEAE were ubiquitous and characterised the spectra. They mostly belong to *Urochloa/Dactyloctenium* cf., Sporoboleae cf., and *Panicum* s.l. Pollen grains > 40 µm, and particularly those > 70 µm (in PS1 and PS2, sub-zone UTB2 *b3*), probably belonging to *Panicum*, *Setaria*, and *Echinochloa* species, were quite frequent. Also at Uan Afuda, large-sized pollen was always present, and, on the whole, Gramineae made up almost half of the spectra. Gramineae with a large proportion of the 'panicoid-type' dominated the pollen spectra of Ti-n-Torha as well (Schulz, 1987). This family mainly consists of annual or perennial herbs that have always had primary geobotanical and economic relevance. At Uan Tabu, grasses were the main element of the plant landscape which probably was a wet open grassland. The presence of grasses offered a suitable habitat for both animals and humans. The high values of Gramineae pollen (up to 79% in PS11, UTB2 *b1*) testifies that they were harvested and accumulated in the rockshelter, and that wild cereals were widespread in the area. Grasses were of great importance for food, as well as fodder, bedding, and basketry. Leaves of grasses were possibly also used to make ropes (Maspero, 1999; see an example of twisted cord in figure 5). As these plants covered numerous basic needs, they must have been greatly valued, and may have also been part of the groups' social rules (Palmer and Van der Veen, 2002).

2) TYPHA was ubiquitous and sometimes dominant in the Uan Tabu spectra. However, it was less represented at Uan Afuda and even less significant at Ti-n-Torha. The pollen of *Typha* includes one type with monads (*T. angustifolia*) and two types with tetrads (*T. minima*, and *T. latifolia* type) is the latter being the more prevalent). This genus is ubiquitous throughout both hemispheres, growing in swamps and ponds, preferably in the shallow water confined to the sides of deep or flowing channels (Haslam, 1978: 110). In the Saharan regional transition zone, *Typha angustifolia* L. and *T. latifolia* L. live in oases and near groundwater outlets (White, 1983). At Uan Tabu, this type of pollen was a clear indicator of widely-spread moist environments in the central Acacus, and found particularly near the river flowing in front of the shelter. This plant was certainly used and accumulated in the shelter during the flowering season, as suggested by the pollen found in the shelter. On the other hand, we cannot completely exclude its use during the fruit season, as the fruits were possibly lost during sampling as they are smaller than 2 mm (see sampling methods). Since the highest value of *Typha* was found in PS7/Layer 6/6a, and the posts of a hut were located in the same layer (Garcea, 2001), it is likely that stems of *Typha* with flowering spikes were used to make the roof of the hut. Thus, the hut was probably built around ca. 8000-7800 BC (8800 - 8700 uncal. years bp), when layers 6/6a were deposited. Since *Typha* is well-known as a multi-use plant, it could be have been used for both building material and food. The rhizomes, shoots, stems, fruits, and pollen itself can be actually eaten (Prendergast *et al.*, 2000). Yellow pollen mixtures also have medicinal properties, and could have been mixed with the flour of wild cereals. It was possibly also used as paint to decorate bodies and walls and is perhaps still preserved in the rock art on the walls of the shelter.

3) ARTEMISIA was widespread in the Uan Tabu spectra. At Uan Afuda, it was even more common, whereas it was sporadic at Ti-n-Torha. This genus includes hundreds of species of small fragrant shrubs and herbs which are usually found in dry areas. In the Saharan regional transition zone today, only *Artemisia herba-alba* Asso grows in the Saharo-montane dwarf bushland, mainly on the Ahaggar above 2600 m a.s.l., and *A. campestris* L. is dominant on gentle slopes of wadis at about 1800 m a.s.l. within the Saharo-montane wadi vegetation. Further north, in the Mediterranean/Saharan regional transition zone, the dwarf *Artemisia*

bushland is widespread from Morocco to Tunisia and abundant on the High Plateau in Algeria, being part of a sub-Mediterranean vegetation with *A. herba-alba* and *A. campestris* occurring in mosaic pattern with Gramineae grassland communities. There, *A. herba-alba* dominates in clay soils and depressions where drainage is impeded and water accumulates (White, 1983: 229-230). As regards the Fezzan, Corti (1942) recorded *A. herba-alba*, *A. monosperma* Delile, and *A. judaica* L. At Uan Tabu, the pollen was an important indicator of bushland. This vegetation was not extended, but probably concentrated in patches alternating with grassland. *Artemisia* lived in the wadi or in upper belts and expanded during phases of relatively drier climate conditions. Plants were certainly harvested and brought into the shelter around 8750 ± 100 years bp, as the high pollen values in the samples from Uan Afuda (44%) and Uan Tabu (15% in PS1, UTB2 *b3*) indicate. Moreover, pollen in human coprolites from Uan Afuda proved that flowers were eaten (Mercuri, 1999). It is also possible that their chemical properties were known. In fact, flowers, dried leaves, and aromatic extracts of different species have traditionally been used as perfumes and medicaments, such as antiseptics, sedatives, or even stimulants and tonics. *A. annua* L. was recently recognised as the most effective antimalarial species (Wright, 2002). *A. herba-alba* is an anti-infectious medicament, an emmenagogue, and a parasiticide, but can be toxic when ingested.

4) *CAPPARIS* was also widespread. A few pollen was recorded at Uan Afuda, but was not visible in the Ti-n-Torha diagram. The genus includes about 250 species of climbing plants, shrubs and trees widespread in the tropics and sub-tropics. In the Saharan regional transition zone, the Sahel species *Capparis decidua* Edigev. lives in *Acacia* wadi communities which are bushland or bushed grasslands. Species of *Capparis* were also part of the original oasis vegetation (White, 1983). In the Fezzan, Corti (1942) did not report any species. At Uan Tabu, these plants must have been part of shrub patches mainly distributed near moist places and probably in the wadi valley. Again, high pollen percentages (6% in PS2 and PS5, UTB2 *b3*) strongly suggest the harvesting of *Capparis* for food or medicinal purposes. Unopened flower buds or capers, semi-mature fruits, and young shoots with small leaves are currently used for food, and also as a diuretic, kidney disinfectant, vermifuge, and tonic. It is also used against arteriosclerosis; caper root bark has been traditionally used for dropsy, anaemia, and arthritis (Simon *et al.*, 1984). Shoots and young leaves of *C. decidua* contain a rubefacient and a vesicant. Seeds contain glucocapparin which releases a mustard oil when the plant is crushed (Smidth, 2003 - Botanical Dermatology Database). Immature buds recorded at Uan Afuda confirmed that capers were harvested, and thus probably eaten by hunter-gatherers (Castelletti *et al.*, 1999).

5) CYPERACEAE, including *Cyperus* and *Scirpus* pollen, in addition to undifferentiated types, were widespread. They were very similar to the spectra from Uan Afuda, whereas, at Ti-n-Torha, they were only second to Gramineae, and mainly belonged to *Cyperus conglomeratus* type. This family includes herbs that are generally related to hygrophilous vegetation or to grassland with Gramineae. Only *Cyperus conglomeratus* Rottb. was a perennial species living on sand dunes and ergs (Ozenda, 1958; Goudie *et al.*, 2000). *C. laevigatus* L. is perennial and lives in a halogypsophilous vegetation in wadi beds also in the Fezzan (Corti, 1942). There, other annual or perennial species of *Cyperus* and *Scirpus* grow in moist places. At Uan Tabu, Cyperaceae mainly lived in wet environments and grasslands. Pollen, reaching 9% in PS2, suggested that these plants were used. The records from Uan Afuda, where a spindle shaped-tuber of *Cyperus rotundus* type was recorded (Castelletti *et al.*, 1999), suggested that tubers were probably eaten. Moreover, a few Cyperaceae fruits were found at Uan Afuda, Ti-n-Torha and Uan Tabu (see below). Some fruits were possibly used in a similar way as grass grains (Wasylikowa and Dahlberg, 1999). Many species of Cyperaceae are still used in traditional medicine: tubers of *C. rotundus* are used as an astringent, diuretic, emmenagogue, emollient, stomachic and vermifuge in fighting against skin diseases, fever, ulcers, headaches, infections, and (scorpion) bites. Finally, several species were used for basketry and fodder.

6) *ZYGOPHYLLUM* was frequent. A few pollen was also recorded at Uan Afuda, but was not visible in the Ti-n-Torha diagram. The genus usually includes tropical herbs or shrubs with stinking foliage. In the Sahara, shrubs of *Zygophyllum* species are part of the halogypsophilous vegetation of wadi beds. This vegetation occurs where salt accumulates in depressions with no drainage. Today, due to the severe climate, halophilous vegetation is very rare in the Central Sahara, being almost only present in the mountain massifs. It is more common towards the northern Sahara and the Mediterranean-Saharan transition zone (White, 1983: 222). *Z. simplex* L. and *Z. album* L. presently live in the Fezzan, being more frequently observed near oases or in salty soils (Corti, 1942). At Uan Tabu, these shrubs must have been part of small saline areas located in the surroundings of the wadis where groundwater outpoured to the surface. Remarkable pollen percentages in two spectra (16% in PS10-UTB2 *b1*, and 7% in PS1) suggested plant accumulation. Flower buds are used as capers (bean capers). Dried flowers have a scent of tea and the plant has analgesic and vermifuge prop-

erties. Plants can be toxic for herbivores.

7) *CASSIA* was frequent. A few pollen grains were recorded in dung from Uan Afuda, and more abundant records were identified at Ti-n-Torha. In tropical regions, this genus includes about 30 species. In the Saharan regional transition zone, *Cassia italica* (Mill.) Lam. ex FW Andr. lives in *Acacia* wadi communities (White, 1983). *C. senna* L. lives in the Fezzan (Corti, 1942). At Uan Tabu, these trees/shrubs lived in bushed grassland. The high pollen value of one sample (7% in PS5, UTB2 *b3*) suggested plant harvesting. People in northern Africa and south-western Asia have always considered Senna as a "cleansing" herb, and used it as a natural laxative for centuries. Actually, leaves and pods of Senna have many medicinal uses, e.g. dried fruits are a powerful cathartic in the treatment of constipation. Furthermore, leaves can be crushed into a paste and applied to various skin diseases. Today, *Cassia* species (mainly *C. senna*) are cultivated in Egypt and Sudan. Plants are toxic for herbivores.

8) *RHUS* cf. *TRIPARTITA* was only present in one sample. Pollen was recorded frequently at Uan Afuda, whereas it was not visible in the Ti-n-Torha diagram. The species is a small shrub which lives in the Saharomontane wadi vegetation, where bushes form a dense cover. It is also quite frequent in the wadis of northern Fezzan (Corti, 1942). At Uan Tabu, these plants must have been part of shrub patches. The high pollen percentages observed here (11% in PS1, UTB2 *b3*) and at Uan Afuda (up to 8%) suggests the use of these plants, as well. They have edible fruits, and their bark contains an important tanning substance, still used by Tuaregs for tanning sheep and goat skins. Leaves, wood and roots of other *Rhus* species are also used.

3.1.2 Main Carpological Flora

Although all samples of the sequence provided plant macrofossils, such as small charcoal, wood, barks, twigs, stems, tubers, and flowers, seed/fruit records were only observed in eleven samples from Unit III to I, partly due to stratigraphical differences from each layer, and partly to the coarse meshes which were used in the sampling procedure. In these samples, concentrations ranged between 6 and 560 sf/l, and the mean concentration was 135 sf/l.

A total of 9274 records were counted, belonging to 51 taxa which have not all been identified yet. Among them, 50 taxa were seeds and fruits and one was a twig. The identified records belong to 14 Families of Spermatophyta. A maximum of 31 carpological taxa were recorded in MS1/Layer 3-Unit I. Paniceae, mainly *Brachiaria*, *Urochloa*, and *Setaria* types, reached their maxima (almost 100% and 439 sf/l in MS5/Layer 7-Unit I). Groups of Paniceae (in MS1/Layer 3 and MS5/Layer 7) and Boraginaceae - type I (in MS2/Layer 4) were recorded in Unit I.

The only identified trees were *Tamarix* cf. *aphylla* (319 twigs) and *Acacia tortilis* ssp. *raddiana* (7 seeds in the deposit). Herbs and shrubs prevailed with Gramineae (8103 fruits representing 17 types including Gramineae) and Boraginaceae (588 fruits; 3 types). Additional taxa are represented by *Citrullus colocynthis*, *Bifora* type, *Torilis* type, *Tribulus terrestris*, *Schouwia purpurea*, *Picris* cf., *Asphodelus* type, Polygonaceae cf., and Cyperaceae.

Four main features characterised macrofossil spectra in the deposit: 1) records had a good/very good state of preservation, being almost entirely in a desiccated state, and only less than 5% remains were charred; 2) Gramineae and Boraginaceae were prevalent in the carpological spectra; 3) concentrations were variable, mainly depending on human action, and generally increased from the bottom to the top of the sequence; 4) floristic diversity also increased from the bottom to the top of the sequence.

3.1.3 Prevalent types

These pollen types were diffused in the spectra, being elements of the local plant landscape. They had hundreds of records in at least one sample, suggesting plant accumulation and use at the archaeological site.

1) *TAMARIX APHYLLA* (L.) Karst. consisted of twigs which were only present and abundant in the upper layers (Unit I). Pollen of *Tamarix* was also recorded in the upper layers. Charcoal was very abundant in the anthracological record from Unit III to I (Neumann and Uebel, 2001). A few twigs, wood and charcoal were also found at Ti-n-Torha and Uan Afuda, where more wood of *Tamarix* sp. was recorded. Actually, it was the most important woody species in the deposit, together with a few seeds of *Acacia tortilis* that were present at the top of the sequence. It has been even suggested that the name of the rockshelter, 'Tabu', could have the same root as Tabarcat or Tabaracar, which in Temajâq language means *Tamarix aphylla* (Corti, 1942; Mercuri, 2001). This species is native of Africa and the Middle East. It occurs in saline habitats, especially near springs and along streams and rivers. In the Fezzan, it preferably grows as an element of the wadi vegetation and is the most diffused tree, together with *Acacia tortilis* ssp. *raddiana*. Like at present, the wood was certainly used for fuel and possibly to make artefacts and to give a special flavour to drinks (Corti, 1942).

2) The PANICEAE tribe included thirteen identified morphological types, and 8075 records of charred and uncharred florets, spikelets, lemmas, paleas, and grains. The most important ones are discussed below.

3) *BRACHIARIA* types, including types A and B,

were the most abundant records i The genus is not quoted in the Floras of Fezzan, Libya and Central Sahara; six species are known in Egypt and thirty-nine in the Flora of Tropical East Africa. It includes species mainly growing in valley alluvium, swamp margins and damp places in deciduous bushland or wooded grassland. *Brachiaria deflexa* and *B. ramosa* (L.) Stapf, which are known to interbreed, have the most similar descriptions to our fossils. They are annual species that grow in moist sandy wadis in Egypt, and in deciduous bushland or at the margins of the riverine forests from Senegal to Yemen and southward into South Africa. *B. deflexa* is a wild cereal harvested in the Sahara and sub-Sahara (Harlan, 1995). Other *Brachiaria* species are cultivated for forage in tropical Africa (Clayton and Renvoize, 1982).

4) *UROCHLOA* type was the second most abundant record. *Urochloa* was a common record also at Ti-n-Torha. This genus is not quoted in the Floras of Fezzan, Libya and Central Sahara either. Nine species are listed in the Flora of Tropical East Africa. They preferably grow in sandy soils of deciduous bushland or wooded grassland, sometimes following cultivation as a weed. *Urochloa panicoides* P. Beauv. and *U. oligotricha* (Fig. & De Not.) Henr. have the most similar descriptions to our fossils. *U. panicoides* is annual and is distributed from Senegal to Yemen and southwards into South Africa. *U. oligotricha* is perennial and is distributed in Eastern Africa from Ethiopia to South Africa. No uses as human food are known at present, but *U. oligotricha* is cultivated as forage.

5) *UROCHLOA/BRACHIARIA* type had florets/spikelets with intermediate characteristics between *Brachiaria* and *Urochloa* types. This intermediate record was also found at Uan Afuda and Ti-n-Torha. The occurrence of such a type is to be expected because the boundary between the two genera is quite problematic and confused by a number of intermediate or anomalous forms. Therefore, the two genera include species that interbreed, as they are genetically closely related, such as *B. ramosa* and *U. panicoides* (Clayton and Renvoize, 1982: p. 603).

6) *CENCHRUS*, including *C.* cf. *BIFLORUS* Roxb., and *C.* cf. *CILIARIS* L., was also quite abundant. Few records of *Cenchrus* were found at Uan Afuda and Ti-n-Torha. No species are quoted in the Flora of Fezzan, but about twenty-two species are distributed in tropical and warm temperate regions of Africa. *C. ciliaris* is still living in Libya today. *C. biflorus* is absent from the flora of Libya but is a common plant in the Southern Sahara. where it is confined to the sandy soils of low-rainfall wooded grassland. (White, 1983: p. 216). *C. biflorus* is introduced in Egypt and distributed in tropical Africa, Arabia and India. It is an annual species harvested and known by the name of Cram-Cram.

7) *SETARIA* type, *PANICUM* sp. and *ECHINOCHLOA* sp. were quite frequent in the Uan Tabu deposit, as well as at Uan Afuda and Ti-n-Torha. *Panicum turgidum, Echinochloa colona* (L.) Link, and *E. haploclada* (Stapf) Stapf are similar to our fossils, but other species must be examined. *Setaria* type was recorded at Uan Afuda and Ti-n-Torha; a few *Panicum* sp. and several *Echinochloa* sp. were recorded at Ti-n-Torha. Moreover, impressions of *Setaria* and *Panicum* grains were observed in the pottery from Uan Afuda (Magid, 1999). The present distribution and habitats of the species are: a) *Setaria verticillata* P. Beauv. is quoted for the Central Sahara and the Fezzan, and twenty-two species are described in the Flora of Tropical Africa; *Setaria* species grow in a wide range of habitats and include some minor crops; b) *Panicum turgidum* presently grows in the Wadi Teshuinat, and is common in the *Acacia-Panicum* wadi communities of the Sahara desert; *Panicum* species live in moist, clayey or sandy soils, grassland or bushland, and include several wild cereals harvested in the Sahara and sub-Sahara, such as *Panicum turgidum* and *P. laetum* Kunth; c) only *Echinochloa colona* lives in Libya and in the Central Sahara, and twelve species of the genus *Echinochloa* in tropical and warm temperate regions of Africa; these herbs grow in moist soils, marshes and flood-plains. In particular, *E. colona* is a semi-aquatic plant which grows in muddy or swampy places. *E. haploclada* lives near stream banks, dry river beds and alluvial flood plains. Several species of *Echinochloa* are wild cereals which are still harvested today in the Sahara and sub-Sahara, such as *E. colona* and *E. stagnina*.

8) *DIGITARIA* and *PENNISETUM* types were only represented by a few records. Some involucres of the *Pennisetum elatum/setaceum* type were recorded at Ti-n-Torha. These genera include species that lived in the Central Sahara, Libya and tropical East Africa, in a wide range of habitats from deciduous bushland, grassland, swamps, and weedy places. They include wild and cultivated cereals harvested in the Sahara and sub-Sahara (Harlan, 1989).

9) Finally, plant remains identified as PANICEAE included broken spikelets and detached lemmas and paleas that were undeterminable or with intermediate characteristics between *Brachiaria*, *Urochloa* and *Setaria*. Thus we can infer that members of Paniceae were certainly harvested and accumulated in the shelter. They include many wild cereals which are harvested even today, mainly as animal and human food. As already mentioned, these food plants could have had an important role in sustaining social relations (Palmer and Van der Veen, 2002).

10) BORAGINACEAE were the other prevalent family including very few records of *Lithospermum* and more abundant records of two types of pyramidal mericarps. They were also frequent and abundant in the Late Acacus layers at Uan Afuda and Ti-n-Torha. Even though generic determination is still uncertain, interesting information can be inferred by the state of preservation of these records: a) charred fruits were found in four layers; b) a group of uncharred fruits was present in MS2/Layer 4 of the Late Acacus horizon; c) fruits were found cracked but not split along natural lines in many layers. The same cracked fruits were observed at Uan Afuda and also in Egypt, at Nabta Playa, dating to 8000 years ago (Wasylikowa and Dahlberg, 1999). This indicates that Boraginaceae must have been exploited and used during the Early Holocene. They were probably harvested because some species of *Arnebia* (*A. hispidissima* [Lehm.]DC., *A. decumbens* [Vent.]Coss. et Kral.) could have been used to extract dye from the roots and several species of *Lithospermum*, *Heliotropium*, *Echium* and other genera have medicinal properties (Ceruti et al., 1993) and could have been harvested for this purpose. This implies that these plants must have had a special consideration among the people and must have also been used for ritual purposes. This latter hypothesis is explained by the unusual abundance of *Echium* pollen which was observed in a dung layer excavated in the inner part of the Uan Afuda cave (Mercuri, 1999; di Lernia and Mercuri, 2001).

3.2 Changes along the Early Holocene sequence: flora, vegetation, climate and plant uses

In the sequence, the pollen and seed/fruit spectra show major changes that were determined by both climatic-derived developments of the landscape and human influence (table 2; figure 4). This not only reflects the general evolution of the environmental context, but also the number of people and their different activities.

Four main features characterised the 'plant assemblage zones, i.e. pollen and carpological zones (table 2): 1) badly-preserved pollen and seeds/fruits were prevalently found in Unit III, and well-preserved records increased towards the top, mainly found in Unit I. On the whole, a small number of charred remains were recorded, being almost absent in the lower levels, and more frequent in the upper levels where broken mericarps of Boraginaceae were also abundant; 2) Gramineae, particularly Paniceae, were prevalent in the pollen and carpological spectra; 3) in general, pollen and seed/fruit concentrations increased from the bottom to the top of the sequence; sterile samples were present everywhere, probably because plant remains had been destroyed by fire in the ash-rich soils; pollen clusters and groups of Paniceae and Boraginaceae were present in the top zone; 4) floristic diversity increased from the bottom to the top of the sequence.

3.2.1 Early Acacus - low human influence (Pollen UTB2 a + Macro UTB-mc 2; Unit III: Layers 20 to 14; from 9810 ± 75 to 8880 ± 100 years bp = from 9550-9100 BC to 8300-7650 BC).

At the beginning of the Early Holocene, the vegetal cover was rich but the landscape must have been quite monotonous, that is, an open grassland with low floristic diversity. Paniceae, prevalently *Brachiaria* and *Urochloa*, and also Cyperaceae, lived in a grassland vegetation, similar to that presently growing in the Sahel. Hydro-hygrophilous plants, mainly *Typha*, grew together with patches of *Tamarix* and *Capparis* near the wadi, or the springs and moist places that were spread along the large Wadi Teshuinat. The local river flow in front of the shelter was stronger than in the tributary wadi Kessan, where Uan Afuda is located. During the Early Acacus, *Cupressus*, *Olea*, and *Quercus ilex* were associated with a Sub-Mediterranean forest type of vegetation (White 1983: p. 226), probably growing in the higher belts of the Tadrart Acacus.

Humans sporadically accumulated a few plants in the rockshelter, especially Paniceae and also *Zygophyllum*, for food. As a quite similar situation of limited plant gathering was also observed at Uan Afuda, it seems reasonable to assume that human pressure on the environment was not high yet. Hunting/gathering groups were probably less numerous than in the periods following and only seasonally frequented the rockshelter (Garcea, 2001, 2004). They exploited the most suitable and available plants in the area, the caryopses of wild cereals of *Brachiaria* and *Urochloa*, being the most prevalent. At ca. 8000 BC (8900 uncal. years bp), the environment started to change and the use of plants consequently changed, as well. Food processing of fruits began to be practised, resulting from the new needs and demands of the inhabitants.

The artefactual assemblage included a lithic toolkit showing the use of a variety of raw materials, namely sandstone, quartz, quartzite, chert, flint, and schist. Microliths were a considerable component in the techno-complex, and endscrapers, perforators, and burins were dominant. Polished stone tools were scarce and pottery was extremely rare (Garcea, 2001).

3.2.2 Late Acacus - a transitional phase (Pollen UTB2 b1 + Macro UTB-mc 3; Unit II: Layers 13 to 9; from 8850 ± 110 to 8800 ± 100 years bp = from 8300-7650 BC to 8200-7550 BC).

During this short period of time, the plant landscape began to develop into a more articulated mosaic of various biotopes, which must have been reasonably richer in this mountainous area than in the lowlands.

The grassland, with a prevalence of Paniceae and Cyperaceae, was still the main type of vegetation in the Wadi Teshuinat. It is possible that it was a seasonally swampy grassland. Wet environments were well-diffused, with *Scirpus* growing near *Typha* in shallower waters of the wadi with a slower flow. *Citrullus colocynthis* has been considered part of the original oasis vegetation (White, 1983: 219) and could have lived near these moist places in the same belt as *Ficus* and *Populus*. Shrubby halophytic vegetation mainly constituting *Zygophyllum* developed in small salt depressions of the wadi. Drought-resistant shrubs, such as Boraginaceae, Chenopodiaceae, and Compositae, but excluding *Artemisia*, constituted small patches of bushland that was more widespread than in the previous period. Psammophilous herbs, such as *Neurada procumbens*, lived in restricted areas of very arid sandy ergs, possibly more diffused towards the lowlands. Moreover, *Celtis* and *Cupressus* could be have been part of a Mediterranean forest located further north where these trees were present in sub-humid and humid belts, or even in the higher belts of the Tadrart Acacus (White 1983: 149).

From these results we can deduce that humid environmental/climatic conditions were still prevalent, but dryer conditions were advancing. A mosaic of environments and a distribution of vegetation in belts became evident.

Paniceae were still largely exploited, and also more drought-resistant wild cereals, such as *Panicum* and *Setaria*, were frequently harvested. Humans experimented with the exploitation of new plants, but plant accumulation was still low, suggesting that only small groups frequented the rockshelter. Nevertheless, human pressure on the environment increased and food processing was widely practised. In fact, this horizon evidenced a large production of polished grindstones that may have been used for various purposes, including wild cereal and other wild seed grinding. The plentiful production of polished stone tools at Ti-n-Torha East was also interpreted as being for the function of wild cereal grinding (Barich, 1992).

As for retouched stone technology, raw procurement was mainly reduced to sandstone. Type variability was also limited to generic formal tools on a macroflake with a prevalence of notches, denticulates, and sidescrapers. This suggested that activities requiring a longer time for processing, preparing, and manufacturing food and secondary products became part of the subsistence pattern (Garcea, 2001, 2003). The use of pottery was still scarce at the beginning of the late Acacus horizon (Garcea, 2001).

3.2.3 Late Acacus Acacus - a high botanical knowledge (Pollen UTB2 b2/b3 + Macro UTB-mc 4; Unit: I Layers 8 to 1; from 8830 ± 75 to 8600 ± 80 years bp = from 8250-7650 BC to 7940-7490 BC).

The plant landscape became even more diverse and articulated than previously as many new species entered the mountains from northern and southern regions. The grassland added Paniceae and other grass species. The herbaceous plant cover seemed to be reduced in favour of a more extended bushland or wooded bushland. This chiefly consisted of large shrubby patches of *Artemisia* and *Zygophyllum* mainly distributed near moist places and in the wadi valley. High seasonality with an alternation of *Artemisia* and *Zygophyllum* in depressions should be hypothesised. Wet environments were generally reduced, especially in the northern part of the Tadrart Acacus. However, permanent water springs with *Potamogeton* and moist biotopes with *Typha, Scirpus, Juncus*, and *Echinochloa* were still present in front of the site and along the Wadi Teshuinat. *Tamarix* and *Ficus* also lived in the same moist places. These trees and shrubs strongly suggest that, during this phase, a Saharo-montane wadi vegetation type (which today also includes *Artemisia* and *Rhus*) grew at 900 m a.s.l. in the Wadi Teshuinat. This vegetation today lives at around 1800 m a.s.l., in the moister wadis of the high Saharan mountains, that is, from 1800 m on the Ahaggar, southern Algeria, and between 1600 and 2300 m a.s.l. on Tibesti, in Chad (White, 1983). This could indicate that temperatures in the Tadrart Acacus were lower than today, although more data are needed as the above-mentioned high mountains lie at lower latitudes of 20°-21° N, i.e., around 3° farther south than the Tadrart Acacus. In the highest belts of the Tadrart Acacus, there probably was a sub-Mediterranean forest with *Pinus, Cupressus, Olea, Pistacia*, and *Quercus ilex*, which represent a type of vegetation that today grows further north in the Mediterranean/Saharan transitional zone (White, 1983).

Besides these types of vegetation distributed in the different belts, more Saharan and Sahelian elements characteristic of the 'desert savannah' were evident during this phase. Today, *Acacia* undiff., *Capparis, Cassia, Maerua*, and *Salvadora persica* grow in the *Acacia-Panicum* community, wherever the annual rainfall exceeds 30 mm (White 1983: 219). Neumann and Uebel (2001) also emphasise the diffusion of woody tropical elements in the charcoal record from Uan Tabu (*Leptadenia pyrotechnica* and *Salvadora persica*; table 2), suggesting that temperatures higher than in the past allowed these plants to enter into the mountains. Another southern biological marker, *Cenchrus biflorus*, which today indicates the boundary between Southern and Central Sahara, considerably increased at the end of this phase. Its shift to the north confirmed that drier and hotter conditions than in the previous periods had advanced. A psammophilous vegetation, including shrubs of *Calligonum* and *Moltkiopsis ciliata*, became more diffused and was closer to the site. Humid environmental/climatic conditions were still present in the area, even though drier conditions advanced. During dry climates, the mountains seemed to offer refuge to both plants and humans.

Humans exploited and experimented with the increased available plant resources. *Brachiaria* and *Urochloa* were still the most exploited grasses, evidencing that the cultural transmission of their use was a long-lasting tradition and that these plants were of continuous value to humans (Mercuri, 2001). Besides these wild cereals, *Echinochloa*, *Digitaria* and *Pennisetum* were also selected. A form of nurturing was also suggested for *Ficus* species at Uan Afuda, due to the abundant syconia recorded in the cave (Castelletti *et al.*, 1999). The extensive harvesting of Paniceae, *Typha*, Boraginaceae, *Artemisia*, *Capparis*, *Cassia*, and *Rhus*, together with many other plants, the higher frequency of food processing, and the spread of plants from trampled areas show that the site was more frequented, and that human pressure on the landscape was considerably stronger than before. The distribution of plant material in the Uan Tabu deposit and its similarities with plant macrofossils from other sites in the Tadrart Acacus evidence both a good regional cultural knowledge of plants, and the skills of transmitting this knowledge from one generation to the next (Mercuri, 2001).

This cultural phase developed a more complex interaction between humans and plants, providing human groups with their highest 'botanical knowledge'. The climatic history of the region, with progressively increasing dryer conditions, may explain why any process of 'care' for grasses or other plants, even if advanced, had to be interrupted.

The decrease in lithic diversity has been related to a decrease in the mobility of the latest hunter/gatherers, which must have contributed to increased human pressure on the available food resources (Garcea, 2001, 2003). This final part of the Late Acacus saw the production of new types of artefacts, many of them obtained from vegetal resources. Remains of vegetal twisted cords (figure 5) and wooden tools (figure 6) were found near the wooden hut that was still partly preserved at the time of the excavations (Garcea, 2001). The use of vegetal resources for a variety of purposes, that is, tool manufacturing, construction, conservation, winnowing, storage, matting, bedding, etc., confirms the inhabitants' vast knowledge of the potential of their environment and their great craftsmanship skills.

Figure 5: Late Acacus: a twisted cord form the hut area.

A considerable increase in pottery production can be linked with the need for preserving and storing goods. Pottery may also have had social purposes for inter-group relations. The Late Acacus pottery from Uan Tabu was not made locally but was probably imported from the Tassili n'Ajjer, some 70 km southwest from the Tadrart Acacus. In fact, petrographic analyses indicated that this pottery included granite in the pastes, which is not a locally available rock, but is present in the Tassili and farther in the Tibesti (Garcea, 2001; Livingstone Smith, 2001). Polished bone tools and ostrich eggshell beads were also typical productions of the Late Acacus horizon. Moreover, colouring materials, such as ochre and limonite, were often found and may have been used for rock art. Other utensils, such as pots, stone tools, and grindstones exhibited traces of pigment, suggesting they might have been used for preparing the colouring materials for the rock paintings depicted on the wall of the shelter. The earliest rock paintings at Uan Tabu were attributed to the Round Head phase (Mori, 1998). Rock art completes the cultural framework of the Late Acacus as it represents a visual means of marking territory, in the sense of expressing the right of access, land ownership and self-identification (David and Lourandos, 1998). Such a cultural acquisition suggests that a strong sense of self-identity and territoriality was developing among these groups, probably due to demographic growth and the ensuing pressures (Garcea, 2001).

4 Final remarks

Since the beginning of the Holocene, first the climate and at a later period humans were responsible for the

Figure 6: Late Acacus: a wooden perforator.

many enviromental changes in the Tadrart Acacus. In the whole Sahara region, during the Early Holocene, climatic conditions were much moister (Prentice et al., 2000) and temperatures were lower than today. In the range of time covered by the Uan Tabu cultural sequence, the climatic change in the area showed a gradual trend from wet to dry conditions, as is also attested to at Uan Afuda and in the south-western Fezzan (Cremaschi and di Lernia, 1998).

At around 9800 years bp, the main vegetation of the Wadi Teshuinat was grassland, possibly a seasonally swampy grassland. Later, around 8800 years bp, a Saharo-montane type of vegetation expanded in the wadi. New floristic elements gradually appeared in the mountains, and the vegetation became a mosaic of grassland, bushland, and wooded grassland that grew around the site, with forest on the higher belts. The Early Holocene vegetation of the Tadrart Acacus was quite similar to that in the high Saharan mountains (Ahaggar and Tibesti) and in the present Aïr Mountains, Niger (18°N), as Schultz (1987) has already correctly observed. This type of vegetation is characterised by a Sahelian enclave within a Saharan environment, with some relict Mediterranean species, e.g., *Salvadora persica*, *Acacia* species, *Panicum laetum*, *Cenchrus biflorus*, *Dactyloctenium aegyptium*, *Pennisetum violaceum*; whereas in the drier Sahel-Saharan transition zone, species include *Panicum turgidum*, *Aerva javanica*, and the trees/shrubs of *Maerua crassifolia* and *Leptadenia pyrotechnica*. Although annual precipitations are usually less than 100 mm, the bare rock surfaces of the massifs and plateaus concentrate large volumes of run-off water into the wadis and temporary pans, where thick covers of Sahelian elements often grow.

In the cultural sequence, archaeological and plant records show changes which run quite parallel to each other. From the Early to the Late Acacus phases, behavioural changes towards increased plant harvesting and processing were observed, confirming that the plant world was of great importance in the cultural evolution of these hunter/gatherers. Fruits of Boraginaceae and pollen of *Typha*, which are two of the most represented plants in the archaeobotanical record, indicate that the Late Acacus people had a wide knowledge of these plants and their attributes (see also Mercuri, 1999). Even though we can not confirm if they were consciously aware of all the medicinal properties of their plants, we have evidence that they selected plants with different qualities in the various seasons and used them for specific purposes. Not only wild cereals, but many other available plants in the area were a relevant part of the culture and competence of the local people. Data show that during the Late Acacus, hunter-gatherers had already developed the means to transfer their cultural knowledge about the plants in the area from one generation to the next. We can assuredly say that such a process was already in existence since the arrival of people to the mountains but archaeobotanical data still do not give the answer as to when this process actually started.

A final remark concerns the possible symbolic significance of plant harvesting. In this pre-pastoral archaeological context, food seems to have always been the main aim of plant accumulation. Nevertheless, the Late Acacus records suggest that other needs must have induced humans to collect plants. Fodder, bedding, colouring, building, crafting, medical, and many other purposes were fulfilled, proving that plants must have played a crucial role in their everyday life. Plants were also part of their spiritual practices. For example, it is well known that food had been frequently used in offerings to deities and in ritual contexts (Palmer and van der Veen, 2002; van der Veen, 2003).

Though other hypotheses can be proposed, the accumulation of seeds and fruits suggests that votive or ritual ceremonies involving plants may have been practised at Uan Tabu. Although this hypothesis needs more data, the plant records could support the existence of a link between the archaeological sequence and the rock art of the Uan Tabu rockshelter. (di Lernia and Mercuri, 2001).

5 Acknowledgments

We would like to thank the Head of the Department of Antiquities of Libya, Dr. Ali Khadduri, and his staff for their support in carrying out fieldwork and laboratory analyses. We also would like to express our gratitude to the Director of the Interuniversity Centre for Research on the Ancient Sahara and Arid Zones, Prof. Mario Liverani, and to the Director of the Joint Italo-Libyan Mission for Prehistoric Research in the Sahara, Dr. Savino di Lernia. Both directors are attached to the University of Rome "La Sapienza", Italy.

6 References

Accorsi, C.A., Bandini Mazzanti, M., Forlani, L., Mercuri, A.M. & G. Trevisan Grandi, 1999. An overview of Holocene Forest Pollen Flora/Vegetation of the Emilia Romagna Region – Northern Italy. *Archivio Geobotanico* 5, pp. 3-37.

Ayyad, S.M. & P.D. Moore, 1995. Morphological studies of the pollen grains of the semi-arid region of Egypt. *Flora* 190, pp. 115-133.

Ali, S. & S.M.H. Jafri (eds.), 1976-1977. *Flora of Libya*. Al Faateh University, Tripoli.

Barakat, H. & A.G. Fahmy, 1999. Wild grasses as 'Neolithic' food resources in the eastern Sahara: a review of the evidence from Egypt. In: M.Van der Veen (ed.), *The exploitation of plant resources in Ancient Africa*. Kluwer Academic/Plenum Publishers, New York, pp. 33-46.

Barich, B.E., 1992. The botanical collections from Ti-n-Torha/Two Caves and Uan Muhuggiag (Tadrart Acacus, Libya) – An archaeological commentary. *Origini* 16, pp. 109-123.

Bartolomei, P. & A. Rizzo, 2001. Radiocarbon dates of charcoal samples from the Holocene sequence. In: E.A.A. Garcea (ed.), *Uan Tabu in the settlement history of the Libyan Sahara*. Arid Zone Archaeology, Monographs 2, All'Insegna del Giglio, Firenze, pp. 63-68, 237-251.

Bonnefille, R. & G. Riollet, 1980. *Pollens des savanes d'Afrique orientale*. CNRS, Paris.

Bottema, S., 1992. Prehistoric cereal gathering and farming in the Near East: the pollen evidence. *Review of Palaeobotany and Palynology* 73, pp. 21-33.

Castelletti, L., Castiglioni, E., Cottini, M. & M. Rottoli, 1999. Archaeobotanical analysis of charcoal, wood and seeds. In: S. di Lernia (ed.), *The Uan Afuda Cave Hunter-Gatherer Societies of Central Sahara*, Arid Zone Archaeology Monographs 1, All'Insegna del Giglio, Firenze, pp. 131-148, 239-253.

Ceruti, A., Ceruti, M. & G.Vigolo, 1993. *Botanica Farmaceutica e Veterinaria*. Bologna.

Clayton, W.D., Phillips, S.M. & S.A. Renvoize, 1974. Gramineae (Part 2). In: R.M. Polhill (ed.), *Flora of Tropical East Africa*. A.A. Balkema, Rotterdam.

Clayton, W.D. & S.A. Renvoize, 1982. Gramineae (Part 3). In: R.M. Polhill (ed.), *Flora of Tropical East Africa*. A.A. Balkema, Rotterdam.

Corti, R., 1942. *Flora e vegetazione del Fezzan e della regione di Gat*. Reale Società Geografica Italiana, Firenze.

Cremaschi, M., 1998. Late Quaternary geological evidence for environmental changes in south-western Fezzan (central Sahara, Libya). In: M. Cremaschi, S. di Lernia (eds.), *Wadi Teshuinat. Palaeoenvironment and Prehistory in south-western Fezzan (Libyan Sahara)*. All'Insegna del Giglio, Firenze, pp. 13-48.

Cremaschi, M. & S. di Lernia, 1998. The geo-archaeological survey in central Tadrart Acacus and surroundings (Libyan Sahara). Environment and cultures. In: M. Cremaschi, S. di Lernia (eds.), *Wadi Teshuinat. Palaeoenvironment and Prehistory in south-western Fezzan (Libyan Sahara)*. All'Insegna del Giglio, Firenze, pp. 245-298.

Cremaschi, M. & L.Trombino, 2001. The formation processes of the stratigraphic sequence of the site and their palaeoenvironmental implications. In: E.A.A. Garcea (ed.) *Uan Tabu in the Settlement History of the Libyan Sahara*. Arid Zone Archaeology Monographs 2, All'Insegna del Giglio, Firenze, pp. 15-23.

David, B. & H. Lourandos, 1998. Rock art and socio-demography in northeastern Australian prehistory. *World Archaeology* 30, pp. 193-219.

Davis, P.H. (ed.), 1978. *Flora of Turkey and the East Aegean Islands*,Vols. 4, 6, 9. Edinburgh.

Di Lernia, S. & E.A.A. Garcea, 1997. Some remarks on Saharan terminology. Pre-pastoral archaeology from the Libyan Sahara and the Middle Nile Valley. *Libya Antiqua* N.S. 3, pp. 11-23.

Di Lernia, S. & A.M. Mercuri, 2001. Penned Barbary sheep at the Uan Afuda Cave (Central Sahara), around 8000 yrs bp. Suggestion from pollen in dung. In: A. Guarino (ed.), *Proceedings 3rd International Congress Science and Technology for the safeguard of Cultural Heritage in the Mediterranean Basin (9-14 July 2001, Alcalá de Henares, Spain), I*. CNR-Progetto Finalizzato Beni Culturali, Roma, pp. 909-915.

El Ghazali, G.E.B., 1993. A study on the pollen flora of Sudan. *Review of Palaeobotany and Palynology* 76, pp. 99-345.

Faegri, K., Kaland, P.E. & K. Krzywinski (eds.), 1989. *Textbook of Pollen Analysis*. John Wiley & Sons, London.

Fiori, A., 1912. *Boschi e piante legnose dell'Eritrea*. Istituto Agricolo Coloniale Italiano, Firenze.

Garcea, E.A.A., 2001. *Uan Tabu in the settlement history of the Libyan Sahara*. Arid Zone Archaeology Monographs 2. All'Insegna del Giglio, Firenze.

Garcea, E.A.A., 2003. Cultural convergences of northern Europe and North Africa during the Early Holocene? In: L. Larsson, H. Kindgren, K. Knutsson, D. Loeffler, A. Åkerlund (eds.), *Mesolithic on the Move. Papers presented at the Sixth International Conference on the Mesolithic in Europe, Stockholm 2000*. Oxbow Books, Oxford, pp. 108-114.

Garcea, E.A.A., 2004. Modern Behaviour and Cultural Complexity in the Upper Pleistocene and Early Holocene in Western Libya. *Beiträge zur Allgemeinen und Vergleichenden Archäologie* 24.

Goudie, A.S., Colls, A., Stokes, S., Parker, A., White, K. & A. Al-Farraj, 2000. Latest Pleistocene and Holocene dune construction at the north-eastern edge of the Rub al Khali, United Arab Emirates. *Sedimentology* 47, pp. 1011–1021.

Grimm, E.C., 1991-93. *Tilia version 2.0*. Illinois State Museum, Research and Collections Centre, Springfield.

Harlan, J.R., 1989. Wild grass seeds as food sources in the Sahara and Sub-Sahara. *Sahara* 2, pp. 69-74.

Harlan, J.R., 1995. *The living fields: our agricultural heritage.* Cambridge University Press, Cambridge.

Haslam, S.M., 1978. *River plants. The macrophytic vegetation of watercourses.* Cambridge University Press, Cambridge.

Hubbard, R.N.L.B., 1992. Dichotomous keys for the identification of the major Old World crops. *Review of Palaeobotany and Palynology* 73, pp. 105-115.

Jafri, S.M.H. & A. El-Gadi, 1977-1978. *Flora of Libya.* Al Faateh University, Tripoli.

Livingstone Smith, A., 2001. Pottery manufacturing processes: reconstruction and interpretation. In: E.A.A. Garcea (ed.), *Uan Tabu in the Settlement History of the Libyan Sahara.* Arid Zone Archaeology Monographs 2, All'Insegna del Giglio, Firenze, pp. 111-150.

Lowe, J.J., Accorsi, C.A., Bandini Mazzanti, M., Bishop, A., Forlani, L., Van der Kaars, S., Mercuri, A.M., Rivalenti, C., Torri, P. & C. Watson, 1996. Pollen stratigraphy of sediment sequences from crater lakes (Lago Albano and Lago di Nemi) and the Central Adriatic spanning the interval from Oxygen isotope Stage 2 to the present day. *Memorie dell'Istituto Italiano di Idrobiologia* 55, pp. 71-98.

Magid, A.A., 1999. Plant impressions in pottery from the Early Holocene occupation of the cave site of Uan Afuda (Libyan Sahara): a preliminary study. In: S. di Lernia (ed.), *The Uan Afuda Cave Hunter-Gatherer Societies of Central Sahara.* Arid Zone Archaeology Monographs 1, All'Insegna del Giglio, Firenze, pp. 183-187, 239-253.

Maspero, A., 1999. Spinning and plaiting. In: S. di Lernia (ed.), *The Uan Afuda Cave Hunter-Gatherer Societies of Central Sahara.* Arid Zone Archaeology Monographs 1, All'Insegna del Giglio, Firenze, pp. 183-187, 239-253.

Mercuri, A.M., 1999. Palynological analysis of the Early Holocene sequence. In: S. di Lernia (ed.), *The Uan Afuda Cave Hunter-Gatherer Societies of Central Sahara.* Arid Zone Archaeology Monographs 1, All'Insegna del Giglio, Firenze, pp. 149-181, 239-253.

Mercuri, A.M., 2001. Preliminary analyses of fruits, seeds and few plant macrofossils from the Early Holocene sequence. In: E.A.A. Garcea (ed.), *Uan Tabu in the settlement history of the Libyan Sahara.* Arid Zone Archaeology Monographs 2, All'Insegna del Giglio, Firenze, pp. 189-210, 237-251.

Mercuri, A.M. & G. Trevisan Grandi, 2001. Palynological analyses of the Late Pleistocene, Early Holocene and Middle Holocene layers. In: E.A.A. Garcea (ed.), *Uan Tabu in the settlement history of the Libyan Sahara.* Arid Zone Archaeology Monographs 2, All'Insegna del Giglio, Firenze, pp. 161-188, 237-251.

Mori, F., 1965. *Tadrart Acacus. Arte rupestre e culture del Sahara preistorico.* Einaudi, Torino.

Mori, F., 1998. *The Great Civilisations of the Ancient Sahara.* L'Erma di Bretschneider, Roma.

Neumann, K. & D. Uebel, 2001. The cold Early Holocene in the Acacus: evidence from charred wood. In: E.A.A. Garcea (ed.), *Uan Tabu in the settlement history of the Libyan Sahara.* Arid Zone Archaeology Monographs 2, All'Insegna del Giglio, Firenze, pp. 211-213, 237-251.

Ozenda, P., 1958. *Flore du Sahara septentrional et Central.* C.N.R.S., Paris.

Palmer, C. & M. Van der Veen, 2002. Archaeobotany and the social context of food. *Acta Palaeobotanica* 42, pp. 195-202.

Prendergast, H.D.V., Kennedy, M.J., Webby, R.F. & K.R. Markham, 2000. Pollen cakes of *Typha* spp. (Typhaceae) – 'lost' and living food. *Economic Botany* 54(3), pp. 254-255.

Prentice, C.I., Jolly, D. & BIOME 6000 participants, 2000. Mid-Holocene and glacial-maximum vegetation geography of the northern continents and Africa. *Journal of Biogeography* 27, pp. 507-519.

Reille, M., 1992. *Pollen et spores d'Europe et d'Afrique du Nord.* URA CNRS, Marseille.

Reille, M., 1995. *Pollen et spores d'Europe et d'Afrique du Nord. Supplement 1.* URA CNRS, Marseille.

Reille, M., 1998. *Pollen et spores d'Europe et d'Afrique du Nord. Supplement 2.* URA CNRS, Marseille.

Schulz, E., 1987. Holocene vegetation in the Tadrart Acacus: the pollen record of two early ceramic sites. In B.E. Barich (ed.), *Archaeology and environment in the Libyan Sahara. The excavations in the Tadrart Acacus, 1978-1983.* British Archaeological Reports, Oxford, pp. 313-326.

Sheriff, A.S. & M.A. Siddiqi, 1988. Poaceae. In A.A. El-Gadi (ed.), *Flora of Libya* 145. Al Faateh University, Tripoli.

Telford R.J., Heegaard E. & H.J.B. Birks, 2004. The intercept is a poor estimate of a calibrated radiocarbon age. *The Holocene*, 14: 296-298.

Turrill, W.B. & E. Milne-Redhead (eds.), 1952. *Flora of Tropical East Africa.* A.A. Balkema, Rotterdam (and followings).

Uphof, J.C.Th., 1959. *Dictionary of economic plants.* Cramer, Weinheim.

Veen van der, M., 2003. When is food a luxury? *World Archaeology* 34, pp. 405-427.

Verdcourt, B., 1991. Boraginaceae. In: R.M. Polhill (ed.), *Flora of Tropical East Africa.* A.A. Balkema, Rotterdam.

Wasylikowa, K., 1992a. Exploitation of wild plants by prehistoric peoples in the Sahara. *Würzburger Geographische Arbeiten* 84, pp. 247-262.

Wasylikowa, K., 1992b. Holocene flora of the Tadrart Acacus area, SW Libya, based on plant macrofossils from Uan Muhuggiag and Ti-n-Torha Two Caves archaeological sites. *Origini* 16, pp. 125-152, 157-159.

Wasylikowa, K. & J. Dahlberg, 1999. Sorghum in the economy of the early Neolithic nomadic tribes at Nabta Playa, Southern Egypt. In: M. Van der Veen (ed.), *The Exploitation of Plant Resources in Ancient Africa.* Kluwer Academic/Plenum Publishers, New York, pp. 11-31.

White, F., 1983. *The vegetation of Africa.* Unesco, Paris.

Wright, C.W. (ed.), 2002. *Artemisia.* Taylor & Francis, New York.

Beyond paper: use of plants of the Cyperaceae family in ancient Egypt

P.L. CRAWFORD

Department of Anthropology, State University of New York, United States of America

The culture of ancient Egypt centered on the Nile River and its associated ecosystems. Plants of the Cyperaceae family played a prominent role in the culture and economy within these ecosystems. The manufacture of paper from *Cyperus papyrus* is only one example of Cyperaceae use in the past.

Because the choices people make when selecting raw materials reflect the materials available and imposed by the limits of their environment, plants of the Cyperaceae family and other hydrophilic plants were most likely those most commonly exploited in the everyday aspects of life. And yet these plants may be underrepresented in the archaeological record, especially in domestic contexts where the ideal preservation conditions of the dry desert sands and closed tombs are not present.

This paper examines how Cyperaceae were or may have been used from prehistoric through the Dynastic periods. Some methods of analysis and types of evidence employed to reconstruct past uses are also discussed.

1 Introduction

Plant remains from archaeological sites can tell us about diet, economy, and other important aspects of society. Plant remains can also provide us with insight into how people adapt to their natural environment. In the case of ancient Egypt plant remains are preserved primarily by charring and desic-cation and in the special preservation contexts of burials.

2 Sources of information regarding plant use in ancient Egypt

The presence of plant remains in archaeological context does not necessarily reveal how these plants were used, only that they were present and available for use and brought into the site by human activity. Use, therefore, must be ascertained by means other than just interpretation of context.

Information for the use of Cyperaceae plants in the past is derived from both direct and indirect types of evidence. Direct evidence for plant use includes the presence of the actual remains in archaeological contexts. Some kinds of direct evidence can be more informative than others. For example, the presence of plant parts in coprolites indicates ingestion but does not tell whether the item was used as food or had a ritual or medicinal purpose.

The presence of well-preserved plants in tomb contexts can also give reliable evidence of use, especially when preservation is good. For example, plant materials preserved in vessels suggest food items, while plants used as funeral bouquets and offerings or as parts of mummy garlands, suggest burial ritual use.

Indirect evidence for plant use in Egypt comes from numerous wall paintings and reliefs. Representations from tombs show scenes such as river and marsh environments, gardens, crop processing, banquets, and various rituals or manufacturing activities using plants. Models, also found in tomb contexts, depict domestic activities where plants are being used or processed although the exact species may not be indicated.

Another form of evidence is that of written records. Ancient documents such as the Ebers medical papyrus provide us with information regarding various medical uses as well as the preparation and application of specific types of plants. Late period accounts by historians such as Theophrastus, Pliny, Dioscorides, Herodotus, and Diodorus also provide descriptions of plant use in ancient times in Egypt and elsewhere in the classical world.

Ethnographic studies of groups living in the present or recent past in similar environments can suggest various uses of specific plants and the associated processing procedures. These types of studies are especially useful for understanding how plant materials might be modified, deposited, and selectively preserved (or destroyed) in the archaeological record.

Since the preservation of plants and the organic objects manufactured from plants is selective, these secondary types of evidence are especially informative but are not always supported by the presence of the plant re-mains themselves. Ancient Egypt is a special case, because of its bountiful secondary information regarding plant use and its excellent preservation conditions in tomb contexts. Both situations provide us with the opportunity and means to fill in the large gaps regarding the relationship between the presence of actual plant remains in the archaeological record and the relative importance of a specific group of plants in the past.

3 General characteristics of Cyperaceae

Plants of the Cyperaceae family are called sedges and are described as erect, grass-like herbaceous plants with heights varying from less than half a meter to more than five meters. As monocots they have narrow linear leaves of varying widths which are sometimes reduced to sheaths only. The stems are three-sided or cylindrical, not jointed, and are generally solid (pith-filled) rather than hollow like grasses. The flowers are small, arranged in single or many-flowered spikelets borne in clusters on the stem or as terminal inflorescences (figure 1). The fruit is a small one-seeded nutlet (achene), three-sided, flat, or lenticular in shape. Cyperaceae have starchy rhizomes that are tuberous in some species such as *Cyperus rotundus* and *Cyperus esculentus* (Täckholm, 1974: pp. 770-790; Feinbrun-Dothan, 1986: pp. 346-375).

Cyperaceae have various salt tolerances and grow in dense clumps or stands in wet or moist environments such as river and canal banks, irrigation ditches, marshes, and standing or brackish water. They are found in association with other moisture-loving plants of the Typhaceae, Juncaceae, and Gramineae families (Täckholm, 1974; Zahran & Willis, 1992).

According to Täckholm (Täckholm, 1974) the genera found in modern Egypt are *Cyperus*, *Scirpus*, *Eleocharis*, *Fimbristylis*, *Carex*, *Cladium*, *Fuirena*, and *Schoenus*. Plants of the *Cyperus* and *Scirpus* genera figure most prominently in the archaeological record of Egypt. *Cyperus papyrus*, the plant from which paper was made and perhaps the most important economic plant of this family, was thought to be extinct in modern times until reported by El-Hadidi in 1968 in the Wadi al Natrun (El-Hadidi, 1971; Zahran & Willis, 1992: pp. 332-333).

4 Cultural importance of Cyperaceae in ancient Egypt

The Cyperaceae family of plants appears to have been of great importance to ancient Egyptians. As a primary component of the environment of the delta marshes, Nile riverbanks, irrigation canals and flooded fields, it figured pro-minently in the cultural ecology of the society. Its importance is attested by early hieroglyphic representations suggesting that Cyperaceae use was well-established before writing was developed. The symbol for a 'stand of papyrus' represents the syllable 'HA' and is a component of the word meaning 'north' (figure 2, top). Papyrus symbolically represents Lower Egypt (the Delta) in iconography while the lotus plant is the symbol for Upper Egypt (the river). The hieroglyphic for a 'sedge plant' (figure 2, middle) represents the sound 'SW' a component of the word for 'south', and also indicates 'Upper Egypt' when combined with a representation of a bee to indicate the 'king of Upper and Lower Egypt'. Also important in the past, the common reed (*Phragmites communis*), a moisture-loving member of the grass family, is represented as the 'flowering reed' and the sound for 'I' or 'Y' in the basic alphabet (figure 2, bottom) (McDermott, 2001: pp. 22, 38, 90).

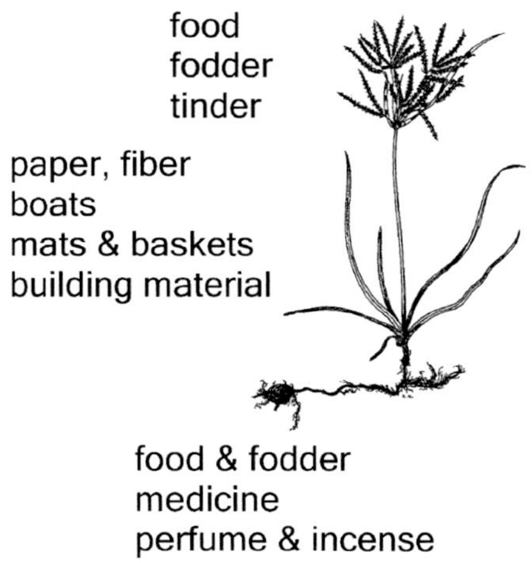

Representations of sedge plants also figure prominently in the art and architecture of Ancient Egypt. The fan-like inflorescence of the papyrus plant is commonly represented in wall paintings and reliefs. As an architectural element, the closed papyrus umbels are depicted on the column capitals of monumental buildings such as the temple of Amun at Luxor and the temple of Amenophis at Thebes (Prisse d'Avennes, 2000; Malek, 2003: p. 166). The bundled stems are represented as supports in stone columns, mimicking the original use of bundled papyrus as roof supports (Hepper, 1992: pp. 182-184). The papyrus inflorescence is depicted in the capitals of the Ramesseum at

Figure 1. Wild nut grass (Cyperus rotundus) as a typical example of Cyperaceae, indicating various uses (after Feinbrun-Dothan, 1986: figure 472).

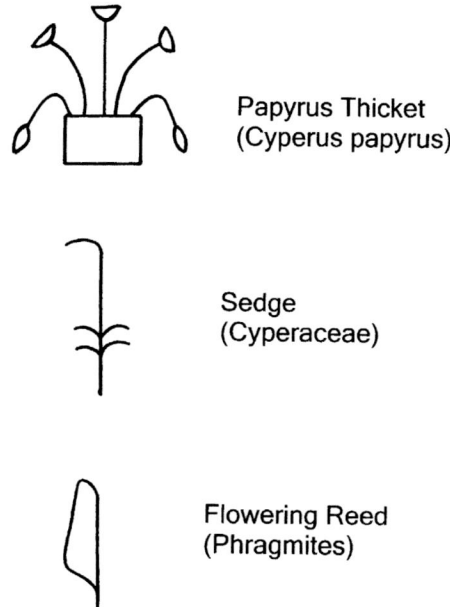

Figure 2. Plants used as symbols in ancient Egyptian hieroglyphics.

Thebes among others. It is also a decorative element in household items such as furniture, cosmetic boxes and utensils, jewelry, and pottery (Prisse d'Avennes, 2000).

Papyrus processed into paper was used as early as the first dynasty and Old Kingdom (Lucas, 1962: p. 163ff). Beyond paper, however, what did Cyperaceae family of plants contribute to the livelihood of ancient Egypt?

5 Evidence for Cyperaceae use in ancient Egypt

5.1 Food
Some of the earliest evidence of Cyperaceae remains comes from Wadi Kubbaniya dating to 19,000-17,000 BC (Hillman *et al.*, 1989: pp. 162-142). Charred rhizomes and tubers of yellow nut grass (*Cyperus rotundus*) were reported. It is suggested that the roasting of these bitter tubers to process them for consumption facilitated their preservation in the archaeological record. Ethnographic observations suggest an alternative method of preparing nut grass tubers, that is, by soaking, drying, and grinding into a powder or pounding into a gruel. This method of processing would not have been as likely to leave an archaeological signal. The presence of items such as mortars and pestles in related contexts further suggests the processing of tubers when the plant parts themselves might not otherwise be present. Also present at Kubbaniya are tubers of club rush, *Scirpus maritimus* or *S. tuberosus* (Hillman *et al.*, 1989: p. 192). Tubers of yellow nut grass and tiger nut (*Cyperus esculentus*) have been reported from the Neolithic through the pharaonic periods (Täckholm & Drar, 1950: pp. 60-69; Darby *et al.*, 1977: pp. 619ff, 649; Germer, 1985: pp. 245-246; Hillman *et al.*, 1989; Manniche, 1989: p. 98; Zohary & Hopf, 2000: p. 198). Their use as food is attested by their presence in stomachs of predynastic burials. They are reported at Neolithic Merimde (Caton-Thompson & Gardner, 1934), New and Middle Kingdom Memphis (Murray, 1993: pp. 165-168), Second Intermediate Period Tell el-Maskhuta (Crawford, 1994: pp. 124-126) and at Ptolemaic El-Hibeh (Wetterstrom, 1984) among other sites.

Evidence for the use and processing of tiger nuts is depicted in 18th Dynasty wall paintings in the tomb of Rekhmire in Thebes (figure 3) (Germer, 1985: p. 245; Manniche, 1989: p. 42). Accounts by Theophrastus (Theophrastus, IV 8.2) also indicate that *Cyperus* tubers were used as food (Manniche, 1989: p. 98). The tubers and rhizomes of many Cyperaceae, especially in the genera *Cyperus* and *Scirpus*, are today consumed raw, boiled, or roasted. Tiger nut is cultivated today for many uses and in some areas is considered a famine food (Simpson & Inglis, 2001).

Other parts of Cyperaceae plants besides the tubers were probably consumed in ancient Egypt. Nutlets of club rush (*Scirpus maritimus* or *S. tuberosus*) were found in fecal matter also at Wadi Kubbaniyah (Hillman *et al.*, 1989: pp. 196-198). The seeds appear to have been roasted before consumption facilitating their preservation in the archaeological record. Seeds of *Cyperus*

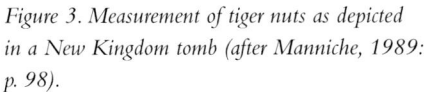

Figure 3. Measurement of tiger nuts as depicted in a New Kingdom tomb (after Manniche, 1989: p. 98).

conglomeratus were found in a basket that dated to prehistoric times. Whether they were collected as food for humans or for other purposes such as medicine is unknown. Ethnographic accounts indicate that seeds of Cyperaceae plants are eaten roasted or raw, but usually ground into a paste and baked (Simpson & Inglis, 2001). Use as a paste cannot be proved in the past since such processing would not leave archaeological remains.

Seeds from Cyperaceae plants used as fuel or as food for animals whose dung is used as fuel, given the proper temperature conditions, would be preserved in the archaeological record by charring. Such evidence of possible use has been reported at Kom el Hisn (Moens & Wetterstrom, 1988) and Tell el Maskhuta (Crawford, 1994: pp. 124-129, appendix 3) where Cyperaceae seeds are present in high quantities compared to the other weed species present. At Amarna, Cyperaceae seeds are reported in deposits of debris associated with animal pens and habitation in the workmen's village. Renfrew speculates that Cyperaceae plants were used here as animal fodder or as bedding (Renfrew, 1985). Most of the plants of the Cyperaceae family are used today as animal fodder. All parts of the plants are consumed by livestock. In some cases, tubers and rhizomes only are fed to pigs and cattle (Simpson & Inglis, 2001: pp. 292-293).

Roots and stems of papyrus and other sedges were also edible (Darby et al., 1977: pp. 645, 649). According to historical accounts the stalks of the papyrus plant were eaten. The lower half meter of the stem was either baked in a hot vessel (Herodotus, II.92), boiled or roasted or chewed raw like sugar cane (Dioscorides, I.115; Theophrastus, IV.8.4,5; Diodorus, 1.80.5; Manniche, 1989: pp. 99-100). Ethnographic information suggests that the roots, rhizomes, and tubers of many of the Cyperaceae species are edible though some need further processing to remove bitterness or toxins (Simpson & Inglis, 2001). Any kind of processing that involves fire or heat increases the possibility that identifiable plant remains would be preserved in the archaeological record.

5.2 Medical Uses

Uses that might not yield obvious archaeological remains fall in the category of medicinal items. Cases in which plants used medicinally are ingested may not result in archaeological preservation. Information about medicinal uses comes from ancient sources such as the Ebers papyrus, historical accounts, and ethnographic observations.

Decoctions from tubers and rhizomes of Cyperaceae were used for colic, indigestion, or cough. Ashes of the papyrus plant were put into wine to induce sleep (Pliny, XXIV.51), used as a tooth powder or a component of poultices (Manniche, 1989: p. 100). Such ancient uses are supported by ethnographic accounts where species of *Cyperus* are used as a component of a wash for rashes. The roots of *C. articulatus* are chewed for headaches. Decoctions of the tubers of *C. obtusiflorus* are used as a cough remedy. *C papyrus* leaves are applied to treat edema while decoctions of the roots and flowers treat heart palpitations. *C. longus* is used as a diuretic and to treat rheumatism. The tubers of *C. rotundus* are used to treat widely diverse conditions such as colic and indigestion, cough, heart ailments. Most genera of sedges are used in some aspect of medicine; for example, as a sedative, diuretic, carminative, stimulant, purgative, tonic, pain relief, or aphrodisiac (Täckholm & Drar, 1950; Kokwaro, 1976: pp. 233-234; Boulos, 1983; Boulos & El-Hadidi, 1984: pp. 58-63; Simpson & Inglis, 2001).

Sedges used externally for medicinal purposes are more likely to leave evidence than ingested plants. Bandages, compresses, poultices, and salves were made from plant materials. Papyrus was combined with other materials in a bandage to bind stiff limbs. Other external uses of papyrus were as an eye compress or as a poultice to draw out infections and cure callosities (Manniche, 1989: pp. 99-100). Tiger nuts (*C. esculentus*) also yield oil as a liniment for stiffness (Germer, 1985: p. 245). Similar external uses are applied today. For example, the tubers of *Cyperus articulatus* are used as a poultice to reduce swelling or as a remedy for snakebite. They are also burned to fumigate the body during sickness (Simpson & Inglis, 2001: pp. 283-285). Modern Africans plait the flower stems of *Fimbristylus ovata* into wrist bangles for rheumatism. Plant parts are also worn as a protective charm against various illnesses (Kokwaro, 1976: pp. 233-234).

5.3 Papyrus boats

Uses of plant parts that do not include not food or medicine are more likely to leave evidence in the archaeological record. Such uses are in the manu-facture of material goods commonly used in everyday circumstances or in burial and religious contexts.

The nature of the Egyptian landscape, because of its focus around the Nile River, demanded the presence of some type of watercraft. One of the earliest uses of *Cyperus papyrus* was for the manufacture of reed-bundle rafts. Wall reliefs in the 5th Dynasty tomb of Ptahhotep at Saqqara depict the construction of these rafts. People are shown harvesting stalks of papyrus and tying them into bundles. These bundles being lashed tightly together to make light rafts or boats suitable for a short distance transport of people or goods across and up and down the crocodile-infested Nile. The shallow,

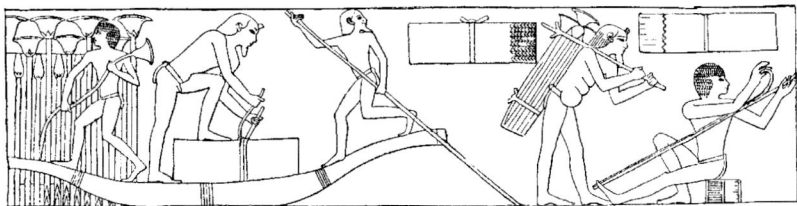

Figure 4. A Theban tomb painting of men in a papyrus boat collecting and preparing the papyrus plant.

raft-like watercraft are illustrated in fishing, fowling or hunting scenes in the marshes. Such craft were used in time periods ranging from early dynastic to as late as Ptolemaic period (McGrail, 2001: p. 21).

In various representations, reed-bundle boats are distinguished from wooden craft by the presence of vertical lines depicting the lashings that bind the papyrus bundles together (figure 4). The pith-filled thick stems of the papyrus plant provide air pockets that make the boats light and easily float-able and less susceptible to water logging. Stems of reeds or grasses are hollow and more brittle and would not provide the same sturdiness and buoyancy as Cyperaceae. These other plant materials, however, were probably used as well. Boats made from these easily acquired raw materials would have been inexpensive, easily replaceable and accessible to all (Digby, 1954: pp. 730-736; Baines & Malek, 2000: pp. 68-69; McGrail, 2001: pp. 14-54). Although there are multiple representations of papyrus boats in Egyptian wall paintings, no remains of papyrus craft have actually been found in ancient context. Wooden boats are represented in tomb models and are found as sacred craft for travel to the underworld after death at the sites of Giza, Abydos, Dahshur and others (McGrail, 2001: pp. 37-39). Wood was used for larger river craft and trade vessels, but was a scarce commodity in ancient Egypt and not likely to have been used for everyday river commerce (Miegs, 1998).

Reed bundle water craft are common in Africa and southern Iraq, as well as parts of the New World (Hodges, 1970: pp. 48, 92-96; McGrail, 2001: p. 21). In the New World these boats are made of reeds (*Phragmites communis* and *Scirpus riparius*), rushes (*Typha angustifolia*) and Palm (*Maurita vinifera*) (McGrail, 2001: p. 404). The same or similar plants could easily have been used in Ancient Egypt as convenient but perhaps not as efficient substitutes for papyrus.

5.4 Fibers, mats and basketry
Fiber manufacture, mat weaving, and basketry are early closely related technologies. Archaeological remains of products of these technologies have been found in sites dating from the Neolithic and earlier (Crowfoot, 1955: pp. 413-455). The raw materials used in these technologies include plants such as, flax, halfa grass, rushes, sedges, grasses, palm and reed. The term 'reed' can refer to several plants such as 'common reed' (*Phragmites communis*), *Arundo donax* or to Cyperaceae (Grant, 1955: pp. 449-450).

Cordage was made of plant material sometimes combined with animal hair or hide. Everyday items such as nets, traps, bindings, etc. were made of cordage (Hepper, 1992: pp. 176-177). Palm fiber was generally used for rope making. Other materials such as papyrus, however, have similar tensile strength properties (Lucas, 1962: p. 161). Rope made from the pith of papyrus was found inside the Step Pyramid of Djoser at Saqqara as well as in the Tura Caves near an old stone quarry (Lucas, 1962: p. 161; Hepper, 1992: pp. 176-177). Papyrus stems, being processed and twisted into fibers to make rope, are depicted on the walls of the tomb of Ptahhotep in Saqqara (Forbes, 1956: pp. 61-62). Accounts by Theophrastus (Theophrastus, IV: 8,4) and Pliny (Pliny, XIII: 22) state that papyrus was used for rope-making (Ryan, 1988). Papyrus was also made into twine or fiber that could actually be weaved into cloth for sails, clothing, and other uses (Herodotus, II.96; Pliny, XII.72; Theophrastus, IV: 8,4; Forbes, 1956: pp. 61-62; Manniche, 1989: p. 99). Palm is still used for fiber in modern times but many plants of the Cyperaceae family are also currently used, especially the genera *Cladium*, *Cyperus* and *Scirpus* (Simpson & Inglis, 2001).

Weaving technology is applied to the manufacture of mats from plant materials. Mats can be weaved by hand or constructed on a ground loom. Loom construction of mats is depicted in the 12th Dynasty tomb of Khety (Forbes, 1956: p. 179). There is a wooden model of a ground loom in the tomb of Meketre (Fagan, 2001: pp. 162-163). Early matting was made from flax, halfa grass, grass or "reeds', palm ribs and fiber, rushes (*Juncus* sp.), reed-mace (*Typha* sp.), and sedges (Täckholm & Drar, 1950; Lucas, 1962; Hepper, 1992: p. 176).

Mats have been found in archaeological context used as wrappings for predynastic burials. The mats found in these burials are described as being made of 'reeds', but without specific identification of the actual plant material used is not known (Lucas, 1962). An early dynastic coffin at Tarkhan described as a 'reed' hamper was found later to be made of *Phragmites* (Germer, 1987: p. 245). Mats made of the leaves and stems of *Cyperus articulatus* as well as *C. alopecuroides*, and *C. papyrus,* have been found in archaeological contexts ranging in time from the predynastic to Roman period (Germer, 1985: pp. 240-250).

Matting can be used as floor covering and as parts of furniture such as chair seats. It was also probably used as part of structures as walls and roofing. Houses of prehistoric villagers living along the river could have been constructed of reed mats or bundles. If boats of bundled papyrus were com-mon at this time then it would have been a small leap of technology to use papyrus or other easily obtained similar plants for house construction. 'Structures' made from mats or reed bundles tied together were also used as shelters for portable sacred images, or as temporary shelter in fields, on boats, or as canopies (Kemp, 1991: pp. 91-100). Huts constructed of rush matting are depicted on 3,000 BC seals from Mesopotamia and are found in modern Africa as well as in the Euphrates Delta. (Hodges, 1970: pp. 38-39, 48) No actual remains of reed or rush structures have been found however. The plants used for construction therefore cannot be identified as Cyperaceae or otherwise.

Mat making today is also done on ground looms using a variety of plants. Foxtail sedge (*Cyperus alopecuroides*) is cultivated in the Fayum for the manufacture of mats and chairs (Boulos & El-Hadidi, 1984, p. 5) and used as a border plant in the Delta. *Cyperus articulatus, C. esculentus, Scirpus articulatus, S. australis*, and *S. lacustris*, to name only a few, are also used for making mats in modern Egypt (Boulos & El-Hadidi, 1984: p. 59; Simpson & Inglis, 2001).

Related to the technology of mat making, sandal makers used similar techniques and plant materials. Palm, rush, grass, reed, and sedges were used to make different parts of the sandals. Both the rind and stem of papyrus have been identified as parts of sandals from the New Kingdom (Täckholm & Drar, 1950: p. 99; Germer, 1985: p. 249; Feindt, 2001).

Basketry provides the technology for the manufacture of many common household items. Although not preserved well in many domestic contexts, baskets are commonly preserved in tombs and sometimes in the dry desert sands of Egypt. In some cases basket remains can be identified from im-pressions on clay pots (Crowfoot, 1955: p. 418).

The technology involved is much like weaving but without the use of a loom and without the need for extensive preprocessing of the raw materials. In ancient Egypt baskets and containers were usually made from palm fronds (both date and doum), as they are in modern times (Lucas, 1962: p. 156). The use of palm was also noted by Theophrastus (Theophrastus,

IV: 2,7). Other basketry materials included flax, halfa grass, cotton grass, rushes, reeds, sedges, and other plant stems (Hepper, 1992: p. 175).

Early well-preserved baskets made of grass are found in the Fayum from the Neolithic Period where in-ground granaries and storage pits were lined with baskets (Caton-Thompson & Gardner, 1934: pp. 43, 44, 46, 89; Crowfoot, 1955: p. 418). Baskets are also common finds in tombs. Cyperaceae plants associated with ancient Egyptian baskets include *Scirpus lacustris, S. lito-ralis, Cyperus dives,* and *Cyperus papyrus* (Germer, 1985: pp. 444-445; Hepper, 1992: pp. 174-175). Hepper speculates that the basket in which Moses was found was that of papyrus sealed with pitch (Hepper, 1992: p. 175). Papyrus boxes were made of thin slices of papyrus pith supported on a framework of reeds (Hepper, 1992: p. 175). Sieves from Roman periods were made of stems of *Cyperus schimperianus* (Forbes, 1954: p. 182, figure 24; Germer, 1985: p. 245). Although palm is still an important raw material for baskets, plants of the Cyperaceae family such as *Cyperus articulatus* are also used in modern times (Simpson & Inglis, 2001: pp. 283-285).

5.5 Other uses
Beyond the most obvious uses of Cyperaceae in ancient Egypt, there are many others, such as fuel, perfume, and various funerary or ritual uses.

The dried stems and leaves of sedges provide a quick burning fuel or tinder. The roots or rhizomes of Cyperaceae are slow burning and could have been used as a substitute for wood (Manniche, 1989: p. 99). In the past, such uses would have allowed preservation by charring in domestic contexts. The rhizomes of many plants of the Cyperaceae family have a sweet, delicate aroma, making them useful as incense, perfume, or for the production of scented unguents or solutions. Cyperaceae rhizomes have been found in the hand of a Theban mummy and also in a toilet case (Germer, 1987: pp. 246-247). Rhizomes of *Cyperus articulatus* have been found in a garment box, presumably as perfume to scent clothing, and in a leather bag at 12th Dynasty Kahun perhaps for a similar purpose of scenting the body (Germer, 1987: pp. 244-245). *Cyperus rotundus* rhizomes were mixed with fats to make scented unguent cones worn on the head by ancient Egyptians (Manniche, 1989: pp. 52-53). In modern Egypt, the rhizomes of *Cyperus longus, C. articulatus, C. esculenttus, C. rotundus* and several species of *Scirpus* are fragrant and used as perfume or to give scent to clothing or hair. The rhizomes of *Cyperus* and *Scirpus* plants are burned as incense making them candidates for preservation in the past (Täckholm & Drar, 1950: pp. 77ff;

Boulos & El-Hadidi, 1984: pp. 59, 62; Manniche 1989: p. 58; Simpson & Inglis, 2001).

Uses of Cyperaceae in tomb contexts include that of stuffing material for mummies. Stems and leaves of *Scirpus tuberosus* were used as crocodile stuffing (Germer, 1985: p. 243). Old pieces of papyrus paper were often recycled in mummy wrappings. Also in tomb contexts, *Cyperus papyrus* and lotus (*Nymphaea*) are popular components of funeral bouquets and garlands (Manniche, 1989: pp. 22, 99-100). In the tomb of Tutankhamen papyrus strips form a base for garlands on the mummy (Manniche, 1989: p. 27). Many other plants of the Cyperaceae family have been found as parts of bouquets or mummy garlands. Split culms of *Scirpus inclinatus* and a stem with fruit of *Eleocharis palustris* have been identified in Greco-Roman and Roman funeral garlands (Täckholm & Drar, 1950: p. 34; Germer, 1985: pp. 242-243; Germer, 1987: p. 246).

As an example of Cyperaceae used in a non-funerary ritual context, tubers of *C. esculentus* were ground into a paste and added to honey and flour as a component of tribute loaves/festival breads made as special temple offerings (Wilson, 1988: p. 18; Manniche, 1989: pp. 42-43).

Modern uses of Cyperaceae that may also be valid for the past, include insect repellents, poisons, water clarification, and soap making (Simpson & Inglis, 2001).

6 Methods of analysis

In spite of the many examples noted, the Cyperaceae family may be under-represented in terms of identified archaeological remains. Many materials from early excavations possibly composed of Cyperaceae parts were simply described as being made from "reeds" or "rushes", non-specific terms that can be applied to plants from the grass (Gramineae), rush (Juncaceae and Typhaceae), or sedge families. Since organic material is not readily preserved, except under special circumstances, many uses of Cyperaceae have left no visible trace. Stems, roots and tubers, and leaf materials that have decomposed or are components of ash are difficult to retrieve and identify. Before flotation was routinely used as a retrieval method, the extremely small seeds (nutlets) and plant parts were not extracted from the soil. Some uses can only be inferred from the ethnographic models and the ancient illustrations and texts give us information regarding use but do not necessarily identify plant remains. For some of these "elusive" remains there are methods of analysis that have been and continue to be used in the identification of plants remains. Such methods include the use of both light microscopy and electron microscopy applied to materials that are badly preserved, fragmentary, or not visible to the naked eye.

An example of the use of microscope analysis on macroremains is the study by N.M. Waly on the organic materials from the Gabalein area of Egypt (Waly, 1999: pp. 261-272). The examination of thin sections of plants making up household items revealed that foxtail sedge (*Cyperus alopecuroides*) was used to make thin ropes, matting, and part of a basket. Other plant materials included flax, halfa grass, bitter rush, date palm, and common reed (Waly, 1999: pp. 263-265). Similar methods were used by Feindt (Feindt, 2000) to examine the various plant materials used in the manufacture of sandals in ancient Egypt.

Methods using phytolith and pollen analysis can give us information about the presence or use of plant materials when no visible remains are apparent in soil samples or as residues in containers or storage facilities. Phytolith analysis of sediments from Tel Miqne in Israel revealed the presence of common reed (*Phragmites*) and *Arundo* grass in various contexts of the site (Ollendorf, 1987). Sediments sampled from contexts where organic materials such as reed mats may have decomposed, such as floors, silos, or the lining of storage pits would most likely produce phytolith material. Phytolith analysis can be applied to ash remains as well as sediments to identify what kinds of fuels were used when the structure of the original plant material is no longer intact. To facilitate the identification of sedge phytoliths in sediments an extensive reference collection of Cyperaceae plants has been assembled (Ollendorf, 1992).

Pollen analysis can give us similar information. In cases where pollen is consumed as food, analysis of coprolites would yield specific dietary information. Hillman *et al.*(Hillman *et al.*, 1989: pp. 223-225) suggest that the pollen of reedmace (*Typha*) was appreciated as a food resource in prehistoric Kubanniya.

Various types of chemical analyses can help determine the nature of materials processed. For example, the grinding stones found at Wadi Kubbaniya were tested using laser microprobe analysis and pyrolysis mass spectrometry to reveal the presence of cellulose or other starches consistent with tuber processing, as opposed to proteins or fats found in nuts and other fruits. Although these analyses did not identify the plant species, they did allow the excavator to determine the type of plant materials processed (Hillman *et al.*, 1989: pp. 190-191; Jones, 1989: pp. 260-266). Plant products producing specific chemical fingerprints may allow us to identify specific plant species in future analyses.

7 Summary and conclusion

The documented and possible uses of plants of the Cyperaceae family in ancient Egypt have been discussed. Many uses of this readily available resource cannot be detected archaeologically, but depend on the documents and representations unique to Dynastic Egypt for elucidation. This circumstance, therefore, is a lesson in cases of 'presence vs. absence' in the archaeological record. Preliterate past societies such as those studied in many parts of Africa do not provide documentation to aid interpretations of the plant remains, present or absent. The absence of remains of plants that otherwise figure prominently in the environment and therefore in the cultural ecology of a society must be considered a consequence of the taphonomic pro-cess, not as a cultural 'blind spot'. Indirect sources of information such as ethnography and ethnoarchaeology must be considered in order to explain and/or fill the gap in the archaeological record. Methods of analysis that address the less visible aspects of plant remains must also be employed.

8 Acknowledgments

I would like to thank my husband, John Shea, for his advice and great assistance in the computer preparation of this document and Carola Borries for assistance with the translation of German. I would also like to thank my colleagues in the IWAA for their helpful comments and interest in my presentation.

9 References

Baines, J. & J. Malek, 2000. *Cultural atlas of Ancient Egypt*, revised edition. Check-mark Books, New York.

Boulos, L., 1983. *Medicinal plants of North Africa*. Reference Publications inc, Algonac, Michigan.

Boulos, L. & M.N. El-Hadidi, 1984. *The weed flora of Egypt*. The American University in Cairo Press, Cairo.

Caton-Thompson, G. & E.W. Gardner, 1934. *The Desert Fayum*. Royal Anthropological Institute, London.

Crawford, P., 1994. Man-land relationships in the Wadi Tumilat of Egypt at Tell el-Maskhuta: A Paleoethnobotanical Perspective. PhD thesis, Boston University. Ann Arbor, Michigan, University Microfilms.

Crowfoot, G.M., 1955. Textiles, basketry, and mats. In: C. Singer, E.J. Holmyard & A.R. Hall (eds), *A History of Technology*, Volume I. Clarendon Press, Oxford, pp. 413-455.

Darby, W.J., Ghalioungi, P. & L. Grivetti, 1977. *Food: the gift of Osiris*. Academic Press, New York.

Digby, A., 1954. Boats and ships. In: C. Singer, E.J. Holmyard & A.R. Hall (eds), *A History of Technology*, Volume I. Clarendon Press, Oxford, pp. 730-743.

Diodorus, 1968. *Diodorus of Sicily*, translated by C.H. Oldfather. Loeb Classical Library. Harvard University Press, Cambridge, Massachusetts.

Dioscorides, 1959. *The Greek herbal of Dioscorides*, translated by J. Goodyer, R.T. Gunther (ed.). Oxford University Press, Oxford.

El-Hadidi, M.N., 1971. Distribution of *Cyperus papyrus* and *Nymphaea lotus* in inland water of Egypt. *Mitteilung Botanische Staatssammlung* 10, pp. 470-475.

Fagan, B., 2001. *Egypt of the Pharaohs*. National Geographic, Washington.

Feinbrun-Dothan, N., 1986. *Flora Palaestina*, IV. Israel Academy of Sciences and Humanities, Jerusalem.

Feindt, F., 2001. Botanische Untersuchung Altägyptischer Sandalen. *Mitteilung AltÄgypten* 30, pp. 284-309.

Forbes, R.J., 1956. *Studies in Ancient Technology*, Volume IV. E.J. Brill, Leiden.

Germer, R., 1985. *Flora des pharaonischen Ägypten*. Verlag Philipp von Zabern, Mainz am Rhein.

Germer, R., 1987. Ancient Egyptian plant-remains in the Manchester Museum. *Journal of Egyptian Archaeology* 73, pp. 245-246.

Grant, J., 1955. A note on the materials of ancient textiles and baskets. In: C. Singer, E.J. Holmyard & A.R. Hall (eds), *A history of technology*, Volume I. Clarendon Press, Oxford, pp. 447-451.

Hepper, F.N., 1992. *Illustrated encyclopedia of bible plants*. Inter Varsity Press, Leicester.

Herodotus, 1946. *Herodotus I-II*, translated by A.D. Godley. Loeb Classical Library. Harvard University Press, Cambridge, Massachusetts.

Hillman, G., Madeyska, E. & J. Hather, 1989. Wild plant foods and diet at Late Paleolithic Wadi Kubbaniya: The evidence from the charred remains. In: A.E. Close (ed.), *The prehistory of Wadi Kubbaniya*, Vol. 2. Southern Methodist University Press, Dallas, pp. 162-242.

Hodges, H., 1970. *Technology in the Ancient World*. Alfred A. Knopf, New York.

Jones, C.E.R., 1989. Archaeochemistry: fact or fancy? In: A.E. Close (ed.), *The pre-history of Wadi Kubbaniya*, Vol.2. Southern Methodist University Press, Dallas, pp. 260-266.

Kemp, B., 1991. *Ancient Egypt: anatomy of a civilization*. Routledge, London.

Kokwaro, J.O., 1976. *Medicinal plants of East Africa*. East African Literature Bureau, Nairobi.

Lucas, A., 1962 *Ancient Egyptian materials and industries*, 4th edition, revised by J.R. Harris. Edward Arnold, London.

Malek, J., 2003. *Egypt, 4000 years of art*. Phaidon Press, New York.

Manniche, L., 1989. *An ancient Egyptian herbal*. University of Texas Press, Austin.

McDermott, B., 2001. *Decoding Egyptian hieroglyphs*. Chronicle Books, San Francisco.

McGrail, S., 2001. *Boats of the world*. Oxford University Press, Oxford.

Meiggs, R., 1998. *Trees and timber in the ancient Mediterranean World*, second edition. Clarendon Press, Oxford.

Moens, M.-F. & W. Wetterstrom, 1988. The agricultural economy of an Old Kingdom town in Egypt's West Delta: insights from the plant remains. *Journal of Near Eastern Studies* 47, pp. 159-173.

Murray, M., 1993. Recent archaeobotanical research at the site of Memphis. In: W.V. Davies & R. Walker (eds), *Biological anthropology and the study of ancient Egypt*. British Museum Press, London, pp. 165-168.

Ollendorf, A.L., 1987. Archaeological implications of a phytolith study at Tel Miqne (Ekron), Israel. *Journal of Field Archaeology* 14, pp. 453-463.

Ollendorf, A.L., 1992. Towards a classifcation scheme of sedge (Cyperaceae) phytoliths. In: G. Rapp Jr. & S.C. Mulholland (eds), *Phytolith systematic, emerging issues*. Plenum Press, New York, pp. 91-111.

Pliny, 1938-1956. *Natural History*, translated by H. Rackham & W.H.S. Jones. Loeb Classical Library. Harvard University Press, Cambridge, Massachusetts.

Prisse d'Avennes, E., 2000. *Atlas of Egyptian art*. Cairo, The American University in Cairo Press.

Renfrew, J., 1985. Preliminary report on the botanical remains. In: B. Kemp, *Amarna Reports II*. Egyptian Exploration Society, London, pp. 175-190.

Ryan, D.P., 1988. Cyperus papyrus. *Biblical Archaeologist* 51, pp. 132-140.

Simpson, D.A. & C.A. Inglis, 2001. Cyperaceae of economic, ethnobotanical, and horticultural importance: a checklist. *Kew Bulletin* 56, pp. 257-360.

Täckholm, V., 1974. *Student's flora of Egypt*, 2nd ed. Cairo University Press, Cairo.

Täckholm, V. & M. Drar, 1950. *Flora of Egypt Vol. II*. Bulletin of Faculty of Science No. 28, Fouad I University Press, Cairo.

Theophrastus, 1980. *Theophrastus. Enquiry into plants and minor works on odours and weather signs*, translated by Sir Arthur Hort. Loeb Classical Library. Harvard University Press, Cambridge, Massachusetts.

Waly, N.M., 1999. The selection of plant fibers and wood in the manufacture of organic household items from the El-Gabalein area, Egypt. In: M. van der Veen (ed.), *The exploitation of plant resources in ancient Africa*. Kluwer Academic/Plenum Publishers, New York, pp. 261-272.

Wetterstrom, W., 1984. The plant remains. In: R.J. Wenke (ed.), *Archaeological investigations at El-Hibeh 1980: Preliminary Report*. Undena Publications, Malibu, California, pp. 50-79.

Wilson, H., 1988. *Egyptian food and drink*. Shire Publications, Bucks.

Zahran, M.A. & A.J. Willis, 1992. *The vegetation of Egypt*. Chapman and Hall, New York.

Zohary, D. & M. Hopf, 2000. *Domestication of plants in the Old World*, 3rd edition. Oxford University Press, New York.

Plant remains from the intact garlands present at the Egyptian Museum in Cairo

R. Hamdy

Botany Department, Faculty of Science, Cairo University Herbarium, Egypt

1 Introduction

All kinds of flowers, leaves or leaflet strips as well as fruits of cultivated plants and wild flowers have been used in ancient Egypt for decoration purposes. Very often they have been found in connection with burials. These floral elements were originally used for ornamentation purposes and at a later stage they also acquired a symbolic and religious meaning. A special kind of decoration is that of garlands which were used as draping on mummies.

Kamal (1891: p. 332) pointed out that the Ancient Egyptians started to ornament their mummies from the 12th Dynasty. Obviously, no botanical records have been found so far dating before the 18th Dynasty in Egyptian museums and it seems, therefore, that this kind of adornment was only of a temporary nature (table 1). Just as Osiris appeared with garlands when leaving the Court of judgment at Heliopolis, Ancient Egyptians similarly ornamented their mummies with garlands wishing them eternal life as they appear in front of Osiris (Keimer 1931: pp. 205-206; Tirard 1970: p. 70). The Ancient Egyptians ornamented their statuettes, musicians and their horses (Kamal l.c.), as well as their sacrificed calves to distinguish them from others and to symbolize their sacredness (Nazir 1972: p. 85).

Garlands were also produced during the Greek and Roman period. During the Greek periods garlands played an important role in traditions. They were used to accompany all the mummies as they expressed the life-giving power of renewal or rebirth, making them essential in funeral rites. Garlands were put in graves to express victory over death (Lindsay 1965: p. 268). For Greeks, garlands were also essential in their festivals. They were hung on doors and placed on altars and at shrines to attract good spirits and to hold on to them. Also as symbols of new life, they were presented to the victors of games and wars. They are made of gold or silvery leaves and they were presented to the benefactors of towns. In Roman times, on the contrary, garlands were only used for decorating honored mummies. The tradition of using plants in funeral contexts is still practised, for example in displaying flowers on funeral cars and on altars in cemeteries.

A garland can be defined as a rope of folded leaves, joined together with thin strips of Date palm leaflets or Papyrus culms which are sewed as clasps for flowers or parts of flowers. The garlands from Egyptian mummies can be categorized into three forms: collars, fillets and wreaths. The collars are made of slices of papyrus culms which are sewed together to make a backing for the rope of individual petals, flowers, fruits or leaves. The most famous intact floral collar was the one found decorating the innermost coffin around the golden face of Tutankhamen. Fillets are made by sewing individual petals into a backing whereas wreaths are a closed form of garlands which are used to crown the head.

The technique of how garlands were made, was described by Manniche (1989):

1. Make a string by twisting together fibers of palm leaflets.
2. Take a quantity of Lotus petals and Persea leaves. Leave some 20 inches of the string free at either ends to use for tying the collar.
3. Fold a Persea leaf one third of the distance from the top. Fold once more one-third further down and fasten the leaf over the string. Fold the remaining third of the leaf to make a neat edge. This will be the front of the collar.
4. Insert a Lotus petal in the Persea leaf so that about half the petal is visible.
5. Stitch with Date palm fiber.
6. Take the next Persea leaf and arrange as the first but overlapping slightly. Insert another Lotus petal, and continue until the required length of the garland has been achieved.
7. For a broad collar prepare another garland of Lotus and Persea and fasten to the first so that the upper row overlaps slightly.

In total, 54 garlands and fragments of garlands are currently kept in the Egyptian Museums. They include:

A   Five intact garlands found with identified mummies kept in the Egyptian Museum, Cairo.
B   Thirteen larger fragments found with identified mummies kept in the Agricultural Museum, Giza.
C   Twenty three smaller fragments associated with identified or unidentified mummies. These fragments are kept in the Egyptian (Cairo) and Agricultural (Giza) Museums.
D   Twelve fragments were recovered from Doush Necropolis and which are kept in the Archaeobotany Laboratory (ABL) of Cairo University (CAI).
E   The intact garland found with Amenophis I, Hamdy (in preparation).

Previous studies on the botanical composition of garlands have revealed 41 plant species (table 1). Garlands kept in Berlin's Museum have been previously studied by Germer (1988). A distinction is made by Germer between garlands of the New Kingdom (XVIII-XXIth Dyn.) and those found on the Royal mummies of Kings and Queens and those which were found on non-royal persons. She also distinguished between traditional garlands, made from only a few plant species and newer garlands, which consist of more plant species from the Graeco-Roman period.

2 Materials and methods

The present study deals with the five intact garlands found with identified mummies kept in the Egyptian Museum, Cairo. The identified mummies belong to the New Kingdom or the Graeco-Roman period. Each garland is briefly described and the botanical constituents of the different rows, indicated with 'R', could mostly be identified to the species level by consulting fresh plant specimens or herbarium specimens from the following herbaria: B, CAI, CAIM and BMBD (abbreviations according to Holmgren & Holmgren, 1998). The identification is based on morphological features as well as anatomical structures (magnification: (2-40 X). Latin and English plant names are presented for each of the identified plants. Finally, explanatory diagrams (figures) and/or photographs (plates) for individual garlands are provided.

3 Description of the garlands

The mortuary garland of Queen Meryet Amun
Locality: Deir El Bahari
Period: New Kingdom (XVIIIth Dyn.)
No. of the mummy: 55150.
It is a compound garland, which was found on the mummy (plate 1 and figure 1a). It consists of two subsidiary garlands (no. I and II). Both garlands are formed of rows of folded Willow (*Salix mucronata*) leaves joined together with strips of Date palm (*Phoenix dactylifera*) leaflets.

Garland no. I consists of nine rows (R1-9; figure 1bI) placed on the chest of the mummy. The upper six rows are ornamented with flower heads of the Nile acacia (*Acacia nilotica*) (figure 1c), the lower three rows are ornamented with petals of the Blue lotus (*Nymphaea caerulea*) (figure 1d).

Garland no. II is placed on the middle part of the mummy and consists actually of two series of garlands tied at the shoulder by linen threads (figure 1bII). The first series (i) has nine rows ornamented with petals of Corn poppy (*Papaver rhoeas*) (figure 1e) and the second series (ii) has another nine rows ornamented by petals of the Blue lotus (*N. caerulea*). This examination confirms the previous identification by Winlock (1932).

The garland of Iriphi the virgin
Locality: Unknown
Period: Late, but precise period unknown.
No. of the mummy: 21/16- 11/13.
This compound garland was found on the mummy of Iriphi the virgin (plate 2 and 3). It consists of three subsidiary garlands which were placed above each other (figure 2a). Each garland consists of several horizontal rows joined together with longitudinal strips of Date palm (*Phoenix dactylifera*) leaflets (figure 2b). The distal garland (I) consists of 18 rows (R1-18), the middle one (II) is made up of 13 rows (R1-13), while the upper garland (III) has ten rows (R1-10). Each row consists of leaves of Persea (*Mimusops laurifolia*), joined with strips of Date palm (*P. dactylifera*) leaflets (figure 2b). In addition, there is a wreath made of leaves of Olive (*Olea europaea*) joined with linen threads (*Linum usitatissimum*) around the head of the mummy (plate 4 and figure 2c).

The mortuary garland of Harsiesis
Locality: Near Edfu, Hassaia.
Period: Late, but precise period unknown.
No. of the mummy: 21/16-11/14.
This garland was found on the mummy (plate 5 and 6). It is formed of six horizontal rows (R1-6; figure 3a) joined together with longitudinal strips of Date palm (*Phoenix dactylifera*) leaflets (figure 3b). Each row consists of leaves of Persea (*Mimusops laurifolia*), joined together with strips of Date palm (*Phoenix dactylifera*) leaflets (figure 3c).

The Mortuary garland of a young woman
Locality: Fayum-Fag El Gamous.
Period: Ptolemaic (220 B.C).
No. of the mummy: 127c.

Table 1: List of plants cited in publications used in garland's formation (plant names follows Boulos 1999, 2000, 2002 and 2005).

Species	Part used	Archaeological Context	Period	Location	Publication
Acacia nilotica (L.) Delile	Flower heads	Mummy of Amenhotep I.	New Kingdom	Deir El Bahari	Schweinfurth (1883)
		Part of a garland.	Roman	Hawara	Germer (1987)
Alcea rosea L.	Petals	Mummy of Ahmose I.	New Kingdom	Deir El Bahari	Schweinfurth (1883)
		Mummy of Amenhotep I.	New Kingdom	Deir El Bahari	Schweinfurth (1883)
Apium graveolens L.	Leaves	On the chest of the mummy of nobel Kent.	New Kingdom	Cheikh Abd el Qurna	Schweinfurth (1886)
		Pectoral garland, Tutankhamun mummy.	New Kingdom	Deir El Bahari	Newberry (1927)
Celosia argentea L.	Leaves	Within a wreath.	Roman	Hawara	Newberry (1889)
		Part of a garland.	Roman	Hawara	Germer (1987)
Centaurea depressa M.B.	Flower heads	Mummy of Nzi-Khonsou.	Late	Deir El Bahari	Schweinfurth (1883)
		Tomb excavation by Schiaparelli.	Late	Cheikh Abd el Qurna	Schweinfurth (1886)
		Within a wreath, Tutankhamun mummy.	New Kingdom	Deir El Bahari	Newberry (1927)
		Pectoral garland, Tutankhamun mummy.	New Kingdom	Deir El Bahari	Newberry (1927)
		Floral collaret, Tutankhamun mummy.	New Kingdom	Deir El Bahari	Newberry (1927)
		In a pottery jar, Tutankhamun tomb.	New Kingdom	Deir El Bahari	Winlock (1941)
Consolida orientalis (Gray) Schr.	Flowers	Mummy of Ahmose I.	New Kingdom	Deir El Bahari	Schweinfurth (1883)
Convolvulus hystrix Vahl	Flowers	Within a wreath.	Roman	Hawara	Newberry (1890)
Cordia myxa L.	Leaves	Part of a garland.	Roman	Hawara	Germer (1987)
Cressa cretica L.		Within a wreath.	Roman	Hawara	Newberry (1890)
Cyperus papyrus L.	Culms	Tomb excavation by Schiaparelli.	Late	Cheikh Abd el Qurna	Schweinfurth (1886)
		On the chest of the mummy Kent.	New Kingdom	Cheikh Abd el Qurna	Schweinfurth (1886)
		Within a wreath, Tutankhamun mummy.	New Kingdom	Deir El Bahari	Newberry (1927)
		Pectoral garland, Tutankhamun mummy.	New Kingdom	Deir El Bahari	Newberry (1927)
		Floral collaret, Tutankhamun mummy.	New Kingdom	Deir El Bahari	Newberry (1927)
		In a pottery jar, Tutankhamun tomb.	New Kingdom	Deir El Bahari	Winlock (1941)
Epilobium hirsutum L.	Flowers	Within a wreath.	Roman	Hawara	Newberry (1889)
Euphorbia forsskalii J. Gray		Within a wreath.	Roman	Hawara	Newberry (1890)
Glebionis coronaria (L.) Tzvelev	Flower heads	Part of a garland.	Roman	Hawara	Germer (1987)
Hedera helix L.	Leaves	Within a wreath.	Roman	Hawara	Newberry (1890)
Helichrysum conglobatum (Viv.) Steud.	Flower heads	Part of a garland.	Roman	Hawara	Germer (1987)
Heliotropium bacciferum Forssk.	Twigs	Within a wreath.	Roman	Hawara	Newberry (1890)
Hordeum vulgare L.	Grains	Around the neck of the mummy Kent.	New Kingdom	Cheikh Abd el Qurna	Schweinfurth (1886)
Iris sibirica L.	Petals	Within a wreath.	Roman	Hawara	Newberry (1890)
Jasminium sambac (L.) Ait.	Flowers	Within a wreath.	Roman	Hawara	Newberry (1890)
Mandragora officinalis L.	Fruits	Floral collaret, Tutankhamun mummy.	New Kingdom	Deir El Bahari	Newberry (1889)
Mimusops laurifolia L.	Leaves	Mummy of Ramses II.	New Kingdom	Deir El Bahari	Schweinfurth (1883)
		Tomb excavation by Schiaparelli.	Late	Cheikh Abd el Qurna	Schweinfurth (1886)
		Tomb excavation by Schiaparelli.	Late	Dra Abu El Nagga	Schweinfurth (1886)
		Mummy of Ramses II.	New Kingdom	Deir El Bahari	Roubet (1981); Roubet & Layer-Lescot (1985)
Myrtus communis L.	Leaves	Within a wreath.	Roman	Hawara	Newberry (1889)
Narcissus tazetta L.	Flowers	Within a wreath.	Roman	Hawara	Newberry (1889)
		Part of a garland.	Roman	Hawara	Germer (1987)

Species	Part used	Archaeological Context	Period	Location	Publication
Nymphaea caerulea Savigny	Sepals, petals and flowers.	Mummy of Ahmose I.	New Kingdom	Deir El Bahari	Schweinfurth (1883)
		Mummy of Amenhotep I.	New Kingdom	Deir El Bahari	Schweinfurth (1883)
		Mummy of Ramses II.	New Kingdom	Deir El Bahari	Schweinfurth (1883)
		On the chest of the mummy Kent.	New Kingdom	Cheikh Abd el Qurna	Schweinfurth (1886)
		Wreath, pectoral garland and floral collaret of Tutankhamun.	New Kingdom	Deir El Bahari	Newberry (1927)
		Mummy of Ramses II.	New Kingdom	Deir El Bahari	Roubet (1981); Roubet & Layer-Lescot (1985)
Nymphaea lotus L.	Petals	Mummy of Ramses II.	New Kingdom	Deir El Bahari	Schweinfurth (1883)
		Within a wreath.	Roman	Hawara	Newberry (1890)
		Mummy of Ramses II.	New Kingdom	Deir El Bahari	Roubet (1981); Roubet & Layer-Lescot (1985)
		Part of a garland.	Roman	Hawara	Germer (1987)
Olea europaea L.	leaves	Within a wreath, Tutankhamun mummy.	New Kingdom	Deir El Bahari	Newberry (1927)
		Pectoral garland, Tutankhamun mummy.	New Kingdom	Deir El Bahari	Newberry (1927)
		Floral collaret, Tutankhamun mummy.	New Kingdom	Deir El Bahari	Newberry (1927)
		In a pottery jar, Tutankhamun tomb.	New Kingdom	Deir El Bahari	Winlock (1941)
Origanum majorana L.	Twigs	Within a wreath.	Roman	Hawara	Newberry (1889)
Papaver rhoeas L.	Petals	Mummy of Nzi-Khonsou.	Late	Deir El Bahari	Schweinfurth (1883)
		Within a wreath.	Roman	Hawara	Newberry (1890)
Phoenix dactylifera L.	Leaflet strips	Mummy of Ramses II.	New Kingdom	Deir El Bahari	Schweinfurth (1883)
		Floral collaret of Tutankhamun.	New Kingdom	Deir El Bahari	Newberry (1927)
		In a pottery jar, Tutankhamun tomb.	New Kingdom	Deir El Bahari	Winlock (1941)
		Part of a garland.	Roman	Hawara	Germer (1987)
		Mummy of Ramses II.	New Kingdom	Deir El Bahari	Roubet (1981); Roubet & Lescot (1985)
Picris asplenioides L.	Flower heads	Mummy of Nzi-Khonsou.	Late	Deir El Bahari	Schweinfurth (1883)
		Floral collaret of Tutankhamun.	New Kingdom	Deir El Bahari	Newberry (1927)
Pluchea dioscoridis (L.) DC.	Flower heads	Within a wreath.	Roman	Hawara	Newberry (1890)
Pseudognaphalium luteoalbum (L.) Hill. & B.L. Burtt	Flower heads	Within a wreath.	Roman	Hawara	Newberry (1889)
Punica granatum L.	Leaves and flowers	Part of a garland.	Roman	Hawara	Germer (1987)
Reseda odorata L.	Flowers	Within a wreath.	Roman	Hawara	Newberry (1890)
Rosa sancta Rich.	Flowers	Part of a garland.	Roman	Hawara	Keimer (1943)
		Part of a garland.	Roman	Hawara	Germer (1987)
Salix mucronata Thunb.	Leaves	Mummy of Ahmose I.	New Kingdom	Deir El Bahari	Schweinfurth (1883)
		Mummy of Amenhotep I.	New Kingdom	Deir El Bahari	Schweinfurth (1883)
		Mummy of Nzi-Khonsou.	Late	Deir El Bahari	Schweinfurth (1883)
		In a tomb excavation by Schiaparelli.	Late	Cheikh Abd el Qurna	Schweinfurth (1883)
		Pectoral garland of Tutankhamun and floral collaret.	New Kingdom	Deir El Bahari	Newberry (1927)
		On a mummy of a lady.	Roman	Hawara	Keimer (1931)
		On a mummy of a lady.	Roman	Arsinoe	Keimer (1931)
		Mummy of Meryet Amun.	New Kingdom	Deir El Bahari	Keimer (1931).
Scirpus inclinatus (Del.) Asch. & Schw.	Culms	Part of a garland.	Roman	Hawara	Keimer (1943)
		Part of a garland.	Roman	Hawara	Germer (1987)
Sesbania sesban (L.) Merr.	Flowers	Mummy of Ahmose I.	New Kingdom	Deir El Bahari	Schweinfurth (1883)
Silene coeli-rosa (L.) Godr.	Fruits	Within a wreath.	Roman	Hawara	Newberry (1889)

Species	Part used	Archaeological Context	Period	Location	Publication
Tilia europaea L.	Leaves	Within a wreath.	Roman	Hawara	Newberry (1890)
Withania somnifera (L.) Dun.	Fruits	Within a wreath.	Roman	Hawara	Newberry (1889)
		Floral collaret, Tutankhamun mummy.	New Kingdom	Deir El Bahari	Newberry (1927)
		In a pottery jar, Tutankhamun tomb.	New Kingdom	Deir El Bahari	Winlock (1941)
		Part of a garland.	Roman	Hawara	Germer (1987)

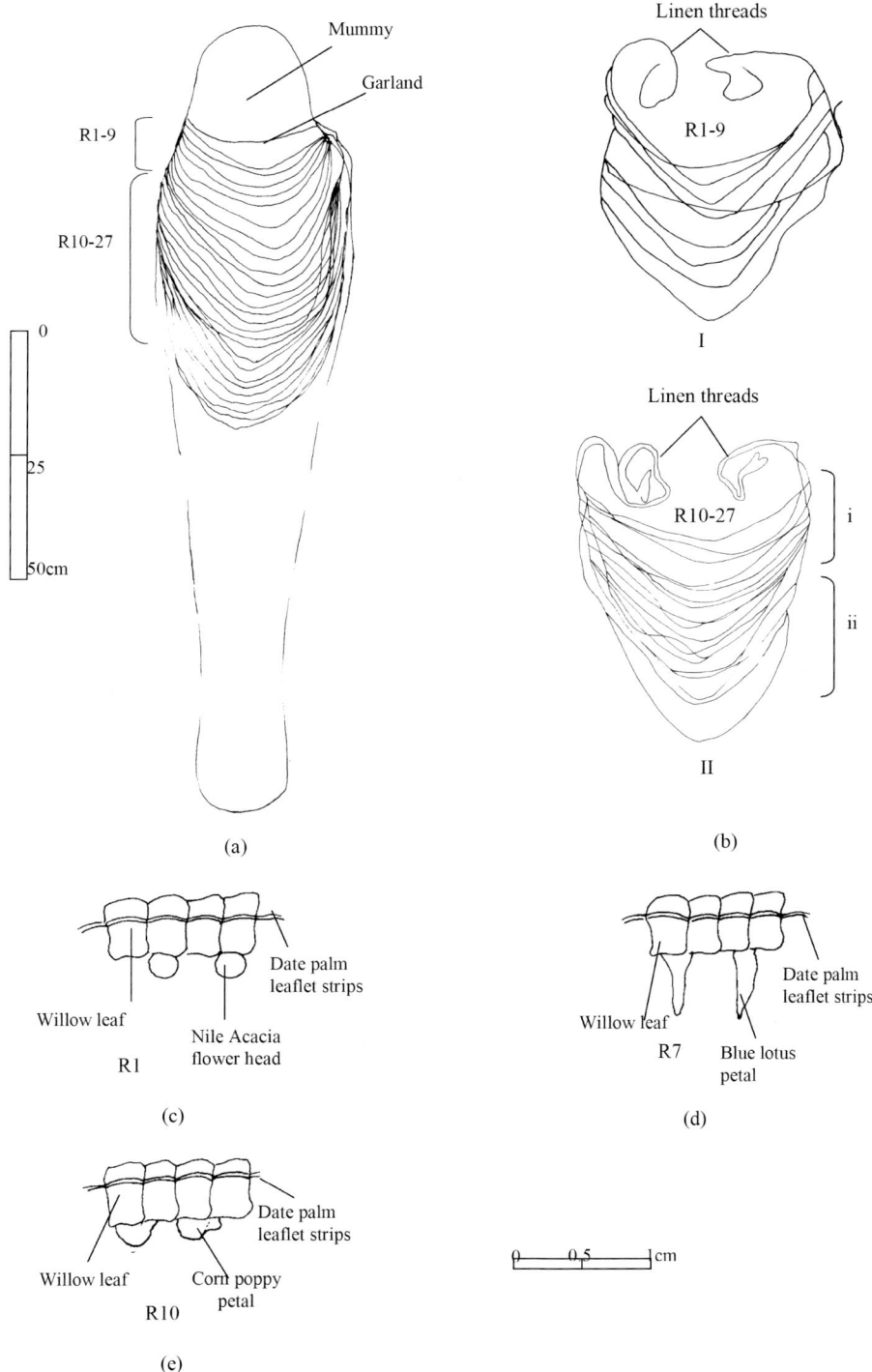

Figure 1: Drawing presenting Queen Meryet Amun's garland showing: (a) the arrangement of the two subsidiary garlands I and II of the compound garland, (bI) the subsidiary garland I consisting of nine rows (R1-9), (bII) the subsidiary garland II consisting of 18 rows (R10-27) in two series (i and ii), and (c, d and e) the enlarged fragments of some rows of the garlands I and II.

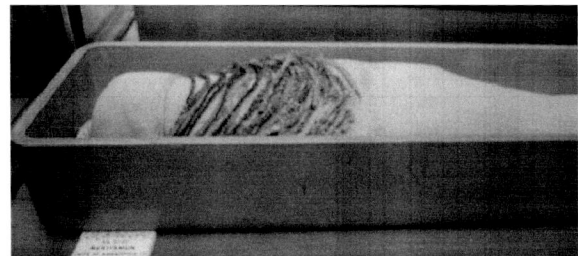

Plate 1: The mummy of Queen Meryet Amun kept in a rectangular sarcophage and covered with a garland (Egyptian Museum, Cairo).

Plate 2: Mummy, cartonnage and cuve of the coffin of Iriphi, the virgin (21/16-11/13). The body is ornamented with a compound garland consisting of three subsidiary garlands. It has also a head collar with a decorative face gilt.

Plate 3: Magnified part of the mummy showing some rows of the upper subsidiary garland III.

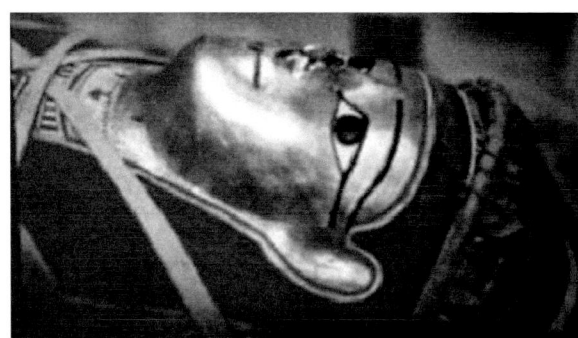

Plate 4: Magnified part of the mummy showing the head crowned with the wreath made of Olive leaves.

This garland is found around the mummy of a young woman (20 years old), who was the daughter of the Egyptian High Priest *Hm-ntr*, son of Ra (plate 7). The mummy was found in the north cemetery; area 18, chamber # 3 (Griggs et al. 1993). The flower garlands wrapped around the body contain at least four different kinds of flowers and a small spray of flowers is placed in the linen wraps over the heart.

The garland is formed of seven rows (R1-7; figure 4a) R1 and R4 are formed of strips of Date palm (*Phoenix dactylifera*) leaf axes, fastened together with thin strips of leaflets of the same plant which serves as clasps for intact Rose flowers (*Rosa* sp.) (plate 8 and figure 4b). The rows R2, R3 and R5 are formed of Papyrus (*Cyperus papyrus*) culms fastened together with thin strips of the culms of the same plant which serves as clasps for detached Rose flowers (*Rosa* sp.) (plate 9 and figure 4c). R6 is formed of strips of Papyrus (*C. papyrus*) culms fastened together with thin strips of Date palm (*P. dactylifera*) leaflets, which serves as clasps for Rose flowers and a white ray floret of a composite head (figure 4d). R7 is formed of strips of Papyrus (*C. papyrus*) culms fastened together with thin strips of culms of the same plant and serves as clasps for Withania nightshade (*Withania somnifera*) fruits. Also some capsules of Rose of heaven (*Silene coeli-rosa*) were found in the sarcophagus. In addition, an olive branch (*Olea europaea*) was placed on the chest of the mummy (figure 4e).

The mortuary garland of a Roman mummy
Locality: Unknown
Period: Roman
No. of the mummy: 25/8-19/2.

This garland was placed on the chest of the mummy. It consists of 12 rows (R1-12) (plate 10 and 11, and figure 5a), joined together on both sides with strips made of culms of a sedge plant ornamented with Myrtle (*Myrtus communis*) fruiting branches and *Senecio glaucus* ssp. *coronopifolius* flower heads, all tied with strips of Date palm (*Phoenix dactylifera*) leaflets. The rows R1, R2, R6, R7 and R10 were formed of Myrtle (*M. communis*) branches, sedge culms, *S. glaucus* ssp. *coronopifolius* flower heads and artificial flowers made of sedge culms, all joined together with strips of Date palm (*P. dactylifera*) leaflets (figure 5b). The rows R3, R4, R5, R8, R9, R11 and R12 were formed of sedge culms and artificial flowers made of the same culm fruits of Withania nightshade (*Withania somnifera*), joined together by strips of Date palm (*P. dactylifera*) leaflets (figure 5c).

4 Concluding remarks

The study reveals the identification of four new mummies; Iriphi the virgin, Harsiesis, Ptolemaic young woman and a Roman mummy kept in the Egyptian museum, Cairo and the re-examination of the last one (Queen Meryet Amun mummy) which was found in

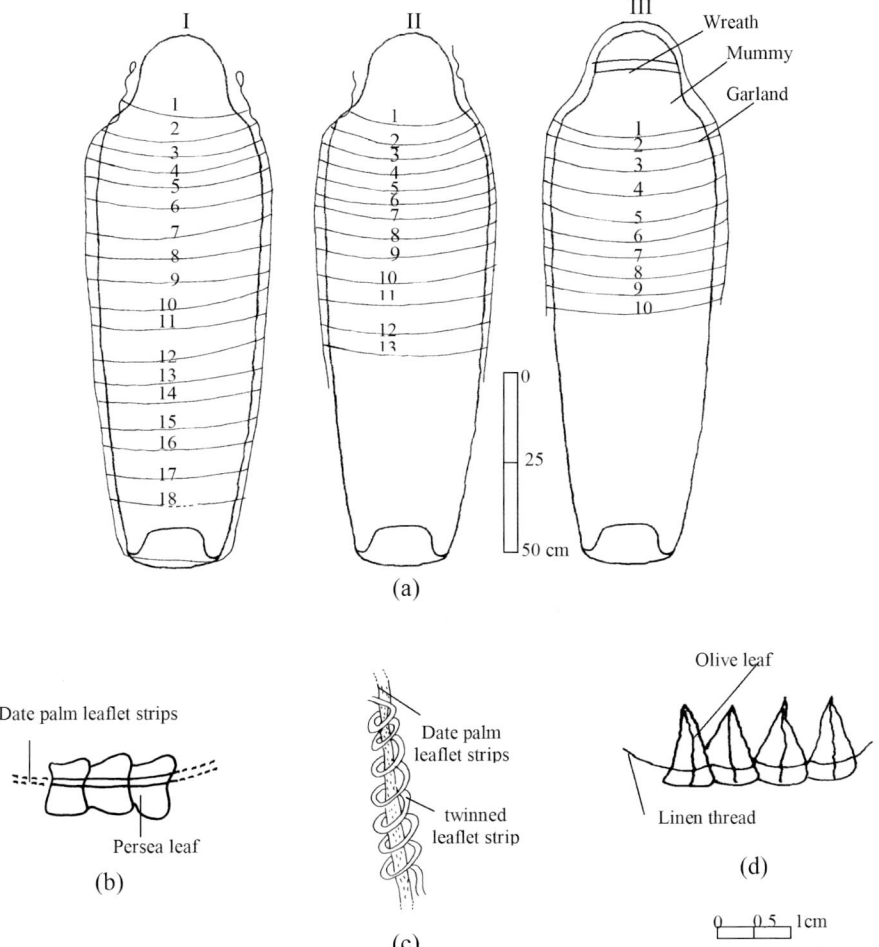

Figure 2: Diagram presenting of Iriphi the virgin compound garland showing: (a) the arrangement of the three subsidiary garlands (I, II and III), (b) the enlarged fragment of a horizontal row of the garland, (c) the enlarged fragment of a longitudinal strip made of Date palm leaflets and (d) the enlarged fragment of the wreath.

the section of Royal mummies. Each of the examined garlands consists of rope-like rows, which were placed horizontally on the mummy and joined vertically by lateral ropes (leaflet strips). This construction is also seen in the other known (fragments of) garlands.

Two main types of garland designs were observed. The rows of garlands during the Late Pharaonic period agree in character with that described by Manniche (1989). Each row is formed of folded leaves, joined together with thin strips of Date palm or Papyrus. Rows were ornamented with accessories such as flowers or part of flowers. The rows of garlands during Ptolemaic or Roman periods are different. Each is formed of a long axis of Date palm leaflets or Papyrus culms which were bound together with strips of the same plant and associated flowers were fixed to the longitudinal axis by binding strips.

The mummies of Iriphi the virgin and Harsiesis were registered in the museum catalogue with uncertain date. By comparing the cartonnage covering these mummies with similar cartonnages found with mummies belonging to the Late period (XXVII-XXXth Dyn.), it became clear that they show identical characters. For that reason, it is suggested that the mummies of Iriphi the virgin and Harsiesis date back to the Late period.

It is concluded that the garlands used in the ornamentation of mummies of upper-class people were rich in their floral constituents (table 2). Garlands from noble class people (Iriphi the virgin and Harsiesis) were comparatively poor in their constituents. Even painted structures were used to replace natural flowers for the ornamentation of this category.

The leaflet strips of Date palm (*Phoenix dactylifera*) are used for binding all the representative garlands. This can be attributed to its common occurrence in all these periods. The most common were Persea (*Mimusops laurifolia*), Olive (*Olea europaea*) and Withania nightshade (*Withania somnifera*). The leaves of the first two were essential in making up garlands and the fruit of

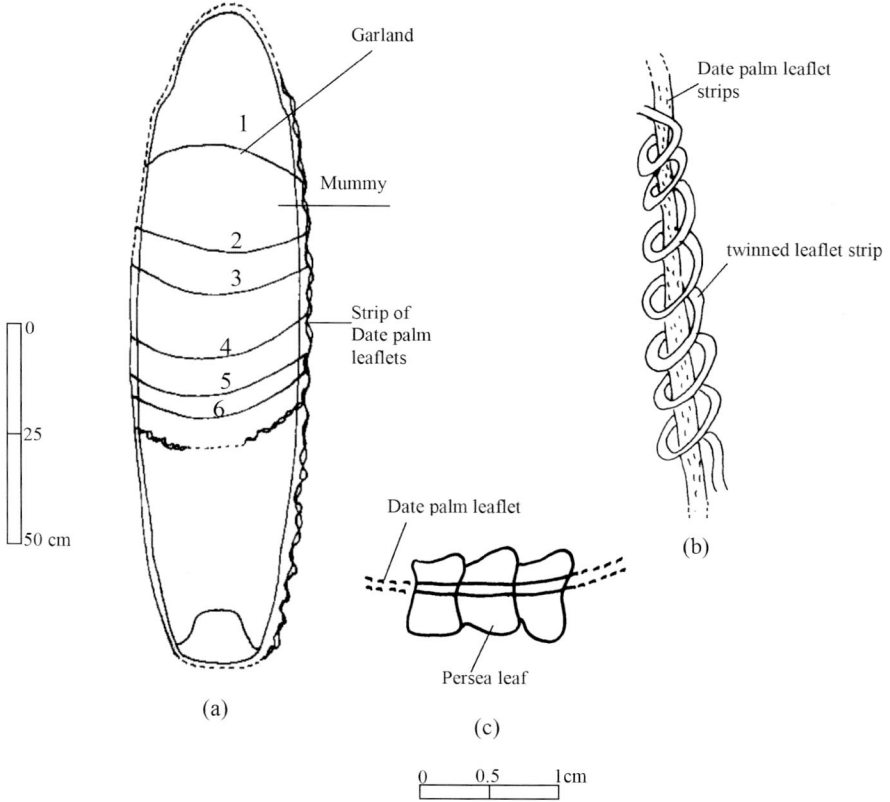

Figure 3: Diagram presenting of Harsiesis garland (G. 21/16-11/14) showing: (a) the arrangement of the six rows (R1-6) of the garland, (b) the enlarged fragment of a longitudinal strip made of Date palm leaflets and (c) the enlarged fragment of a horizontal row of the garland.

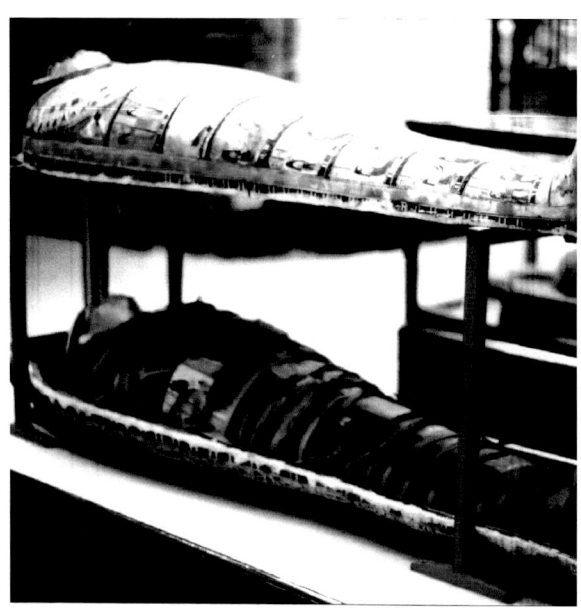

Plate 5: Anthropoid coffin with the mummy of Harsiesis (21/16-11/14, about 1.62 m long) placed in a sarcophage (below) with lid (above). The mummy is ornamented with a garland (Egyptian Museum, Cairo).

Plate 6: Enlarged upper part of the mummy (21/16-11/14) showing some rows of the garland. The mummy has gilt cartonnage and a bead garment.

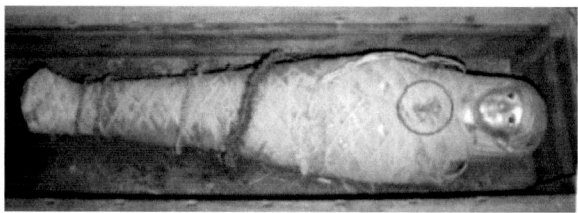

Plate 7: Top view of the mummy of a young woman (127c) showing four floral rows (R4, R5, R6 and R7) of a garland and a branch of Olive (in a circle) placed on the chest of the mummy.

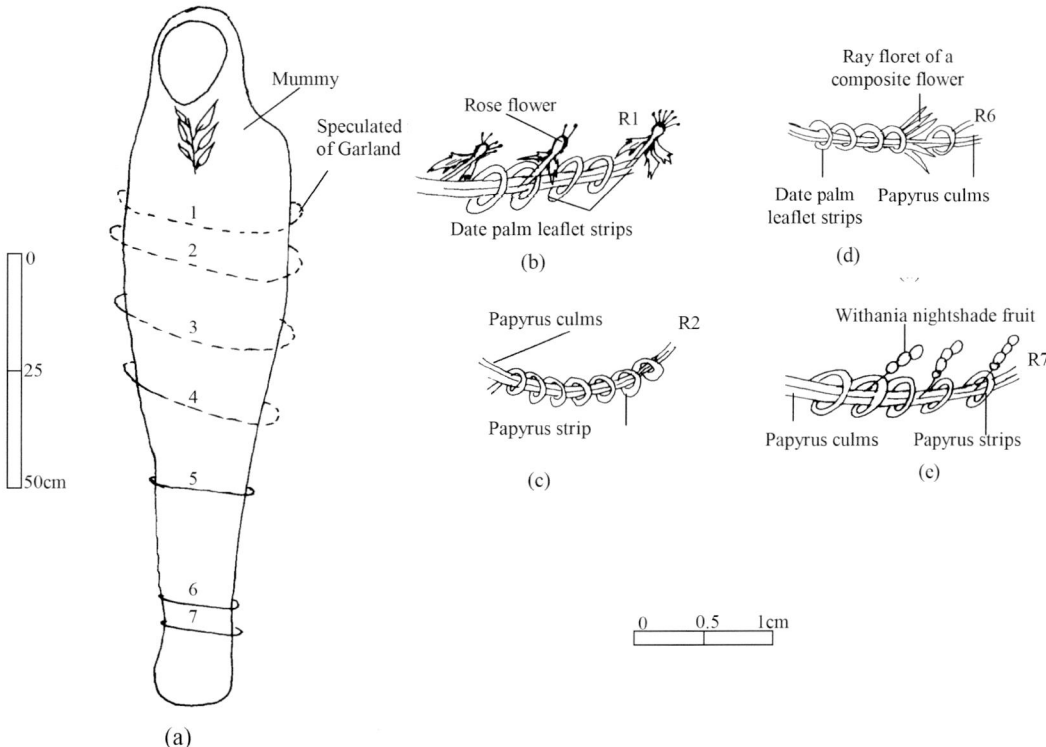

Figure 4: Diagram presenting of the mummy garland on the young woman showing: (a) the arrangement of the seven rows (R1-7) of the garland, (b) the enlarged fragment of R1 of the garland, (c) the enlarged fragment of R2 of the garland, (d) the enlarged fragment of R6 of the garland and (e) the enlarged fragment of R7 of the garland.

Plate 8: First floral row (R1) of the garland G127c. A few detached Rose flowers are shown below the scale.

Plate 10: Roman mummy (25/8 − 19/2, about 1.55 m long) in a rectangular sarcophagi and covered with the rows of a garland (Egyptian Museum, Cairo).

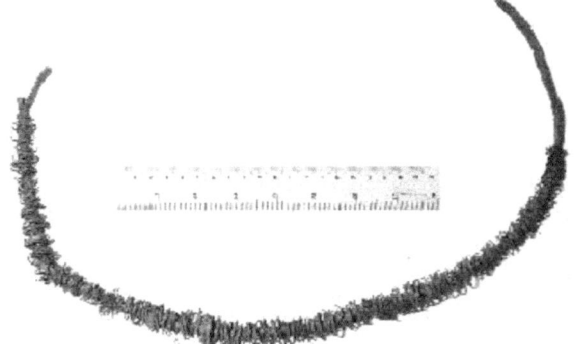

Plate 9: Second floral row (R2) of the garland (G127c) consisting of Papyrus culms.

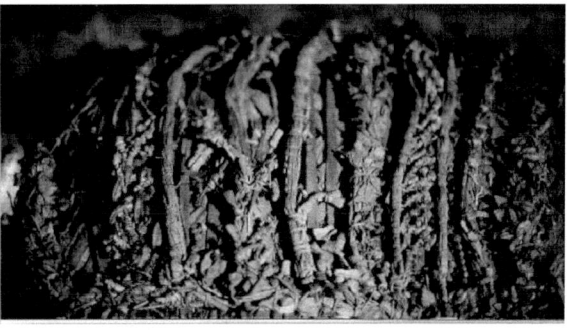

Plate 11: Magnified part of the Roman mummy (25/8 − 19/2) showing several rows of the garland.

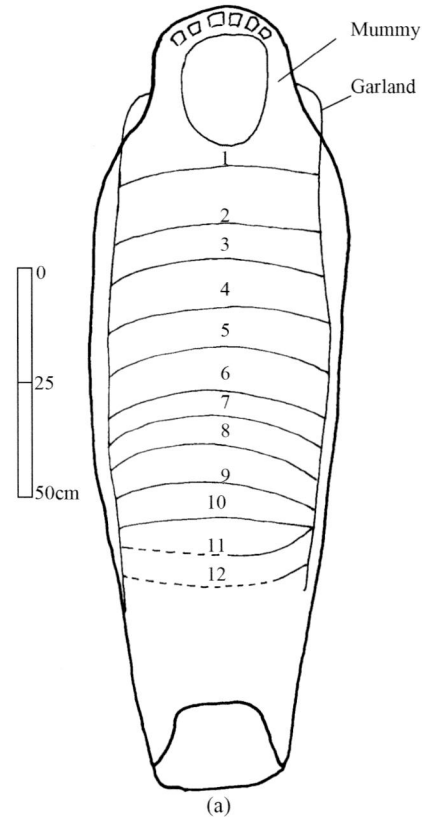
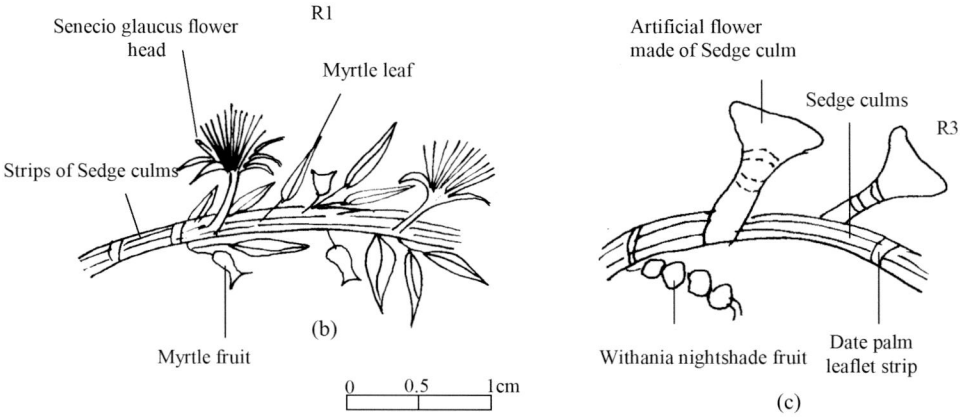

Figure 5: Diagrammatic presentation of the Roman mummy garland showing: (a) the arrangement of the 12 rows (R1-12) of the garland, (b) the enlarged fragment of R1 of the garland and (c) the enlarged fragment of R3 of the garland.

the last one was used for ornamenting Graeco-Roman garlands. The others are less common and represented only once in the investigated garlands.

It could be possible, by using the monthly calendar, to indicate the season of the year in which the mummy was laid to rest. Table 2 shows that Nile acacia (*Acacia nilotica*), *Rosa* sp., *Senecio glaucus* ssp. *coronopifolius* and Blue lotus (*Nymphaea caerultea*) are summer flowers, while *Papaver rhoeas* and Rose of heaven (*Lychnis coeli-rosa*) are autumn flowers.

In addition to the many indigenous plants also some exotic plants have been used for making garlands. Examples are Myrtle (*Myrtus communis*), *Rosa* sp., and Rose of heaven (*Lychnis coeli-rosa*), which are considered to be introduced in Egypt during the Graeco-Roman period. The introduction of exotic plants clearly increased the floral diversity of the garlands during this period.

Table 2: The different parts of 14 species used in the ornamentation of the five intact garlands (b = branch; cu = culm; fh = flower head; fl = flower; fr = fruit; l = leaf; lb = leafy branch; p = petal; st = leaflet strip).

Species	Meryet Amun	Iriphi the Virgin	Harsiesis	Young woman	Roman mummy
Acacia nilotica	fh				
Cyperus papyrus				cu	
Mimusops laurifolia		l	l		
Myrtus communis					lb
Nynphaea caerulea	p				
Olea europaea		l		lb	
Papaver rhoeas	p				
Phoenix dactylifera	st	st	st	st	st
Rosa sp.				fl	
Salix subserrata	l				
Carex/Cyperus					cu
Senecio glaucus subsp. coronopifolius					fl
Silene coelirosa				fr	
Withania somnifera				fr	fr
Number of plant species	5	3	2	6	5

5 Acknowledgments

The author would like to extend her deepest gratitude to the staff of the Egyptian Museum-Cairo for their help in facilitating the study these materials, especially Mr. Nasry Iskander, director of the restoration and conservation laboratory. A special word of thanks to René Cappers for his help in making it possible for me to attend this workshop and to Eng. Essam A. Gawad for his great assistance in writing and formatting the manuscript.

6 References

Boulos, L., 1999. *Flora of Egypt: 1*. Al Hadara, Cairo.
Boulos, L., 2000. *Flora of Egypt: 2*. Al Hadara, Cairo.
Boulos, L., 2002. *Flora of Egypt: 3*. Al Hadara, Cairo.
Boulos, L., 2005. *Flora of Egypt: 4*. Al Hadara, Cairo.
Germer, R., 1987. Ancient Egyptian Plant remains in The Manchester Museum. *Journal of Egyptian Archaeology* 73: pp. 72-73.
Germer, R., 1988. Katalog der Altägyptischen Pflanzenreste der Berliner Museen. *Ägyptologische Abhandlungen* Band 47: pp. 1-90.
Griggs, C.W., M.C. Kuchar, S.R. Woodward, M.J. Rowe, R.P. Evans, N. Kanawati & N. Iskander, 1993. Evidences of a Christian population in the Egyptian Fayum and genetic and textile studies of the Akhmim Noble Mummies. *Brigham young University Studies* 33 (2): pp. 215-243.
Holmgren, P.K. & N.H. Holmgren, 1998. *Index herbariorum*. New York Botanical Garden.
Kamal, A., 1891. *The arts, the habits, the industries and the origin of the Ancient Egyptians* (In Arabic). Cairo.
Keimer, L., 1931 L'Arbre | TR T ≡-↓] est-il réellement le saule Égyptien (Salix safsaf Forsk.)? *Bulletin de l'Institut Français d'Archéologie Orientale* 31, pp. 177-237.
Keimer, L., 1943. La Rose Égyptienne. *Études d'Égyptologie* 4, pp. 1-28.
Lindsay, J., 1965. *Leisure and pleasure in Roman Egypt*. London, Frederick Muller.
Manniche, L., 1989. *An Ancient Egyptian Herbal*. Austin, University of Texas Press.
Nazir, W., 1972. The Egyptian rural traditions and customs (In Arabic). *Choose for the farmer* 73. Cairo.
Newberry, P.E., 1889. On the vegetable remains discovered in the cemetry of Hawara. In: W.M.F. Petrie (ed.), *Hawara, Biahmu and Arsinoe*. Field & Tuer "The Leadenhall Press", London, pp. 46-53.
Newberry, P.E., 1890. The ancient botany. In: W.M.F. Petrie (ed.), *Kahun, Gurob and Hawara*. Kegan Paul, Trench, Trübner and Co., London, pp. 46-50.
Newberry, P.E., 1927. Report on the floral wreaths found in the coffins of Tut-Ankh-Amen. In: H. Carter (ed.), *The tomb of Tut.ankh.Amen* Volume 2. Cassell and Company, London, pp. 189-196.
Roubet, C., 1981. La parure florale de Ramsès II. *Bibliothèque d' étude 88 (Prospection et sauvegarde des antiquités de l' Égypte)*, pp. 151-161.
Roubet, C. & Layer-Lescot, M., 1985. La Parure florale de Ramsès II. In: L. Balout & C. Roubet (eds.), *La Momie de Ramsès II*. Editions Recherche sur les Civilisations, Paris, pp. 158-161.

Schweinfurth, G., 1883 (1882). De la flore pharaonique. *Bulletin de l'institut Égyptien* 2ᵉ Série, No. 3, pp. 51-76.

Schweinfurth, G., 1886 (1885). Les dernières découvertes botaniques dans les anciens tombeaux de l'Égypte. *Bulletin de l'Institut Égyptien* 2ᵉ Série, No. 6, pp. 256-283.

Tirard, H.M., 1970. *The Book of the dead*. London.

Winlock, H.E., 1932. The tomb of Queen Meryet-Armun at Thebes. *The Metropolitan Museum of Art Egyptian Expedition* Vol. 6, pp. 1-100.

Winlock, H.E., 1941. Materials used at the embalming of king Tūt-ʿAnkh-Amūn. *Metropolitan Museum of Art* Papers 10, pp. 17-18.

Food supply along the Theban Desert roads (Egypt): the Gebel Roma^c, Wadi el-Hôl, and Gebel Qarn el-Gir caravansary deposits

R.T.J. Cappers
Groningen Institute of Archaeology, University of Groningen, the Netherlands

L. Sikking
Freelance Archaeobotanist based in Ottawa, Canada

J.C. Darnell & D. Darnell
Yale University, Near Eastern Languages & Civilizations, U.S.A.

1 Introduction

Since 1992, Prof. J.C. Darnell and D. Darnell have studied the ancient routes across the large Qena Bend of the Nile. Crossing the escarpment, these tracks connect Thebes to the northwestern part of the Qena Bend, the terminus for many caravan routes through the Western Desert. This project began as the '*The Luxor-Farshût Desert Road Survey*', and has continued since 1998 as the '*Theban Desert Road Survey*' of Yale University. The survey project is aimed at mapping the ancient routes through this mountainous area of the desert, and studying the archaeological sites and accompanying rock inscriptions and rock art, which date primarily from the earliest Predynastic cultures to the Coptic Period.

During the past several years, numerous sites along these caravan tracks have been discovered, some of which have been studied in detail, including Gebel Roma^c, Wadi el-Hôl, Qarn el-Gir and Gebel Antef (figure 1). The sites provide epigraphic evidence for the use of the route to transport temple grain (J. Darnell 2002; D. Darnell 2002). During the sixth survey season, archaeobotanical samples were collected from Gebel Qarn el-Gir. The sorted samples contained a considerable quantity of decomposed animal (equid) dung, as well as a wide range of other organic material.

The analysis of these initial samples confirmed that many of the botanical remains represented not only digested animal fodder, but also the remnants of foodstuffs both shipped across the road and intended for consumption at the sites, perhaps by both people and animals. It was, therefore, decided to undertake a program of systematic sampling targeting the different strata of the debris deposits in order to reveal what kind of food was eaten by travellers during their trips through the desert. The work of the Theban Desert Road Survey continues at these sites; the following data represent the results of field seasons VI-IX.

2 Materials and methods

From all visible strata of the debris deposits of Gebel Roma^c, Wadi el-Hôl and Qarn el-Gir, subsamples were taken, measuring 10 cm x 10 cm x stratum thickness. These soil samples have been dry-sieved through a stack of sieves of mesh sizes 5.0 mm, 2.0 mm, 1.0 mm and 0.5 mm, although the 0.5 mm sieve residues have not yet been analysed due to lack of time. Sorting of the sieve residues and identification of the plant remains were performed with a dissecting microscope. Additionally, 70 samples were collected from the different strata and sieved through a 3.0 mm sieve and sorted in the field. Of these hand-picked samples, 65 yielded identifiable plant remains.

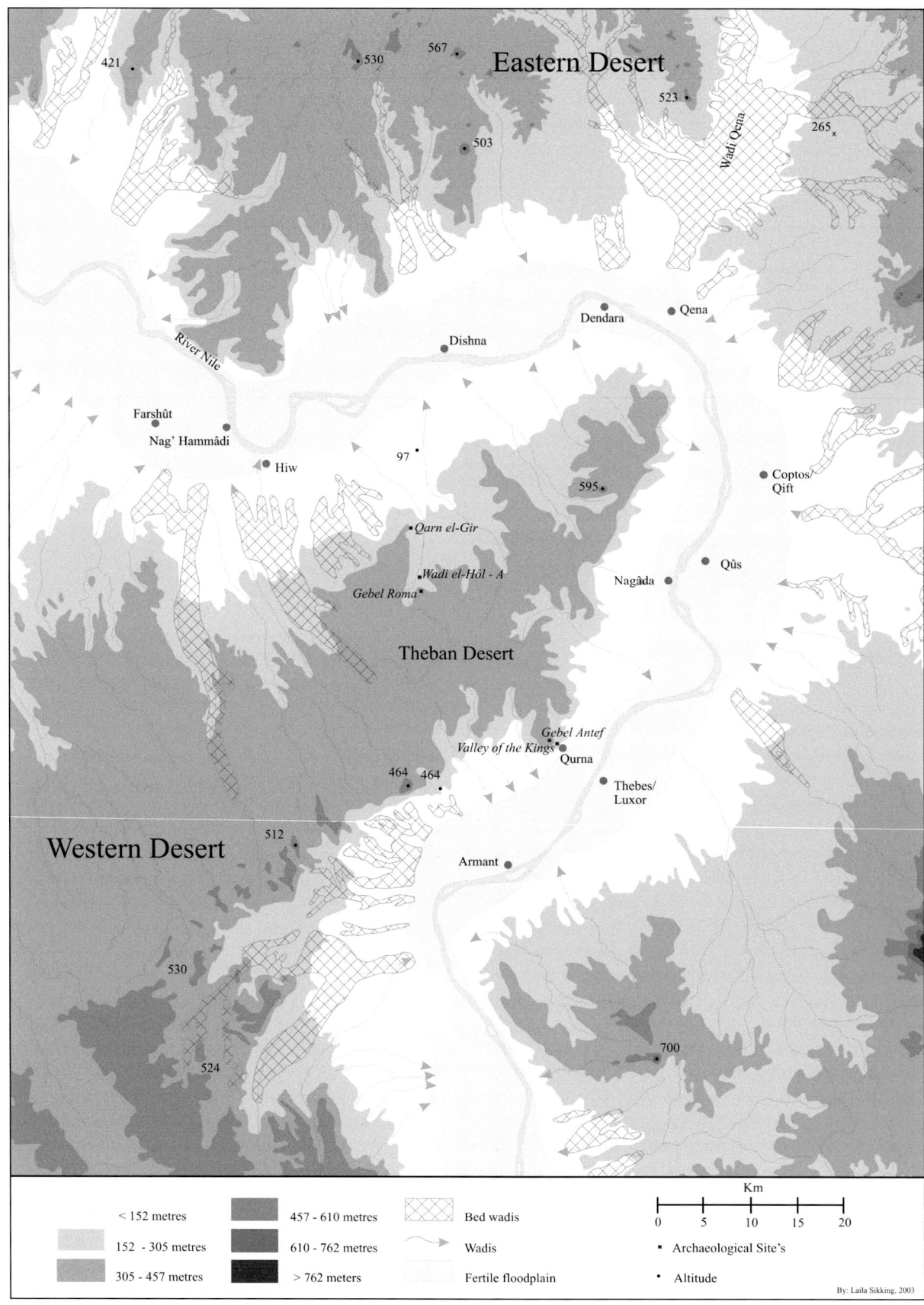

Figure 1: Qena Bend of the Nile with the main archaeological sites.

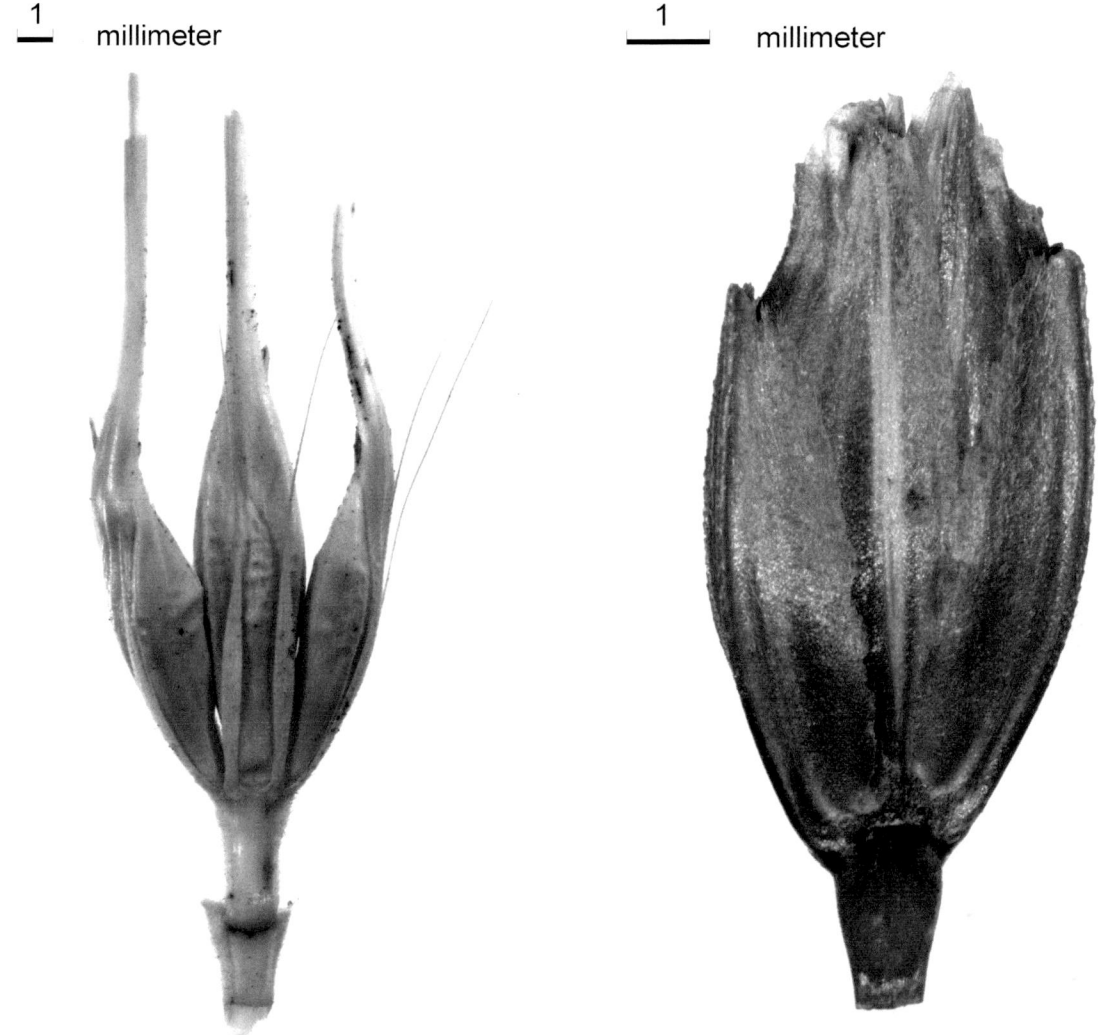

Figure 2: Rachis node of 6-row barley (Hordeum vulgare ssp. Vulgare; recent) bearing three spikelets each consisting of a single grain kernel (left), and rachis node of emmer wheat (Triticum turgidum ssp. dicoccon; from Neolithic K-pit in the northeastern Fayum) bearing one spikelet consisting of two grain kernels (right).

The debris deposit from Gebel Romaᶜ was sampled at three different locations: the Main Trench (18 strata: 17 soil samples and 36 hand-picked samples), the West Trench (9 strata: 9 soil samples and 1 hand-picked sample) and the Southeast Trench (7 strata: 9 hand-picked samples). From the debris deposit at Wadi el-Hôl, two profiles were sampled: the East Trench (6 strata: 6 soil samples and 4 hand-picked samples) and the South Trench (4 strata: 4 hand-picked samples). The large debris deposit at Qarn el-Gir was sampled at a location near the center of the deposit and yielded samples from 17 strata (18 soil samples and 16 hand-picked samples).

The quantitative comparison of 6-row barley and emmer wheat is based on both the number of grain kernels and the number of rachis nodes. In 6-row barley, each rachis node bears three grain kernels, whereas each rachis node of emmer wheat bears two grain kernels (figure 2). Traditionally, the rachis nodes of emmer wheat are labelled either 'spikelet forks' or 'glume bases', depending on the degree of fragmentation and the presence of glume fragments (Cappers et al., 2004). To obtain a solid unit for a quantitative comparison of the rachis fragments with the grain kernels, it makes more sense to ignore the difference in fragmentation and to quantify only the rachis nodes, being the relevant part of the rachis. This implies that a so-called 'spikelet fork' corresponds to one rachis node and a glume base with half a rachis node.

Table 1: Cultivated plants from Gebel Roma^c
Charred plant remains are presented in bold (second row). Identification: (N): uncertain at species level; [N]: uncertain at genus level (MK=Middle Kingdom; IP2= 2nd Intermediate Period; NK=New Kingdom; IP3=3rd Intermediate Period; GR=Graeco-Roman period).

Gebel Roma^c		MK	IP2	IP2 NK	NK	NK IP3	IP3	GR	Mixed / undated	
Number of samples (N=72)		5	6	3	24	3	12	6	13	
CEREALS										
Hordeum vulgare ssp. vulgare	grain	5	195	240	233 **2**	10 **1**	98	6	91 **1**	6-row barley (hulled)
	RN	32	745 **2**	83	1719 **5**	54	390 **2**	16	515	
Triticum	grain						4	2		wheat
	RN		101		42					
Triticum turgidum ssp. dicoccon	grain	2	43 **1**	16	189	140	266	24	50 **1**	emmer
	RN	2	479 **1**	18 **1**	3139 **3**	574	1917	56 **1**	1843	
Triticum turgidum ssp. durum	grain									hard wheat
	RN		(7)		7 (15)	(18)		(1)	4	
Triticum monococcum	grain	**(1)**								einkorn
	RN	**(4)**								
Gramineae tribe Triticeae	grains						2			cerealia
PULSES										
Lens culinaris	seed								[1]	lentil
OIL PLANTS										
Carthamus tinctorius	fruit				1					safflower
Linum usitatissimum	seed		7		19	1	33		4	flax
	fruit	1	5		19		5		7	
TUBERS										
Cyperus esculentus	tuber				1 (4)		(1)			earth-almond
Cyperus rotundus	tuber				[2]					nut-grass
VEGETABLES										
Allium sativum	bulb	1			1		(2)			garlic
Citrullus lanatus	seed						6	1	2	watermelon
Cucumis	seed						1		2	cucumbers
Cucumis melo	seed			5	20		1		4	melon
Cucumis sativus	seed			2	10	1	5			cucumber
FRUIT TREES										
Balanites aegyptiaca	fruit				4					sugar date
Ficus sycomorus	fruit	212	60	7	45	1	5		70	sycomore fig
	syconium						1			
Mimusops laurifolia	seed		1		6		4	2		persea
Olea europaea	fruit		1		8			4	5	olive
Phoenix dactylifera	seed		3	4	24	4	6	6	12	date
Punica granatum	seed		1							pomegranate
Vitis vinifera	seed					1				grape
CONDIMENTS										
Coriandrum sativum	fruit				2		3		1	coriander
Nigella sativa	seed				2		1		1	black cumin

3 Results

3.1 cultivated plants

The cultivated plants are represented by cereals, pulses, oil and fibre crops, vegetables, fruit trees and condiments (Tables 1, 2 and 3). The plant records of Gebel Roma^c, Wadi el-Hôl and Qarn el-Gir are dominated by two cereals: hulled 6-row barley (*Hordeum vulgare* ssp. *vulgare*) and emmer wheat (*Triticum turgidum* ssp. *dicoccon*). The identification of a single charred grain kernel and a few rachis fragments as einkorn (*Triticum monococcum*), originating from stratum 17 of the Main Trench and dated to the 12th-13th Dynasty (Middle Kingdom), is only tentative. This charred grain kernel may have originated from one of the upper spikelets of emmer, which often produces only one grain kernel instead of two. Such grain kernels develop under similar growing conditions as those of einkorn and may, therefore, have a concave shape on both the dorsal and ventral side. A positive identification of einkorn by their rachis fragments is also difficult as they may also indicate small specimens of emmer wheat, which passed through the sieve during crop processing. The small number of fragments does not allow one to distinguish between einkorn and emmer (Cappers *et al*, 2004).

Hard wheat (*Triticum turgidum* ssp. *durum*) is represented by rachis fragments in the Gebel Roma^c and

Table 2: Cultivated plants from Wadi el-Hôl. For further explanation, see table 1.

Wadi el-Hôl		IP2 NK	NK	NK IP3	GR	Mixed / undated	
Number of samples (N=14)		2	2	3	1	6	
CEREALS							
Hordeum vulgare ssp. vulgare	grain	7	4 **3**	7 **2**		4	6-row barley (hulled)
	RN	15	19	46	22	43 **1**	
Triticum	RN			5			wheat
Triticum turgidum ssp. dicoccon	grain	4		17	11		emmer
	RN	25 **1**	382	314 **1**	400	338	
Gramineae tribe Triticeae	culm			4			cerealia
Sorghum bicolor	chaff					1	sorghum
OIL PLANTS							
Linum usitatissimum	seed				2	1	flax
	fruit	1		1	1	3	
VEGETABLES							
Allium sativum	bulb					2	garlic
Citrullus lanatus	seed	1					watermelon
Cucumis	seed					1	cucumbers
Cucumis melo	seed			2			melon
Cucumis sativus	seed					1	cucumber
FRUIT TREES							
Ficus sycomorus	fruit				1	12	sycomore fig
Mimusops laurifolia	seed			1		2	persea
Olea europaea	fruit	3	1	1		1	olive
Phoenix dactylifera	seed		4	1		1	date
Vitis vinifera	seed					2	grape
CONDIMENTS							
Nigella sativa	seed			1			black cumin

Table 3: Cultivated plants from Gebel Qarn el-Gir. For further explanation, see table 1.

Gebel Qarn el-Gir		MK	IP2	IP2 NK	NK	NK IP3	IP3	GR	
Number of samples (N=34)		3	1	1	10	10	8	1	
CEREALS									
Hordeum vulgare ssp. vulgare	grain	14		18	376	314	76		6-row barley (hulled)
	RN	45 **1**	1	49	1509	470	193	3	
Triticum	RN	8			9	1			wheat
	culm					3			
Triticum turgidum ssp. dicoccon	grain	2		10	97	154	87	1	emmer
	RN	18 (4)		25	1260 [1]	2695 (85)	1295	3	
Triticum turgidum ssp. durum	grain								hard wheat
	RN				5	86		3	
PULSES									
Lens culinaris	seed				1	2	1		lentil
Pisum sativum	seed				[1]				
OIL PLANTS									
Linum usitatissimum	seed				20		3		flax
	fruit	1		1	8	2	2		
Ricinus communis	seed					1			castor-oil-plant
TUBERS									
Cyperus esculentus	tuber					1			earth-almond
VEGETABLES									
Citrullus lanatus	seed					5	1		watermelon
Cucumis	seed		1		1	6	2		cucumbers
Cucumis melo	seed				2		3		melon
Cucumis sativus	seed				(1)	1			cucumber
FRUIT TREES									
Balanites aegyptiaca	fruit			[1]					sugar date
Ficus sycomorus	fruit	41	3		32	3 [1]			sycomore fig
Mimusops laurifolia	seed				2				persea
Olea europaea	fruit					1			olive
Phoenix dactylifera	seed				6	5	5	1	date
Vitis vinifera	seed	1			1				grape
CONDIMENTS									
Coriandrum sativum	fruit					1			coriander
Nigella sativa	seed					4			black cumin

Figure 3: Rachis fragment of hard wheat (Triticum turgidum *ssp.* durum) *from Gebel Roma^c (stratum 9).*

the Gebel Qarn el-Gir caravansary deposits. From the New Kingdom onwards, these fragments clearly show the diagnostic features of tetraploid naked wheats (figure 3). The number of positively identified fragments from the 1.0 mm sieve fraction has been adjusted to the whole volume of the samples and represents dozens of specimens. The single chaff fragment of Sorghum (*Sorghum bicolor*) from Wadi el-Hôl is considered to be contamination. The current subfossil record from Egypt shows that this African domesticate only appears during the Roman period.

Pulses are barely present in the samples examined. Only a few lentil seeds (*Lens culinaris*) could be positively identified from Gebel Roma^c and Gebel Qarn el-Gir. The identification of a single pea seed (*Pisum sativum*) from the Gebel Qarn el-Gir deposits, dated to the New Kingdom, remains uncertain.

Oil plants are represented by flax (*Linum usitatissimum*), safflower (*Carthamus tinctorius*) and the castor-oil plant (*Ricinus communis*). The sugar date (*Balanites aegyptiaca*) and black cumin (*Nigella sativa*), which have been categorized as a fruit tree and condiment respectively, may also be included in this group. Flax is the only plant found at all three sites and is represented by both seeds and capsule fragments. A single safflower fruit comes from the Gebel Roma^c deposits, and the only seed of the castor-oil plant was found at Qarn el-Gir. Both these records are dated to the New Kingdom.

Several tubers were found, mainly originating from the Gebel Roma^c deposits. Only two such specimens could be positively identified and turned out to be earth-almond (*Cyperus esculentus*), they were found in the Gebel Roma^c and the Gebel Qarn el-Gir deposits. The majority of the other tubers probably also belong to this plant species, but their identification remained uncertain due to the lack of reference material in the field lab.

Vegetables are well represented by garlic (*Allium sativum*), watermelon (*Citrullus lanatus*), melon (*Cucumis melo*) and cucumber (*Cucumis sativus*). These species are recorded from all three sites, with the exception of garlic, which was not evidenced at Gebel Qarn el-Gir.

Also well represented are edible fruits and seeds from trees, of which the sycomore fig (*Ficus sycomorus*) stands out due to its large number of fruits. In addition to the small fruits, which are also erroneously referred to as 'seeds', one (or possibly two) so-called accessory fruits have been found. The edible false fruit (*syconium*) of a fig consists of many small fruits (*sensu stricto*) surrounded by a hollow, fleshy receptacle, the last part of the flower stalk to which the flowers are attached. The fruits of the sycomore fig are conspicuously better represented as those of grape (*Vitis vinifera*) and pomegranate (*Punica granatum*). The highly valued dates (*Phoenix dactylifera*) are well represented and could be obtained in large quantities from the Nile valley. Also, persea (*Mimusops laurifolia*) is represented by seeds, though in smaller numbers. Today, only a few planted persea trees still grow in the garden of the Agricultural Museum in Cairo, but the archaeobotanical record, dating back to the 3rd Dynasty and comprising seeds as well as leaves, indicates that this tree was more common in ancient Egypt. The sugar date (*Balanites aegyptiaca*) is represented by a few inner parts of the fruit (*endocarps*). The outer part of the fruit, which resembles the shape of the date, has a sugar content of about 45%, while the relatively small seed is quite soft and yields 40-60% balanos oil. Contrary to the date palm and the persea tree, the sugar date is adapted to desert environments, such as the middle and upstream zones of the *wadis* of the Eastern Desert.

Condiments are represented by coriander (*Coriandrum sativum*) and black cumin (*Nigella sativa*). Coriander has been cultivated in Egypt from the Predynastic period onwards and today is still produced on a large scale. The fruits from Egypt are spherical and are easily distinguished from the more elongated fruits from India and Nepal and the smaller fruits from east Asia (figure 4). Although its vernacular name suggests a taxonomic relationship with cumin (*Cuminum cyminum*) and sweet-cumin (anise; *Pimpinella anisum*), both

Figure 4: Recent coriander fruits (Coriandrum sativum) *from Egypt (left), India (middle) and China (right).*

Figure 5: Pressing oil out of black cumin (Nigella sativa) *(Cairo, 1997).*

members of the Carrot family (Apiaceae), the black cumin (*Nigella sativa*) belongs to the Buttercup family (Ranunculaceae). The tasty seeds are used as a condiment in dishes or for making brown-yellowish cumin oil. This oil is still produced in Egyptian spice shops, where seeds are cold-pressed in two turns (figure 5). The Egyptian archaeobotanical record of black cumin is much more restricted. Thus far it has been recorded from Kahun (Middle Kingdom; Germer, 1989), Thebes (Germer, 1988), the Valley of the Kings (18th Dynasty, Germer, 1989) and Mons Claudianus (Roman period, Van der Veen, 1996).

3.2 Wild plants

The wild plants that have been retrieved from the three sites are summarized in Table 4. Some wild plants are conspicuously present and include Acacia and some plants that are considered weeds: *Ambrosia maritima*, cf. *Enarthrocarpus lyratus*, *Lolium temulentum*, *Phalaris paradoxa* and *Sorghum halepense*. Acacia is represented by seeds and thorns, and it is possible that the spiny acacia branches were used in the construction of temporary animal pens (*zaribas*). Although the weed plants are associated with barley, emmer and flax, it seems most likely that they have to be considered weed plants from cereal fields. Cereals can be reaped with sickles close to the ground and in this way many weed plants are harvested along with the crop plants. Flax, on the other hand, is traditionally harvested by uprooting and so only the tallest weed plants are uprooted along with the crops. Moreover, *Lolium remotum*, being a typical weed in flax fields, was not found.

Table 4: Wild plants from Gebel Roma⁽(R), Wadi el-Hôl (H) and Gebel Qarn el-Gir (G). For abbreviations of periods, see table 1.

		MK		IP2	IP2 NK			NK			NK IP3			IP3	GR		? and mixed	
Site		R	G	R	R	G	H	R	G	H	R	G	H	R	R	H	R	H
Number of samples		5	3	6	3	1	3	24	10	2	3	10	3	12	6	1	13	6
Acacia	seed	6		1	2		1	4	5	2	4	14	3	18	1		13	
	thorn	1	1						2	1		3	5				1	2
Aerva javanica	seed				3													
Aizoaceae	fruit							5										
Amaranthus	seed								9									
Ambrosia maritima	fruit	30		16				4				3		5			1	
Asphodelus tenuifolius	seed			2														
Bromus	fruit								9									
Carex	fruit																1	
Chenopodium murale	fruit			6														
Citrullus colocynthis	seed								1					1			1	
Compositae	fruit						1		2									
Cruciferae	seed			8										2				
Cyperaceae	fruit							5		4	2							
Digitaria	fruit			11				3										
Echium	fruit			2											1		1	
Cf. Enarthrocarpus lyratus	fruit	4		14	4	1	1	8		4			1	1			2	
Eragrostis	fruit		1														18	
Gramineae	fruit			13				1				2		1			13	
	RN				5													
Lathyrus	seed			1														
Lathyrus hirsutus	seed								1									
Leguminosae	seed			6													1	1
	hilum						1							2				
	fruit			7														
Lolium temulentum	fruit		4	16		1	4	50	108				1	4			12	
Lupinus digitatus	seed													2				
Panicum turgidum	fruit							4	4					1	1	1	2	3
Phalaris paradoxa	fruit			47	5			81	39			5	1	41			20	
Raphanus raphanistrum	fruit													1	5		3	
Rumex	fruit			8				3	1					1				
Scirpus	fruit										1							
Silene	seed			6				1										
Sorghum halepense	fruit			8				5	6		1	3	1	2		1	9	
Tamarix	branch							1										
Trigonella	seed								1									

Of special interest are two seeds of *Lupinus digitatus*, which were found in stratum 5 of the Gebel Romaᶜ deposit, dated to the 20ᵗʰ-24ᵗʰ Dynasty. These seeds have the typical square shape of lupins. The hilum is clearly exposed at one of the corners (figure 6). The geographical distribution is restricted to northern Africa (Egypt, Libya, Algeria, Morocco and Western Sahara) and northeast tropical Africa (Sudan). In Egypt, it is recorded from fallow fields and sandy desert soils (Boulos, 1999). Due to the new irrigation system that is practised since the construction of the Aswan High Dam, *L. digitatus* has recently become an endangered species in Egypt (Gladstones, 1998). In addition, the subfossil record of this wild legume is numerically limited.

In Abusir, a single seed has been found among emmer chaff from the pyramid of Neuserre, dated to the 5ᵗʰ Dynasty (Germer, 1988). The only other record is that of some seeds found at Hawara in the Fayum, dated to the Graeco-Roman period. Originally, these seeds were identified as white lupin (*L. albus*), but Schweinfurth re-identified them as *L. varius* ssp. *orientalis* (= *L. digitatis*) (Keimer, 1984). It may very well be that the seeds of the tall *L. digitatus* found at Gebel Romaᶜ were also incidentally mixed in with the emmer or barley during the harvesting process. As the large seeds are not easily separated from grain kernels by sieving, it is possible that the Lupinus seeds were transported with the unthreshed harvest as well as the threshed.

Figure 6: Seed of Lupinus digitatus *from the Gebel Roma^c deposit.*

It is remarkable that plants, such as *Avena fatua*, *Beta vulgaris* and *Convolvulus arvensis,* which are considered noxious weeds today in Egyptian cereal fields, have not been found. Judging from their limited presence in the archaeobotanical record of Ancient Egypt, is seems as if these weed plants have only become dominant in more recent times (Cappers, 2006a; Fahmy, 1997).

Some wild plants can be considered as desert plants which grew in the near surroundings of the caravansaries. They include *Aerva javanica*, *Amaranthus* (probably *A. graecizans*), *Asphodelus tenuifolius*, *Citrullus colocynthis*, *Panicum turgidum* and *Tamarix*. It is possible that fresh fruits of the colocinth (*C. colocynthis*), being juicy but extremely bitter in taste, were gathered in the *wadis* and used as fodder for donkeys.

4 Conclusions and discussion

4.1 Representativeness of pulses and cereals

A number of factors, including seed production, seed dispersal, preservation conditions and the sampling procedure, determine the representativeness of the different groups of cultivated plants. Concerning the cultivated plants, the process of seed dispersal can be interpreted in terms of harvesting practices and crop processing. It seems that cereals have a much better chance of ending up in a site than pulses do. This can be explained by the threshing procedure, the economic value of the threshing remains and the loss of seeds during sieving.

Although the harvesting of a pulse crop produces a substantial amount of threshing remains, consisting of pods and vegetative plant remains, the economic value of these remains is low, for which reason they might be used as fertilizer instead of fodder to be brought to the settlement. As a result, the archaeobotanical recovery of pulses is in most cases based on the seeds only. In addition to this, the presence of seeds of pulses in debris deposits is linked to the seed's size. Lentil seeds, for example, are relatively small and some of them will pass through the sieves, as a result of which they might become mixed up with the debris. Peas and chick peas, on the other hand, are relatively large, and spoilage will be rather exceptional.

Both hulled and naked (or, free-threshing) cereals are transported along with their chaff to the site. Hulled cereals, such as hulled barley and emmer wheat, will not produce large quantities of threshing remains. The first stage of threshing concerns the fragmentation of the ears. Spikelets, consisting of grain kernels enclosed by chaff and a rachis fragment, are stored in specialized facilities. Prior to food production, small quantities of the spikelets undergo a second stage of threshing, in which the grain kernels are separated from chaff and rachis fragments. In this way, only relatively small quantities of threshing remains are produced, and it is assumed that such small amounts have only a limited economic value.

Cultivating naked cereals such as hard wheat, on the other hand, will produce large quantities of threshing remains as the threshing of the whole harvest can be easily done in one go. As huge quantities of threshing remains are produced in this way, they found their way into the archaeological record as building material (tempering of clay), fuel (tempering of dung), and fodder. This explains the omnipresence of threshing remains of naked cereals in settlements.

The presence of threshing remains of naked cereals on sites outside production areas, such as the caravansaries along the Qena Bend, depends on the method of transport. If only grain kernels are traded, which makes sense when large quantities are concerned as it reduces both volume and weight, the archaeological recovery of the cereal relies predominantly on the lost grain kernels. Small numbers of such grain kernels could be indicative of substantial transport. However, when naked cereals are transported as whole ears, which was true, for example, at Roman Berenike, the recovery of the cereal is much more likely since large quantities of threshing remains are produced on the site proper (Cappers, 2006b). Judging from the number of rachis fragments of hard wheat attested at Gebel Qarn el-Gir, it is suggested that in addition to the transport of hulled cereals (*viz.*: barley and emmer wheat), naked wheat was also a substantial commodity.

Due to the arid climate, uncharred plant remains are excellently preserved in Egypt and unfavourable preservation conditions are mainly related to wind dis-

Table 5: Fruit weight, seed numbers and energy content of eight fruits evidenced from Gebel Roma^c, Wadi el-Hôl, and Gebel Qarn el-Gir.

	weight (g)[1]	seeds/fruit	kJ/100 g[2]	seeds/100 g	seeds/ 1000 kJ[3]
Citrullus lanatus	1500-4000	c. 200-800[4]	80	5-53	63-663
Cucumis melo	500-1000	c. 500[5]	42[6]	50-100	1190-2380
Cucumis sativus	200-300	c. 200[7]	40	67-100	100-450
Punica granatum	150	c. 250-400[8]	343	167-266	469-752
Ficus carica (dried)	21[9]	c. 700-2200[10]	969	3325-4284	3431-4421
Vitis vinifera (dried)	1[11]	4	286	400	1399
Phoenix datcylifera (dried)	7[12]	1	724	14	19
Olea europea (dried)	2[13]	1	369	50	136

1) Fresh weight of whole fruit. Sources: Kiple & Ornelas 2000; Voedingscentrum The Hague.
2) Amounts per 100 gram portion including peel and seeds. Sources: National Public Health Institute, Nutrition Unit. Fineli; Breedveld *et al.* (1996).
3) Daily intake of energy is c. 10,500 kJ for a man and 8,400 kJ for a woman.
4) Diploid species: Sugiyama & Morishita 2002.
5) Shoemaker.
6) *Cucumis melo* ssp. *melo* var. *cantalupensis* Naudin (Musk melon).
7) Thoa 1998.
8) Kučan 1995.
9) Measurements taken from herbarium specimens collected in Groningen (shop).
   Weight: (15.82-)20.64(-29.88) gram (N=10, s.d.=1.31).
10) Condit 1947 and measurements taken from herbarium specimens (Groningen, small: N=918; 22,62 gram = 1598 fruits, 29,86 gram = 2239 fruits; Iran: N=621).
11) Measurements taken from herbarium specimens collected in Yemen (market; grapes with pips).
    Weight: (0.2-)0.83(-1.45) gram (N=111, s.d.=0.03).
12) Measurements taken from herbarium specimens collected in the Fayum (Egypt).
    Weight: (1.82-)6.69(-12.89) gram (N=78, s.d.=0.27; fruits from 6 different trees).
13) Measurements taken from herbarium specimens collected in the Fayum (Egypt).
    Weight: (0.29-)1.72(-3.04) gram (N=79, s.d.=0.08; fruits from 5 different trees).

persal, browsing animals and damage by insects and the like. It is assumed that the botanical composition that is currently present in debris deposits is only a faint reflection of the plant remains that were deposited over the course of time. Recent observations of debris deposits at Berenike (Eastern Desert), for example, have attested to the presence of the following animals: cattle, pigs, chickens, sheep, goats, donkeys, black rats, dogs, Egyptian vultures and brown-necked ravens (Cappers, 2006b). Small creatures, such as beetles and larvae, also eat organic material. This kind of reduction in volume of deposited botanical remains may occur over thousands of years, as could be observed when special attention was paid to this process while analyzing a Roman debris deposit in Karanis (Arabic: Kôm Aushim). It was found that such small animals are present at least in the upper 10 cm of a debris deposit.

It should be realized that both formation and reduction processes are responsible for the reduction and contamination of the botanical composition of debris deposits. The removal of debris particles by wind blowing is enhanced in arid environments and seems to be a common phenomenon on Egyptian sites. Wind erosion is a selective process and will only affect particles up to a specific weight class. Particles on the surface which are too heavy, will gradually sink down if small particles around them are blown away. Only when such heavy particles have become embedded in a matrix of heavy particles, will they remain in situ. Over the course of time, wind erosion will create a surface layer that consists only of heavy particles, which are mostly of a considerable size. Only when such a soil crust is disrupted, for example by visiting animals, and mixed up with smaller particles underneath, further wind erosion may take place. Small particles are only preserved when sufficient amounts of new trash are deposited.

The stratified deposits of Gebel Roma^c, Wadi el-Hôl and Qarn el-Gir are composed primarily of pottery intermixed with dung and debris, with interspersed "laid" floors of both gypsum plaster and crushed sherds used to create a sealed surface (D. Darnell 2002). Such hard layers could have prevented the materials from blowing around the site in high winds. Furthermore, plant species with small seeds are present in the plant record, indicating that proper sealing of certain levels within the deposits does occur.

The debris deposits at Wadi el-Hôl, located at the edge of the *wadi*, may also have suffered occasional erosion by water currents. The only clear evidence of water damage consists of rain lenses in stratum no. 3 of the main debris deposit at Gebel Roma^c. The poor preservation of organic material in this particular layer may be caused by microbial decay that was possible under temporarily moist conditions.

Finally, also the number of samples that have been processed effects the representativeness. Most of the samples originate from Gebel Roma^c and it is only from this site that members of all cultivated plant groups are present. It can therefore be assumed that the plant record of Gebel Roma^c is also representative of the other sites.

4.2 Representativeness of vegetables and fruits
Fruits with a high water and sugar content are perfect treats during desert trips. Fruits that are especially refreshing for their high water content include watermelon, melon, cucumber, and pomegranate. These juicy fruits can be easily transported fresh over large distances. Fresh figs and grapes, on the other hand, are more vulnerable to damage, and so were probably normally transported in a pre-dried state. Dates and olives have a lower water content and can be transported either fresh or dried.

It is difficult to determine the contribution of each of these fruits to the ancient traveller's diet. One reason is that some fruits are consumed in their entirety, whereas with others, the seeds or the inner part of the fruit is discarded. Examples of the former fruit type are fig and pomegranate. Fruits that are only partly eaten include watermelon, melon, cucumber, date, and olive. The relatively large seeds of the watermelon are present throughout the fruit, whereas those of the melon and cucumber are concentrated in rows and are therefore more easily removed. Grapes are an indeterminate group since some people eat the entire berries, whereas others spit out the seeds, especially when the seeds are relatively large.

This difference in consumption pattern also affects the deposition of the inedible or indigestible seeds and fruit parts. When entirely swallowed, fig, pomegranate, and grape seeds will be excreted with the faeces. The presence of their seeds in the processed samples from the debris deposits indicates that this area would also have been used, at least temporarily, by visitors to defecate. When grape, date and olive seeds are spit out at random spots, their chance of ending up in the archaeobotanical record is strongly reduced. Cucumber and melon seeds may become part of kitchen refuse that will be thrown into the debris deposits.

Another reason that obscures reconstruction of the proportion in which these fruits were consumed, is that seed numbers differ greatly among fruits and are not proportionally related to their possible consumption. Each date and olive, for example, produces only one seed (olive: endocarp), whereas a single fig contains about 700-900 small fruits. A possible solution to this problem lies in converting the seed numbers to fruit weight and the amount of energy (expressed in kilojoules [kJ; 1 kilojoule = 0.24 kilocalories]).

In Table 5 the number of seeds per 100 grams, as well as the number of seeds per 1000 kJ, is presented for watermelon, melon, cucumber, pomegranate, fig, grape, date and olive. Figs and watermelon are two fruits that are clearly remarkable in terms of the number of seeds by weight and the amount of energy contained. With the exception of the watermelon, the number of seeds per fruit in the species retrieved is proportional to both the fruit weight and the amount of energy. Watermelon seeds represent a relatively large amount of energy in comparison with the standardised weight of its fruit.

Some prudence is called for when comparing fruits that occur in large numbers. In such fruits, there may be a considerable variation in the actual number of seeds that are present in a particular specimen. A number of factors, including the flowering sequence and the pollination, cause this variation, which is also expressed in Table 5. Fruits with only a single seed, such as date and olive, can be interpreted unequivocally.

Comparing the seed numbers from Table **5** with those retrieved from Gebel Roma^c, the site which produced the largest number of seeds, it is evident that the fig is the most favourite fruit. It should be mentioned that at Gebel Roma^c the sycomore fig (*Ficus sycomorus*) is found, and that the average number of fruits (s.s.) might be different from that of the fig (*F. carica*). The second most common fruit consumed at Gebel Roma^c is the date. Although its number of seeds per unit of energy is less than that of the watermelon, the recovery of this fruit is enhanced by its relatively large seed size.

Since the number of seeds from the debris deposit of Gebel Roma^c is still relatively small, the interpretation is only tentative. A more balanced interpretation of fruit consumption can be obtained when substantial numbers of seeds are available from a particular site, preferably from deposits which combine kitchen refuge and locations where faeces are concentrated. Such a combination of contexts would take into account the different consumption patterns and would reduce the necessity of uncertain corrections.

5 References

Boulos, L. (1999): *Flora of Egypt.* Volume 1. Cairo, Al Hadara Publishing.

Breedveld, B.C., J. Hammink & H.M. van Oosten (1996): *Nederlandse voedingsmiddelentabel.* Den Haag.

Cappers, R.T.J., T. van Thuyne & L. Sikking (2004): Plant remains from Predynastic El Abadiya-2 (Naqada area, Upper Egypt). In: S. Hendrickx; R.F. Friedman, K.M. Cialowicz, M. Chlodnicki (eds.), Egypt at its Origins. Studies in Memory of Barbara Adams. Proceedings of the International Conference "Origin of the State. Predynastic and Early Dynastic Egypt", Krakow, 28th August - 1st September 2002. *Orientalia Lovaniensia Analecta* 138. Leuven: Peeters Publishers, pp. 297-293.

Cappers, R.T.J. (2006a): The reconstruction of agricultural practices in ancient Egypt: an ethnoarchaeobotanical approach. In: *Palaeohistoria* (47/48), pp. 429-446.

Cappers, R.T.J. (2006b): *Roman foodprints at Berenike. Archaeobotanical evidence of trade and subsistence in the Eastern Desert of Egypt.* Monograph 55, Cotsen Institute of Archaeology, UCLA.

Condit, I.J. (1947): *The fig.* Waltham, Chronica Botanica Co.

Darnell, D. (2002): Station on the Plateau above the Wadi el-Hôl. In: J. C. Darnell with the assistance of D. Darnell (eds.): *Theban Desert Road Survey in the Egyptian Western Desert vol. 1: Gebel Tjajuti Rock Inscriptions 1–45 and Wadi el-Hôl Rock Inscriptions 1–45* (Oriental Institute Publications 119), Chicago, p. 91.

Darnell, J.C. (2002): Opening the Narrow Doors of the Desert: Discoveries of the Theban Desert Road Survey. In: R. Freidman (ed): *Egypt and Nubia - Gifts of the Desert.* London, British Museum Publications, pp. 138-139.

Fahmy, A.G. (1997): Evaluation of the weed flora of Egypt from Predynastic to Graeco-Roman times. In: *Vegetation History and Archaeobotany* 6, pp. 241-247.

Germer, R. (1985): *Flora des pharaonischen Ägypten.* Mainz am Rhein, Verlag Philipp von Zabern.

Germer, R. (1988): Katalog der altägyptischen Pflanzenreste der Berliner Museen. In: *Ägyptologische Abhandlungen* Band 47.

Germer, R. (1989): *Die Pflanzenmaterialien aus dem Grab des Tutanchamun.* Hildesheim, Gerstenberg Verlag.

Gladstones, J.S. (1998): Distribution, origin, taxonomy, history and importance. In: J.S. Gladstones, C.A. Atkins & J. Hamblin (eds.): *Lupins as crop plants. Biology, production and utilization.* Cambridge, Wallington University Press, pp. 23-26.

Keimer, L. (1984): *Die Gartenpflanzen im alten Ägypten.* Band 2. Herausgegeben von Renate Germer. Mainz am Rhein, Verlag Philipp von Zabern.

Kučan, D. (1995): Zur Ernährung und dem Gebrauch von Pflanzen im Heraion von Samos im 7. Jahrhundert v. Chr. In: *Jahrbuch des Deutschen Archäologischen Instituts* 110, pp. 1-64.

National Public Health Institute, Nutrition Unit. Fineli. Finnish food composition database. Release 4. Helsinki 2004. http://www.ktl.fi/fineli/

Prestel, P. (2004): Die Lebenselixiere der Pharaonen. In: G. Graichen (ed.): *Heilwissen versunkener Kulturen. Im Bann der grünen Götter.* Econ Verlag, pp. 22-81.

Shoemaker, B. Melon culture & varieties (WWW).

Sugiyama, K. & M. Morishita (2002): New method of producing diploid seedless watermelon fruit. In: *Japan Agricultural Research Quarterly* 36, pp. 177-182.

Thoa, D.K. (1998) AVRDC-ARC research report, Cucumber seed multiplication and characterization (WWW).

Veen, M. Van der (1996): The plant remains from Mons Claudianus, a Roman quarry setlement in the Eastern Desert, Egypt. An interim report. In: *Vegetation History and Archaeobotany* 5, pp. 137-141.

Growing, gathering and offering: Predynastic plant economy at Adaïma (Upper Egypt)

C. Newton
Centre de Bio-Archéologie et d'Écologie, Université Montpellier, France

1 Introduction

The Late Predynastic site of Adaïma lies in Upper Egypt, in the Nile valley, at the border between the present cultivated plain and the desert, and covers 35 ha. The eastern and western cemeteries and a large settlement area have been excavated since 1990 under the direction of Béatrix Midant-Reynes (CNRS, Toulouse). The occupation of Adaïma extends over 700 years, spanning from the cultural periods Nagada IC (Predynastic) to Nagada IIIC/D (2nd Dynasty). Since preliminary studies of plant macro remains (De Vartavan, 1991, 2002), wood (Dietrich, 2002) and charcoal (Pernaud, 2002), further work has been undertaken by the author, on material both from the cemeteries and from part of the settlement site (the northern terrace), whose occupation was longer and later than that of the sector previously studied. Four charcoal samples from this terrace were radiocarbon dated, resulting in the following uncalibrated dates B.P.: 4765±50, 4445±45, 4440±45 and 4430±45. These dates correspond culturally to the period between the end of Nagada II (Predynastic) and Nagada IIIB/C (Protodynastic, dynasties I and II) (Midant-Reynes, Buchez, 2002).

The features excavated between 1997 and 2000 are mainly pits dug into the substratum of the terrace. In some cases, their floor and walls were lined with a mixture of soil and plant material – pisé in the broad sense. These features were also probably covered with such pisé material supported by twigs or reed-type culms, as imprints on large fragments of pisé found fallen at the bottom of one feature indicate. The archaeological sediment filling the pits is mainly a mixture of sand and silt, and no vertical stratigraphy can be observed in the sector studied. The filling of the pits consists of the settlement's trash, with all kinds of material being found together, and no clear domestic feature is defined. Sampling the pisé *in situ* in the pits eliminated the problem of the antiquity of the plant remains, and was therefore favoured. A specific study on the taphonomical and technical implications of this building material is described elsewhere (Newton, 2004).

The cemetery, excavated under the direction of professor Éric Crubézy (University Paul Sabatier, Toulouse), is divided into two main zones. The western area, excavated up until 1996, is located on a sandy butte separated from the settlement site by a wadi bed (Crubézy et al., 2002). Its continuous use as a burial place spans from Nagada IC until Nagada IIIA/B. It comprises tombs of several degrees of richness, including one which is likely to have been the first tomb (S55). It attests the existence of an elite group and of complex ritual practices, as can be derived from the burial of the bodies and the deposit of offerings (Crubézy et al., 2002; Buchez, 1998). The eastern cemetery is located in and near the wadi bed, and can be divided into two parts. One is mainly the burial site of children aged between 0 to 12 years, and seems to have been in use during the Nagada IIIA/B period (end of the Predynastic) The other part is dominated with a later group from Nagada IIIC/D (2nd dynasty). The graves from the eastern cemetery were found intact, remarkably well-preserved, and rich in archaeological material. The organic remains found are therefore considered to be *in situ*.

The purpose of the study was to have an overview of the plant environment and the plant economy of the site during its occupation. Therefore, all macroscopic plant remains from a variety of provenances were studied. The main results are presented here in a condensed form, with the exception of the charcoal analyses (Newton, 2005). The palaeoethnobotanical topics discussed here will centre on crops and agricultural practices, the importance of wild plants in the economy, and the implications of the plant deposits in funerary contexts.

Table 1: Plant macroremains identified from the settlement site of Adaïma, excluding wood, summarized by excavated square. The figures are the counted numbers of remains per sample, the result of a calculation if only a subsample was sorted. The type of remain is a seed and/or fruit, except otherwise mentioned. An 'x' indicates that the plant remain is not quantified.

Types	Taxa	Type of remain	Square	1030/17	1040/10	1040/16	1040/17	1050/16	1050/17	1060/12	1070/13	1080/13	1080/22	1090/13	4001/16	7001	TOTAL
			Number of samples	5	4	18	34	4	8	4	2	4	1	1	6	1	92
			Volume (l)	27.8	19.5	114.4	137.9	7.8	18.55	10.95	7.25	14.75	2.9	1.1	30.5	4.3	397.7
CROPS																	
Cerealia	*Triticum* sp.					36	36		6	13		3			6	3	103
	Triticum turgidum subsp. *dicoccum*			1	2	1	1	2	7	11	5			4		8	41
		rachis node		345	365	1079	928	90	207	441	142	567	3	155	1	166	4489
		chaff			3	3	1	11								1	19
	Hordeum vulgare			112	82	1197	964	60	186	124	66	6	3	4	6	206	3016
		rachis node		972	428	11720	7735	427	1137	316	104	722	5	122	x	656	24344
		chaff				x	30	1	51	x			x				82
	Cerealia			2	20	973	402	21	43	77	13	103	4	26	96	17	1796
		rachis node		64	10	612	118	24	113	51	10	80		4	2		1088
		chaff					29										29
		culm node				38	7							17			62
Other crops	*Linum usitatissimum*			91	43	1795	371	3	40	32	11	7		21		42	2454
	Lens culinaris			1		2	5				1						9
	type *Vicia* sp.				11	5	7	6	12	19		2				10	72
	Cucumis sativus/melo				2	3	4									14	24
WILD TAXA																	
Edible wild fruit	*Balanites aegyptiaca*								6								6
	Citrullus colocynthis			31	46	492	128	1	248	55	22	56	1	13	4	20	1116
	Ziziphus spina-christi				5	4				15	1						25
Other woody taxa	*Acacia* types				2	33	7		60		2					2	106
		leaflet		16	7	6	66	2								1	98
	Capparaceae type *Cleome* sp.						2										2
	Tamarix spp.	stem/leaf		x	x	x	x		x	x	x	x			x	x	x
WILD/WEED DICOTYLEDONES																	
	Ceruana pratensis			225	73	10073	2406	40	295	111	73	88	5	21	8	97	13515
	Ambrosia maritima			378	77	11036	2743	17	409	55	11	136	1			20	14883
	Asteraceae			1	2	44	7	1		4						1	60
	Cocculus pendulus			1		4	2				1						8
	Coronopus niloticus			42	21	85	81	5		12	3	132				7	389
	Enarthrocarpus cf. *strangulatus*			3	10	200	143	1	11	25		41				8	443
	Brassicaceae				4	58	1		1			9					73
	Fabaceae including *Medicago*, *Trifolium* and *Trigonella* types			6	24	364	146	2	21	18		3			8	2	594
	Fabaceae/Brassicaceae								1						14		15
	Resedaceae														1		1
	Echium rauwolfii			30	17	346	176	10	24	24		13	2	13	6	114	774
	Heliotropium sp.			69	36	93	93		3	14		305	1			1	616
	Boraginaceae				5		5		3			18					31
		leaf				6	4										10
	Boraginaceae/Labiatae			67		298	69		1			9					445
	Labiatae					24	4	1		2		2					33
	Labiatae type 1/cf. *Verbena* sp.			2	2	161	39	1	6	4		4				9	228
	Caryophyllaceae			303	44	2038	1158	588	108	56	45	342	1	4		8	4694
	Portulaca oleracea			146	15	3	2	2				186				6	359
	Amaranthaceae						3									1	4
	Chenopodium murale			106		34	68			23		363					594
	Chenopodiaceae				29	84	28	1	1	2							145
	Raunnculaceae					3	48		3								54
	Hyoscyamus sp.														2		2
	Solanum nigrum			27	2		19		4			20				1	73
	Solanaceae					98	21	1									120
	Apiaceae			23	1	13	1					40					78
	Fumaria sp.			34	5	10	4			14		55				1	122
	type Euphorbiaceae					8	61	5	13	4		11					102
	Malvaceae						1										1
	Rumex spp.			61	46	1779	572	36	100	38	13	57		30	10	89	2830
	Polygonaceae				37	15	10		5			94				42	203
	Polygonaceae/Cyperaceae					16	4										20
	Dicotyledone	leaf				8	15										23
CYPERACEAE																	
	Fimbristylis bisumbellata			53		199	246	12	23	12	6	5				19	576
	Scirpus spp.					1	9					2					12
	Cyperus spp.			287	133	4096	2538	65	216	216	78	543	4	13	8	685	8880
		root/rhizome/tuber			3		1	2									6
	Cyperaceae			60	7	361	234	3	11	6		9	1		4	11	707
		root/rhizome/tuber					1										1
WILD POACEAE																	
	Bromus sp.						1									11	12
	Chloris sp.						3										3
	Crypsis schoenoides			112		1307	865	69	33	2	45	9	2	4		13	2461
	Eragrostis sp.			3		3	1		17	8							31
	wild *Hordeum*			2	21	432	154		28	22		10			20	4	692
		rachis node				3	71										74
	Lolium sp.						72	4	6								82
	Phalaris sp.			5	7	205	75	4	12	12		13	4			20	356
	Panicoid types			24	2	86	1020	69	208	12	16	39		13		4	1494
		chaff						1	1								2
	Poaceae including wild Poaceae types			703	14	4893	3366	79	434	285	38	120	1	36	136	118	10223
		rachis node		16		73	170		11	6		101	1	8			386
		chaff		x		8	34	x	x	x		3			2		47
		culm base		16			3	x		11						1	31
		culm node		19	39	265	228	8	81	119	15	80				47	900
		root/rhizome/tuber				1			8							4	13
Unidentified		varia		25	23	2259	560	20	48	75	1	4	4	5	41	24	3089
TOTAL				4483	1726	59091	28428	1695	4261	2341	721	4412	43	514	358	2531	110601

2 Materials and methods

The plant remains studied from Adaïma are from two distinct: the northern terrace of the settlement area, and the tombs. The samples were collected and treated differently depending on their provenience and nature.

2.1 The settlement area

Apart from 53 samples which were hand-picked during the excavation and will not be used in quantified analyses, 92 of the samples were bulk soil samples. These were machine-floated using an interior 3 mm mesh sieve, and exterior 1 and 0.2 mm mesh sieves. The pisé material lining some of the pits was sampled directly on site from their walls and floor. The pisé, instead of being machine-floated, was hand-floated using a bucket and sieves with 1 and 0.4 mm mesh sizes. They represent 27 of the 92 samples. The resulting samples were dried and then sorted under a low-power microscope at 6 to 15 x magnifications. Some samples were sub-sampled depending on the quantity of material present. In these cases, the numbers of remains used in the tables and figures were calculated to represent the total number of remains per sample; they are therefore not necessarily the actual counted remains.

2.2 The tombs

All the plant remains and artefacts made of plant material were hand-picked during the excavations; in total, 134 of these samples were examined. Moreover, the contents of part of the ceramic vessels found in the graves were investigated; from field season 1999 onwards. They were systematically checked for their contents, emptied and the contents were dry-sieved. The state of preservation did not allow the floatation of these contents, as a test on one sample demonstrated. The 15 vessel contents were examined in the same way as the bulk soil samples from the settlement.

The basketry type samples were swept clean with a small paintbrush and examined without further preparation under a stereomicroscope and a reflection microscope. Specimens from a personal reference collection and studies by Greiss (1957) were used for identification.

Table 1 summarizes the results of identification and counts of plant remains from the settlement area, excluding charcoal. A recapitulative list of the identified taxa with their occurrence is given in table 2.

2.3 States of preservation of the remains

Overall, 38% of the counted remains were preserved by charring. However, the variation in this percentage from one sample to another is high (standard deviation = 25), and in part can be explained by the mode of deposition; the pisé samples and those collected from the fill of the lined pits are richer in desiccated remains (Newton, 2004). The quality of preservation of the charred remains is homogeneous throughout the settlement, whereas that of the desiccated remains differs from one area to another. The fragments present as pisé temper are generally better preserved. In the eastern cemetery, the preservation is not as good. This may be due to the location of the tombs close to the surface and in the wadi bed, where occasional flooding and/or rising of the water table may have leached part of the organic constituents.

3 The settlement

The most abundant and most ubiquitous taxa are the cereals: *Hordeum vulgare* (hulled barley), including both the two-row form (*H. vulgare* ssp. *distichum*) and the six-row form (*H. vulgare* ssp. *vulgare*), and *Triticum turgidum* ssp. *dicoccum* (emmer wheat). A naked wheat, probably *Triticum turgidum* ssp. *durum*, is also present but in small amounts and in fewer contexts. Cereal remains amount to 33% of the remains, of which 27% are chaff fragments. Culm fragments, but for nodes and bases, were not counted.

Another frequent crop is *Linum usitatissimum* (flax/linseed), in the form of seeds and capsule fragments representing 2% of the total number of remains. It is both an oil and a fibre crop, though the use of linseed for its oil has never been attested in ancient Egypt.

Domestic legumes are scarce and badly preserved: *Lens culinaris* (lentil), and *Vicia/Pisum* type (vetch/pea). A few seeds of *Cucumis melo/sativus* and *Cucumis melo* (melon) were also found. The remains are infrequent and the identification is tentative: it could be a wild unidentified species, or a non-sweet domestic variety, such as *C. melo* var. *chate*.

3.1 Cereal identification and variability

Barley was found in the form of hulled and dehusked grains, empty spikelets and, more frequently, rachis fragments. The grains are often in a desiccated and more or less "wrinkled" state that does not allow a morphological analysis. This state of preservation is not due to the recovery technique, since wrinkled grains occur in floated and dry-sieved samples, and both wrinkled, partly wrinkled and full grains occur in a single sample. Whole charred grains can theoretically help to identify the type of barley, 2-row or 6-row, according to the percentage of twisted *versus* straight grains in the samples. This proportion was only calculated for one

Table 2: List of the identified plant remains (excluding wood), and occurrence in the samples taken from the settlement zone on the terrace at Adaïma (number of samples in which the taxon was identified, on a total of 92). In bold: taxa present in at least 50 % of the samples (i.e. 46).

Types/Families	Taxa	Occurrence	Types/Families	Taxa	Occurrence
CROPS			Labiatae/Verbenaceae	Labiatae type 1/cf. *Verbena* sp.	19
Cerealia	*Triticum* sp.	19	Malvaceae	Malvaceae	1
			Menispermaceae	*Cocculus pendulus*	5
	Triticum turgidum subsp. dicoccum	82	Polygonaceae	**Rumex cf. dentatus**	63
	Hordeum vulgare	90		**Rumex sp.**	66
	Cerealia	79		Polygonaceae	18
Fiber/oil	**Linum usitatissimum**	62	Polygonaceae/Cyperaceae	Polygonaceae/Cyperaceae	2
Legume	*Lens culinaris*	6	Portulacaceae	*Portulaca oleracea*	16
	type *Vicia* sp.	18	Raununculaceae	Raununculaceae	3
Fruit	*Cucumis sativus/melo*	5	Resedaceae	Resedaceae	1
EDIBLE WILD FRUIT			Solanaceae	*Hyoscyamus* sp.	2
	Balanites aegyptiaca	1		*Solanum nigrum*	11
	Citrullus colocynthis	58		Solanaceae	12
	Ziziphus spina-christi	7	Tamaricaceae	*Tamarix* spp.	42
OTHER WOODY TAXA			Dicotyledone		7
	Acacia nilotica	2	CYPERACEAE		
	Acacia type	22		Cyperaceae type *Carex* sp.	1
	Acacia/Prosopis	9		*Cyperus rotundus*	3
	Capparaceae type *Cleome* sp.	1		**Cyperus type 1**	51
	Tamarix spp.	42		**Cyperus type 2**	86
WILD/WEED DICOTYLEDONES				*Cyperus* sp.	2
Amaranthaceae	Amaranthaceae	2		**Fimbristylis bisumbellata**	47
Apiaceae	Apiaceae	7		*Scirpus* cf. *maritimus*	4
	Apiaceae A1	1		*Scirpus* sp.	2
	Apiaceae A2	2		Cyperaceae type 1	31
	cf. *Coriandrum* sp.	1		Cyperaceae type 2	8
Asteraceae	**Ceruana pratensis**	84		Cyperaceae type 3	4
	Ambrosia maritima	82		Cyperaceae type 4	1
	Asteraceae	16		Cyperaceae type 5	2
Boraginaceae	**Echium rauwolfii**	64		Cyperaceae	14
	Heliotropium sp.	31	WILD POACEAE		
	type *Lithospermum* sp.	3		*Bromus* sp.	3
	Boraginaceae	5		*Chloris* sp.	1
Boraginaceae/Labiatae	Boraginaceae/Labiatae	18		**Crypsis schoenoides**	69
Brassicaceae	*Coronopus* sp.	40		*Eragrostis* sp.	5
	Enarthrocarpus cf. strangulatus	53		wild *Hordeum*	36
	Brassicaceae	8		*Lolium* sp.	5
Caryophyllaceae	Caryophyllaceae type *Silene* sp.	2		**Phalaris sp.**	47
	Caryophyllaceae type 1	63		*Setaria viridis/verticillata*	2
	Caryophyllaceae type 2	16		*Setaria* sp.	10
	Caryophyllaceae type 4	3		*Panicum* sp./panicoid	14
	Caryophyllaceae	35		Panicoid	38
Chenopodiaceae	*Chenopodium murale*	20		Poaceae type 1	1
	Chenopodiaceae	15		Poaceae type 2	2
Euphorbiaceae	type Euphorbiaceae	16		Poaceae type 3	2
Fabaceae	type *Medicago* sp.	25		Poaceae type 4	2
	type *Trifolium* sp.	6		Poaceae type 5	4
	type *Trigonella* sp.	1		Poaceae type 6	13
	Fabaceae type 1	3		**Poaceae type 7**	60
	Fabaceae	37		**Wild Poaceae**	64
Fabaceae/Brassicaceae	Fabaceae/Brassicaceae	5		**Poaceae**	90
Fumariaceae	*Fumaria* sp.	15	**Unidentified**		79
Labiatae	Labiatae type 2	5	Total		92
	Labiatae	2			

sample, in which grains were numerous enough (*i.e.* 1040/17.26); of the 117 grains, 45 were twisted. This low proportion and the presence of sterile spikelets in some samples indicate a probable mixture of these two types of barley. Moreover, the shape of the charred grains varies from plump to long and thin types, which may be due in part to the distorting effect of charring on dehusked grains.

The rachis fragments are also highly variable in size and morphology. The length of the internode, its more or less tapered shape, the presence or absence of lateral robust rachillas are the main varying characters. They allow the distinction between 6-row and 2-row forms, the latter bearing typical lateral robust rachillas. However, the morphological variability and the differences in preservation do not always allow this distinction. We conclude that probably both types were grown in Adaïma, without being able to decide in what proportions.

Two main factors lead to the morphological variability of both grains and rachis fragments: the probable culture of two barley varieties, *Hordeum vulgare* var. *vulgare* and *H. vulgare* var. *distichum*, defined in relation to modern varieties that may not have existed as such in the past, and the intra varietal variability, due to a genetic diversity and to the varying growth conditions. In addition, an intra individual variability also exists within the spike between proximal, intermediate and distal nodes.

Wheat remains are not very abundant in the Adaïma samples and consist mainly in rachis nodes and glume bases; grains are infrequent. The grains belong to *T. turgidum* ssp. *dicoccum*, although some of them could belong to a *Triticum aestivum/turgidum* ssp. *durum* type. Rachis nodes have a more varied morphology, with einkorn and emmer types partly overlapping. Two types can be distinguished: *T. monococcum/dicoccum* and *T. dicoccum* Schübl. [syn. *T. turgidum* ssp. *dicoccum* (Schrank) Tell.].

The most frequent type of rachis node is *T. monococcum/dicoccum*, with einkorn morphology – glumes slightly diverging or even converging, smooth and glossy, upper disarticulation scar measuring more than half the rachis' width at that height – but emmer size. The glumes are thick and the rachis bears a robust aspect. The recovery of a few top rachis nodes of the emmer type, and the fact that no einkorn type grain was identified in Adaïma, could allow us to consider all the specimens as belonging to emmer wheat. Emmer's morphological variability is high, and its size range is in average inferior to other assemblages studied on early Egyptian sites. Indeed, the rachis nodes are similar in size to those from Nagada/Khattara, Upper Egypt (Wetterstrom, in press: figure 4), and inferior to those from Maadi, Lower Egypt (Van Zeist & De Roller, 1993: p.7) and from Hierakonpolis, Upper Egypt (El-Hadidi *et al.*, 1996: p. 50). Adaïma is the youngest of all studied contexts mentioned; therefore, the difference in size cannot be explained solely by the differences in the sites' geographical location or chronology.

Helbaek also noted the high size variability of emmer grains from the site of El-Omari (4700-4300 B.C., Lower Egypt) and from the underground Djoser complex in Saqqâra; some grains resemble einkorn grains, whereas the largest can reach 7.3 mm in length (Helbaek, 1955: pp. 93, 95). Wheat grains are scarce in Adaïma assemblages, but for a few samples from the cemetery. The largest assemblage of emmer grains was found in ash deposits (S359) in the cemetery area but not directly related to any grave. Their size have a wide range of variation; three drawings illustrate the different size classes observed (figure 1). Measures were taken automatically through an AxioCam digital camera and using a macro routine in the KS400 software (Zeiss), at the Groningen Institute of Archaeology, (the Netherlands). The surface of each grain was measured on whole specimens from one sub sample (S359/57, N=211). The surface measures 10.96 mm^2 on average, ranging from 2.25 to 16.94 mm^2, with a standard deviation of 2.6. These results were compared to measures obtained in the same way on Spanish material (Castro de Bureba, 4th century BC, N=10) by Neef (2000). The average surface of emmer grains is similar (11.64 mm^2), but its range is much narrower (8.04-15.34) (Neef, 2000: p. 228). Indeed, the smaller emmer grains found in Adaïma are not represented in Castro de Bureba material. This is probably due to the archaeological context, and the mode of deposit of the emmer grains; the smaller ????tailgrains are not usually found together with the average size grains, because they are eliminated during the processing of the ears. The fact that the exceptional Adaïma sample also includes smaller grains may reveal that they come from the carbonization of whole ears, at an unprocessed stage.

As for barley, the variability of emmer grain size and rachis node morphology can be related to intra individual variability within the spike, intra specific variability according to growth conditions, and genetic diversity of the crop. Whereas we do not have access to genetic data for Predynastic crops, their growth conditions can be assessed by the study of their weed assemblage (*confer infra*).

3.2 Quantitative comparison between emmer and barley

If we compare only the by-products of emmer and barley, represented by their rachis nodes – without taking into account the kernels because of their scarcity – the general result for all the samples is that barley remains

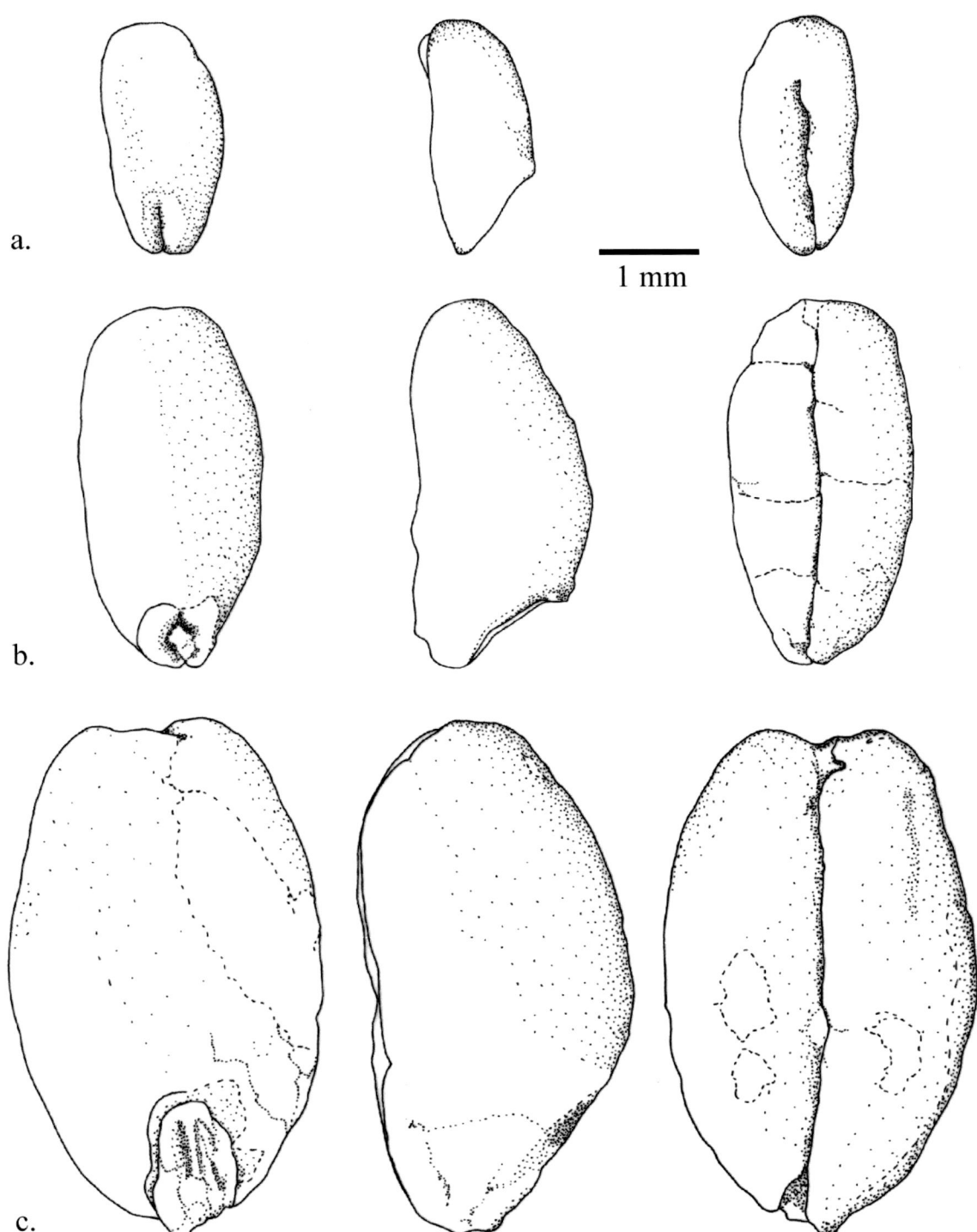

Figure 1: Three emmer (Triticum turgidum ssp. dicoccum) grains from context S359, illustrating the high size variability of the material. All drawings are to the same scale.

are 2.5 times more abundant than emmer remains. If the figures are corrected in terms of grain equivalents, *i.e.* one (six-row) barley rachis node represents three kernels and one emmer rachis node represents two kernels, the fraction equals 4.2. If we consider that two-row barley is predominant and that one rachis node represents one kernel, then it becomes 2.8. In all cases, whether we consider barley as predominantly six-row, predominantly two-row or a mixture of both, its remains are always at least 2.5 times more abundant than that of emmer.. However, this varies greatly from one sample to another, and emmer is more abundant than barley in 9 samples.

The same type of comparison using desiccated remains points to greater differences between barley and emmer remains: with uncorrected data, barley is 9.4 times more abundant than emmer, with data corrected for six-row barley, the latter is 19.3 times more abundant. The variation between samples is even greater for desiccated remains. The samples in which barley is the most predominant are pisé samples, rich in desiccated material but especially in barley threshing remains: barley remains can be as much as 286 times more abundant (191 times with uncorrected data, in pisé sample 1050/17.1A.2). This characteristic bias in favour of barley in pisé may represent a technical choice (Newton, 2004).

The percentages of emmer and barley remains in a sample vary between samples. The statistical study of these two variations show that they are highly significantly independent from one another. The taphonomical implication is that the two cereals do not have the same provenience and/or mode of deposition on the site; interpreting their relative abundance must therefore be closely related to the archaeological context.

3.3 Edible fruits

Remains of fruit from indigenous trees, shrubs and palms are infrequent, except for *Acacia* seeds. The edible fruit comprise B*alanites aegyptiaca* (desert date), *Ziziphus spina-christi* (sidder) and *Hyphaene thebaica* (doam palm). In total, 49 such fruit remains were found, including 9 in hand-picked samples. *B. aegyptiaca* was found twice and *H. thebaica* only once, but *Z. spina-christi* remains occur in 7 bulk samples (table 2); these fruit were probably part of the diet on a regular basis – their fleshy mesocarp is slightly acidic, and may be consumed fresh, dried, as a drink or as a sweet "cake" (Arbonnier, 2000: p. 442).

3.4 Wild herbaceous plants

Seeds and fruit remains from wild herbaceous taxa represent 60% of the counted plant remains in the settlement samples. The most frequent taxa – present in over half the samples – are *Ambrosia maritima*, a Caryophyllaceae with minute seeds, *Ceruana pratensis* (ceruane), *Citrullus colocynthis* (colocynth), *Echium rauwolfii*, *Enarthrocarpus* cf. *strangulatus*, *Fimbristylis bisumbellata*, *Rumex* spp. (dock), several Cyperaceae, and wild Poaceae including *Crypsis schoenoides*, *Phalaris* sp. and Paniceae. In particular, *Echium rauwolfii* achenes, with slightly less than 1% of the remains, are probably over-represented due to their good preservation without charring.

Among the wild herbaceous taxa that may have belonged to the ruderal flora, the separation of (Predynastic) non-segetals and segetals is difficult. Firstly, their ecological range today often comprises both naturally-disturbed grounds, such as riverbanks, and cultivated fields, and secondly, the ecological conditions prevailing during Predynastic times differ from today's, and the number and identity of segetals also differ. In fact, regularly flooded lands are a major source of segetals, because plants are already adapted to disturbed growth conditions, through a particularly efficient sexual reproduction (therophytes) or vegetative multiplication (geophytes) (Jauzein, 1995: pp. 14-17).

3.5 Assemblages of taxa in the samples

Statistical analysis has been used to explore the relation between assemblages of wild taxa and the cereal remains. Firstly, all 92 samples and most wild/weed species were taken into account. Resedaceae, Raununculaceae, Malvaceae and *Cocculus pendulus* were not taken into account because of their scarcity and difficulty of grouping with other taxa. Other taxa were grouped by family or type for simplification: 28 types (figure 2) occurring in at least nine samples, *i.e.* 10% of the samples. A correspondence analysis was carried out on these variables. The results bear evidence for the separation into several wild/weed clusters, with 49% of the discrimination explained by the first three dimensions. The first group is very distinct and comprises Amaranthaceae and Chenopodiaceae, Apiaceae, *Coronopus niloticus*, *Fumaria* sp., *Heliotropium* sp., Polygonaceae (unidentified), and *Portulaca oleracea*. The second group comprises the remaining taxa/groups of taxa, *i.e.* the following: Asteraceae, Boraginaceae/Labiatae, Brassicaceae/Fabaceae, Caryophyllaceae, *Ceruana pratensis*, *Citrullus colocynthis*, *Crypsis schoenoides*, Cyperaceae (including *Scirpus* spp.), *Cyperus* spp., *Echium rauwolfii*, *Enarthrocarpus* cf. *strangulatus*, Euphorbiaceae, *Fimbristylis bisumbellata*, panicoid types (including *Panicum* sp. and *Setaria* spp.), *Phalaris* sp., *Rumex* spp., Solanaceae, *Solanum nigrum*, Trifoliae (FAB), wild barley, and wild Poaceae (including *Bromus* sp., *Chloris* sp., *Lolium* sp., *Eragrostis* sp. and unidentified types).

Secondly, the main crops – *Triticum* sp., *Triticum turgidum* ssp. *dicoccum*, *Hordeum vulgare*, unidentified cereals and *Linum usitatissimum* – were added to the wild/weed taxa and the analysis repeated. The grouping of the wild taxa does not change, and the percentage of discrimination explained by the first three dimension changes to 48% (figure 2). The crops are discriminated by the first two dimensions, but not by the third, and belong to the second group of taxa. The first two dimensions seem therefore to be linked to the crop remains.

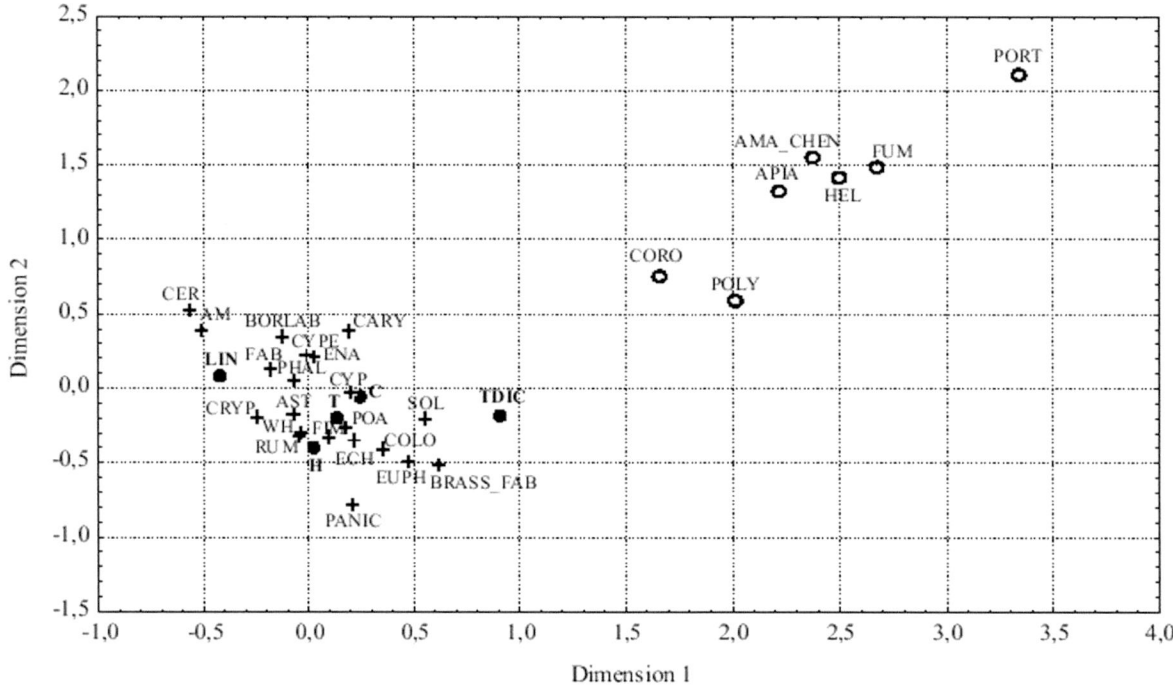

Figure 2: Correspondance analysis plot of wild/weed taxa and crops in the 92 settlement samples, showing two clusters, 1: hollow dot, 2: cross. The crops are linked to clusters 2. Legend: Crops (full dots): Linum usitatissimum *(UN),* Triticum turgidum *ssp.* dicoccum *(TDIC),* Triticum *sp. (T),* Hordeum vulgare *(H), and* Cerealia *(C). Wild taxa: AM =* Ambrosia maritima, *AMA_CHEN =* Amaranthaceae *and* Chenopodiaceae, *APIA =* Apiaceae, *AST =* Asteraceae *(other than* Ceruana pratensis*), BORLAB =* Boraginaceae *and* Labiatae, *BRASS_FAB = unidentified* Brassicaceae *and* Brassicaceae/Fabaceae *types, CARY =* Caryophyllaceae, *COLO =* Cilrullus colocynthis, *CER =* Ceruana pratensis, *CORO =* Coronopus niloticus, *CRYP =* Crypsis schoenoides, *CYP =* Cyperus *spp., CYPE =* Cyperaceae *(other than* Cyperus *spp. and* Fimbristylis bisumbellata, *including* Scirpus *spp.), ECH =* Echium rauwolfii, *ENA =* Enarthrocarpus *cf.* strangulatus, *EUPH =* Euphorbiaceae *types, FAB =* Fabaceae *including* Medicago, Trifolium *and* Trigonella *types, FIM =* Fimbristylis bisumbellata, *FUM =* Fumaria *sp., HEL =* Heliotropium *sp., PANIC =* Panicoid *types, PHAL =* Phalaris *sp., POA =* Poaceae *(including unidentified types,* Bromus *sp.,* Chloris *sp.,* Eragrostis *sp.,* Lolium *sp.), POLY =* Polygonaceae *(except* Rumex *spp.), PORT =* Portulaca oleracea, *RUM =* Rumex *spp., SOL =* Solanaceae, *WH = wild* Hordeum.

The first cluster of wild taxa is characterized by its absence of relation with the crops; it may group ruderal and garden taxa. The two other clusters are related to the crop by-products and may represent in part the weeds associated to the fields of cereals and flax. However, from these results it is impossible to distinguish between emmer- and barley-associated weed assemblages.

3.6 Diversity and ecology of the taxa

The taxa comprised in the first cluster include potentially cultivated plants; Purslane (*Portulaca oleracea*) may have been gathered or grown as a leaf vegetable, Apiaceae include many aromats used in the form of leaves and/or fruit. Amaranthaceae, Chenopodiaceae, Polygonaceae and *Coronopus niloticus* may have been ruderals pastured by domestic animals, and the inclusion of their seeds in the archaeological soil may have originated from animal dung.

Within the segetal taxa cluster, two wild taxa are found frequently in the samples but today not considered as segetals are: *Ceruana pratensis* and *Citrullus colocynthis*. The first is a weed growing on Nile banks and large irrigation canals (Boulos, El-Hadidi, 1994: p. 56), and may bear witness to its use as a broom, in particular in the cereal processing areas (*ibid.*; Henein, 1988: p. 118). This hypothesis would explain the presence of heads and achenes of this plant in association with the cereal processing by-products. *Citrullus colocynthis*, a non-segetal ruderal, may be eaten by donkeys, goats (flowers, green parts and seeds) and humans (processed seeds only), may serve as light fuel, and as a medicine (Benchelah *et al.*, 2000: p. 162; Gast, 2000: pp. 44-45). The presence in the archaeological record of both *Ceruana pratensis* and *Citrullus colocynthis* may therefore be related to their various domestic uses.

Considering only the 28 taxa and taxa groups found relating to crops by the statistical analysis, winter weeds are more numerous in the samples, both in number of taxa (26 taxa) and in number of remains. Their ecological affinities are diverse and probably reflect the growth conditions of the crops they were associated with.

In relation to the status of soil water content, 16 are hydro/mesophilous, one hydro/halophilous, two xerophilous, one xero/halophilous, three indifferent and three undetermined. Thus, it seems that humid soil conditions prevailed in the winter-cultivated fields. It could be an indication of natural irrigation practices, in accordance with the traditional model of flood irrigation, in which sowing occurs during the Fall, once the flood waters have receded. The xerophilous plants (*Echium rauwolfii*, *Panicum* sp., *Citrullus colocynthis*) may in this context either represent the ruderal habitat close to fields, or the existence of drier areas within the irrigated fields.

Considering the three main types of winter-cultivated lands today, as described by Kosinová (1975) *i.e.* rain fed/dry fields, irrigated fields in the Nile valley and irrigated fields in the oases, only one taxon (*Enarthrocarpus* cf. *strangulatus*) is typical of dry winter crops and does not belong to the other associations. Nine taxa are typical of irrigated crops of the Nile valley, and 18 belong to this weed association. The taxa typical of irrigated fields in the oases are also typical of the Nile valley ones. Most of the taxa belong the the weed association growing with the winter crops in Nile valley irrigated fields. It may be concluded that the weed assemblage identified in the settlement samples is characteristic of winter irrigated fields. The fact that the irrigation was probably a flood irrigation, compared to the permanent irrigation system today, may account partly for the differences observed in the composition of the weed assemblage.

4 The cemetery

4.1 Objects

Plant remains recovered from the tombs include desiccated wood fragments, wooden coffins, and wood and monocotyledonous stems and leaves belonging to baskets or mats. These often contain the corpse. Tables 3-4 summarize the new identifications of these plant materials.

The objects identifiedare a harpoon handle and three coffins discovered in N IIC tombs in the western cemetery. The harpoon (S100) was made of *Mimusops laurifolia/Olea* sp. wood (Dietrich, 2002). The types of wood used for the coffins are: *Ficus sycomorus* (S116, Dietrich, 2002), *Faidherbia albida* (S116, this study), and both *Acacia* cf. *nilotica* and *Faidherbia albida* (S100, this study). All the coffin woods belong to taxa that prefer humid growth conditions; they were probably growing in the low lands near the river or in the floodplain. The Nile acacia is renowned for its resistance to termites and xylophagous insects (Sidiyene, 1996: p. 31), whereas *Faidherbia albida* and *Ficus sycomorus* are not very good quality woods (Arbonnier, 2000: p. 389; Broun & Massey, 1929: p. 211).

Medemia argun (the argun palm) was not previously identified as leaflet; the species is known archaeologically only in the form of fruit, in particular as tomb offerings in the Nile valley. It is known in present Egypt from a single location, but its distribution area covers the Nubian desert, a great part of present Sudan, and may have anciently comprised oases in the western desert; fruit were indeed found in domestic contexts from the Persian, Ptolemaic and Roman periods in the Kharga oasis ('Ayn-Manâwir. Barakat, 1999, unpublished report; Newton, 2001). Its presence in the natural vegetation of the Nile valley is doubtful. The matting may therefore have been imported either from a western desert oasis or from Nubia, as Nubian ceramic vases in the cemetery (in tombs S552 and S597, Buchez, *in verbis*).

Other taxa are a halfa (*Desmostachya bipinnata/Imperata cylindrica*), *Desmostachya bipinnata*, Poaceae, *Cyperus* sp., *Cyperus papyrus* and an unidentified Monocotyledone. All these taxa belong to the ruderal flora and bear affinity with humid conditions; they grow on fallow land, near the canals and on the Nile banks, near water sources. They belong to the materials commonly identified for such artefacts in Egypt prior to the Roman period (Wendrich, 2000: p. 255).

Matting or basketry artefacts also include wood pieces (*Tamarix* sp. in three cases), but their poor state of preservation prevents any functional interpretation.

Some leather "purses" in tombs S216 and S103, measuring approximately 5 cm in diameter, contained a stuffing, composed of barley and emmer chaff mixed with decomposed organic matter. These objects attest a type of utilisation for cereal chaff rarely encountered in archaeological context, but already recorded from a Predynastic cemetery (Fahmy, 2003).

4.2 Food offerings and funerary deposits

Plant remains were found in the tombs' sediment, mostly in ceramic vessels. Some are primary deposits of plant products (cereal, fruit, processed matter). Other plant remains such as ashes are considered as secondary deposits, *i.e.* plant material not charred *in situ*.

Ashes

Many ashy deposits were found in the tombs, filling ceramic vessels. Ten samples from eight western

Table 3: Identifications of wood and other plant material from the western cemetery.

Tomb no.	69	100	102	103	116	218	326
Date (Nagada)	IIIA1	IIC	IC-IIC ?	IIC	IIC	IIB	?
Wood		*Faidherbia albida*, *Acacia* cf. *nilotica*	*Tamarix* sp. (matting) and Capparaceae	*Tamarix* sp. (103/8, matting)	16 *Faidherbia albida*, 1 Palmae		
Matting	*Cyperus papyrus*		Superposed monocotyledone culms	Superposed *Cyperus* sp. stems (103/8)		Poaceae culms, string in monocotyledone (218/6 & 8)	Superposed halfa (*Desmostachya bipinnata*/*Imperata cylindrica*) culms

Table 4: Identifications of wood and other plant material from the eastern cemetery.

Tomb no.	469	500	515	520	556	559	597	608
Date (Nagada)	IIIC2/D	IIIA-B	IIIA-B	IIIA-B	IIIA-B	IIIA-B	IIIA-B	IIIA-B
Wood			*Salvadora persica*				*Tamarix* sp. (matting)	
Matting	Superposed halfa *Desmostachy bipinnata*/*Imperata cylindrica*) culms	Monocotyledone stems (basket)		Superposed *Desmostachya bipinnata* culms	Monocotyledone stems	Poaceae leaves	Two layers: (1) superposed *Cyperus* sp. stems (2) monocotyledone stems & halfa culms	Three mats: *Medemia argun* leaflets

cemetery tombs were studied from the 17 that were discovered (Buchez, 1998). Six studied samples come from the eastern cemetery. The context S359 found within the western cemetery is not a tomb in itself, but a compound of ash deposits in the sand (*confer* next paragraph).

The deposits in vases were of two kinds: those consisting entirely of ashes, and those composed of carbonized remains and ashes dispersed in a sandy/loamy sediment. The assemblages of the ash samples, composed of a mixture of fuel, cereal-based food preparation residues and animal remains – in particular fish bones – suggest that the vases were filled with domestic hearth residues (table 5). One sample in particular (S451 plv 97017), not taken into account in the table because of its lack of botanical remains other than charcoal, contained mainly fuel and fish remains; in this case, it might represent a single episode, the preparation of a fish-based meal. The domestic character of the proveniences is highlighted by the nature of the fuel, a mixture of plant remains and wood. The charcoal from some of the samples was identified and reveals a spectrum of taxa similar to that identified in the settlement area: *Tamarix* is the main taxon, followed by *Acacia* and a group of less common taxa (Newton, 2002, 2005). The wood does not seem to have been selected especially for a funerary purpose. The interpretation of these ashes as remains of a funerary meal is a possibility, in particular in the case of S451 plv 97017. However, the diversity of the remains, and in particular the taxonomic diversity of the charcoal, is indicative of the use of a hearth over a period of time, not of a single fuelling. The fact that the fuel comprises not only wood, but also cereal chaff and straw is a further clue against the hypothesis that the ashes come from a ritual fireplace.

The samples richer in sediment have a similar composition to that of the ash samples (table 5). The presence of the ashes in these contexts could be explained in two ways; either a small quantity of ashes was laid in the vases, which were later filled with sediment, or the vase was filled with sediment collected on the settlement, a sediment containing dispersed items typical of domestic activities. The second explanation would reveal a ritual, a symbolic act, that of burying with the deceased some of the soil of his land, representing a link with the living world. To our knowledge, this practice is not attested in ancient Egypt, but the importance of soil is well-known in Pharaonic Egyptian culture.

Plant products

Three types of deposits are distinguished: cereal deposits, fruit deposits and transformed organic matter. Each of them was found either filling a vase or lying on a support of sediment, in such a way that the deposit

Table 5: Identified charred plant remains in ashy deposits of the cemeteries. All the remains are diaspores, except otherwise mentioned. The volume is not mentioned when it is unknown. For S359, the total volume of the deposit is unknown; the sub-samples studied only yield an approximative number of remains. Abbreviations: ++: many fragments present other than the counted remains, vol.: volume, *: charred remains only, w: western cemetery, e: eastern cemetery.

	Sandy/loamy ash deposits							Pure ash deposits								
tomb S	103	421	433	619	618	515	559	318	318	429	430	430	99	584	621	359
sample reference number	1*	#122	#139	2	2	3	4	79	80	#142	#143	#144	5	5	2	
cemetery	w	w	w	e	e	e	e	w	w	w	w	w	e	e	e	w
vol. of deposit (l)	0.021	3.5	8					0.8	1	?	1.7		0.06	2	0.475	3.25
vol., sorted fraction (ml)	21	200	400					55	60	20	?	50	60	180	475	358
Taxa, type of remain/ date of tomb	IIC	IIB	IIC	IIIA-B	IIIA-B	IIIA-B	IIIA-B	IIB	IIB	IIB	IIB, content	IIB, plug	IIC	IIIA-B	IIIA-B	?
Charcoal	xxx	x	x	x	x		2	x	xxx	xxx		xxx	xxx	xxx	x	
Poaceae culm node		1									1	1	3	15		
Cerealia	19		7					62	4	2	1	6	9	23		355++
H. vulgare	3		20	13	1				55	7		12		54		
H. vulgare rachis node	2	4	10	16	2			6	9	6		10	2	416	31	
H. vulgare rachis base														10		
H. vulgare empty spkikelet			7													
T. turgidum subsp. dicoccum		4							4				1			2030++
T. dicoccum rachis node	10		101					14	14	5		14	1	72	1	59
T. aestivum/durum rachis fragment	2													1		
Triticum sp.																253
Collected fruit																
Acacia sp.		4						1	1							
cf. Acacia nilotica leaflet												3				
Wild Poaceae																
Phalaris sp.								1		1						
Bromus sp. type																1
Wild Hordeum		2	7						5							
Wild Poaceae		17	1					3				2	3		2	1
Poaceae		60	18						29			28				xxx
Poaceae spikelet													1	2		
Poaceae rachis fragment									2							
Other wild/weed taxa																
Vicia sp. type			2													
Fabaceae	2							10	6							
Ceruana pratensis		1	1							2						
Arnebia type	6	1										2	1	1		1
Chenopodiaceae type												2				
Enarthrocarpus cf. strangulatus								1				1				
Rumex sp.			1					2	1							
Polygonaceae type		1										1		2	2	
Cyperaceae											1				3	
Dicotyledone leaf, fragment			5													
unidentified		x	4									19	2	8		
Total	47	87	178	57	3	0	2	98	88	64	23	83	17	607	62	2423
Other remains																
bone, heated or D	16	x	xxx	1		x		x	x	8			50	38	15	1
insect cuticle C/D	x	4										1			1	1
sheep/goat faeces C/D	x	4	1									1		2		2
organic matter	x	xxx							x			xxx	x	2		

was presented and therefore visible at the time of offering. The special case of feature S359, consisting in ashy concentrations located in the western cemetery but not directly related to any tomb, is distinctive by the absence of fuel and the almost exclusive presence of emmer grains. Its role toward the cemetery is unknown to us, but a symbolic deposit of grain storage is a possibility.

Table 6: *Summary of the presence of plant products as offerings in the tombs. D: desiccated. "Shaped" processed organic matter refers to whole items, whose limits are identified. "Loaf" refers to flat roundish items.*

Reference	tomb	context	date	type of deposit
Fruit and cereals				
	S100	in vase	N IIC	*Ziziphus spina-christi* fruit, D (De Vartavan 2002)
	S103	in vase	N IIC	*Ziziphus spina-christi*, D (De Vartavan 2002)
	S105	in vase	N IIC	*Ziziphus spina-christi*, D (De Vartavan 2002)
S515/3	S515	in vase	N IIIA-B	decomposed hulled barley grains, D
S516/1	S516	in large bowl	N IIIA-B	hulled barley grains, D
S522/4	S522	in vase	N IIIA-B	*Ziziphus spina-christi* fruit, D
S524/2	S524	in small pot	N IIIA-B	barley grains, D and charcoal
S543/?	S543	in tomb	N IIIA-B	*Ziziphus spina-christi* fruit, D
S543/12	S543	in vase	N IIIA-B	Fabaceae legume, cf. *Faidherbia albida*, D
S552/4	S552	in vase	N IIIA-B	*Ziziphus spina-christi*, D
S559/4	S559	in small bowl	N IIIA-B	decomposed hulled barley grains, D
S617/3	S617	in vase	N IIIA-B	barley hulled grains + *Cyperus* sp. tubers, D
S622/2	S622	in vase	N IIIA-B	decomposed hulled barley grains, D + flaky organic matter
Processed roducts				
S397/95	S397	in tomb	N IIIC2/D	processed organic matter, flaky, D
S398/98	S398	in tomb	N IIIC2/D	processed organic matter, D
S430/141	S430	in vase	N IIC	processed flaky organic matter D + very hard black organic matter
S430/145	S430	in vase	N IIC	processed organic matter, D
ad97/017	S451	?	?	processed organic matter D
S500 #2	S500	in vase	N IIIA-B	processed organic matter D
S508/1	S508	bottom of jar	N IIIA-B	processed organic matter D, "loaf"
S510/2	S510	in cup	N IIIA-B	processed organic matter D, "loaf"
S515/1	S515	in vase	N IIIA-B	processed organic matter D
S532/1	S532	in bowl	N IIIA-B	processed organic matter D, shaped
S536/1	S536 (dog)	in vase	N IIIA-B	processed organic matter D, shaped
S537/2	S537	in long vase	N IIIA-B	processed organic matter D, shaped
S538/1	S538	in vase	N IIIA-B	processed organic matter D, shaped
S543/4	S543	in bowl, on fill	N IIIA-B	processed organic matter D, shaped and small
S543/5	S543	in bowl, on fill	N IIIA-B	processed organic matter, D, containing one *Ziziphus spina-christi* endocarp fragment.
S543/8	S543	in bowl, on fill	N IIIA-B	processed organic matter D, "loaf"
S543/9	S543	in tomb	N IIIA-B	processed organic matter D, shaped (similar to S536/1)
S543/10	S543	in round vase	N IIIA-B	fragments of processed organic matter, shaped
S544/3	S544	in vase	N IIIA-B	processed organic matter D, shaped
S559/10	S559	in vase, on fill	N IIIA-B	processed organic matter D, "loaf"
S559/?	S559	in vase	N IIIA-B	fragments of processed organic matter, shaped
S597/7	S597	bottom of Nubian vase	N IIIA-B	processed organic matter D, shaped
S606/7	S606	?	N IIIA-B	cluster of barley hulled grains' processing by-products, D
S608/8	S608	in bowl	N IIIA-B	fragment of processed organic matter, shaped
S619/1	S619	in vase	N IIIA-B	fragments of processed organic matter, shaped
S621/1	S621	bottom of jar	N IIIA-B	processed organic matter D, shaped
S622/2	S622	in vase	N IIIA-B	processed organic matter, shaped, D + partly decomposed hulled barley grains, D
S625/4	S625	underneath bowl	N IIIA-B	processed organic matter D, "loaf"

All the other cereal deposits come from the children's tombs in the eastern cemetery and date to Nagada IIIA-B. When the cereal was identifiable, it was always found to be hulled barley (*Hordeum vulgare*). In most cases, the grains were found in a desiccated state, the degree of preservation varying: the grain itself may be visible, or only the lemmas and paleas are preserved with or without traces of organic matter on them. In the latter case, the original state of the deposit is unknown; it could consist of hulled grains that deteriorated after deposition, or grain processing (dehusking) residues, such as from beer preparation (Samuel, 2000: pp. 542 *et sq.*).

In one case, hulled grains were mixed with tubers of *Cyperus esculentus* (tiger nut/yellow nutsedge) (S617/3). The association of barley grains with tiger nut tubers may imply a common use as food, either in this form or in a processed form, like bread (Curtis, 2001: p. 118). In any case, it indicates the importance of these tubers in the human diet, at least in that of children.

Ziziphus spina-christi (sidder) fruit remains – either whole fruit with desiccated mesocarp or endocarps only – were found in three tombs from the western cemetery, dated to Nagada IIC (S105, S100, S103) (De Vartavan, 2002), and three tombs from the eastern cemetery, dated to Nagada III A-B (S543, S522, S552). They were located either in the tomb's sediment fill next to the body, or in the filling of ceramic vases, or as fragmentary remains in processed organic material that could be interpreted as a bread-like food ingredient (S543). The occurrence of *Z. spina-christi* in bread is attested in the tomb of Tutankhamun from the 18th Dynasty (De Vartavan, 1990: p. 486; Hepper, 1990: pp. 53, 68). Kamal (1913: p. 241) mentions bread made of *Z. spina-christi* fruit as a funerary offering together with other kinds of bread. The organic matter from S543 could be one of these two types of bread. The presence of the fruit in Adaïma tombs as offerings illustrates the antiquity of this practice before its formalisation in ritual texts, and confirms its importance as food, already shown in the settlement.

A legume, possibly *Faidherbia albida*, was found in a vase from tomb S543 in the eastern cemetery, lying on a thin layer of organic matter comprising cereal chaff and leaves. *F. albida* is attested at Adaïma in the form of charcoal (Newton, 2002, 2005), and may have had medicinal or food uses, as it does now in the Sahara (Sidiyene, 1996; Gast, 2000: p. 89; Arbonnier, 2000: p. 389). A *F. albida* legume was also identified in a Predynastic or early Dynastic tomb at Gebelein (Germer, 1988: p. 47).

Desiccated processed organic matter was frequently found in ceramic vessels of the tombs (table 6). It was either laid on a layer of soil or directly on the bottom of the vessel. Several types were defined based on differences in texture, structure, colour and shape, but they could not be clearly identified. A few items were found in a charred state together with similar desiccated ones (S500); they contain macroscopic elements such as fibres and cereal grains. The latter could be grain that escaped grinding, and the material a cereal-based food, such as bread. In one case *Ziziphus spina-christi* endocarp fragments were found in the desiccated matter (*confer supra*). Some of the items however seem to comprise only chaff, especially lemma and palea. These could be residues of a food preparation, such as beer making. Indeed, the residues separated from the sweet liquid that will ferment and become beer, are composed of chaff, bran and grain fragments (Samuel, 2000: pp. 554-5). Only more detailed analyses (SEM and chemical analyses) could validate or invalidate the hypothesis that some of the material is bread and other, beer residues.

5 Discussion

5.1 Crop husbandry

The results from the study of the weed assemblage in the settlement soil indicate that winter crops were cultivated: emmer wheat, barley, flax, and perhaps legumes. There is evidence for the practice of flood irrigation given that many of the winter weeds prefer or grow well in humid growth conditions and many are known today as weeds in irrigated fields, whether in the Nile valley or in the oases. The sandy plain of Adaïma is located south of the floodplain, and although the water table was higher than it is today and may have allowed a permanent woody vegetation, it probably could not have supported annual crops, which need surface water at the beginning of their life cycle.

5.2 Food: cultivated and wild plants in the diet

During the second half of the Nagada III period, the most important crops in the agricultural system seem to have been cereals, and more particularly barley. Although barley remains are more frequent than emmer's, this may be due in part to the taphonomical bias introduced by the decomposition of the pisé, rich in barley chaff, in the archaeological sediment (Newton, 2004). However, if we suppose that barley and emmer were of similar economic importance, the presence of barley grains as offering in the cemetery and the absence of emmer grains as such, tend to show that barley had a greater cultural status at that time, perhaps related to its production in larger amounts. The fact that these offerings were only found in chidren's tombs also introduces a bias ; the analysis of more adult tomb offerings is necessary in order to have significant data. Moreover, the uses of the two cereals only partly overlap: grinding and milling for consumption in the form of processed grains (breads, porridges) seem to have been used mainly for emmer, whereas the way in which barley grains were consumed is less clear. It is possible that they were processed into a beer-type food, but no proof was found. Their by-products were used as fuel and as tempering in pisé material, but barley was favoured for the latter.

Food also included wild plant products: Cyperaceae tubers – probably *Cyperus esculentus* –, vegetative parts and seeds of purslane (*Portulaca oleracea*) and mallows (*Malva* spp.), maybe colocynth seeds (*Citrullus colocynthis*), and fruits from trees and shrubs – heglig's desert date (*Balanites aegyptiaca*), Christ's thorn (*Ziziphus spina-christi*), figs (*Ficus* sp.), *Salvadora persica*. Their importance in the diet is also evidenced by their presence in tombs as food offerings, in the case of *Ziziphus spina-christi* fruit, one *Faidherbia albida* legume, and *Cyperus* tubers.

Agriculture had a great importance in daily life and in the plant economy, both through its products and by-products. Its place in food, building and foddering/pasturing was probably a central one. Moreover, the various uses of products from the spontaneous vegetation also held a considerable importance in the same plant economy. If we consider agriculture as one of several types of exploitation of the plant environment, it had been integrated in the plant economy at the end of the Predynastic period for already two thousand years. However, the spontaneous vegetation of the Nile and desert still persisted in the plant economy; it was exploited for the manufacture of domestic objects, and increased the variety of meals. In addition, wild plant products may have supplemented the diet in times of shortage, and even been consumed as famine food. The diversity of these floras could compensate for the losses induced by the still random conditions of flood irrigation.

Apart from the winter crops, a year-round horticulture was probably practised, outside the floodplain, on levees within the plain, on the Nile banks during the dry season and/or at the margins of the desert, near the settlement (during flooding) using manual irrigation methods. The plants grown could have been purslane (*Portulaca oleracea*), melon (*Cucumis melo*) and maybe cucumber (*Cucumis sativus/melo*), perhaps food legumes (lentil, vetch) and other vegetables and condiments, such as Apiaceae.

5.3 Plant economy viewed from the grave
Two main interests for studying material from funerary contexts lie in the access it gives to material seldom found in domestic contexts and in the cultural information it yields on the plant world. The items almost exclusively found in tombs are artefacts made of plant materials: wooden objects, matting, padding. They show the use of local materials from the riparian and floodplain vegetation, and in one case of imported mats, probably from Nubia. Some taxa are known only as material in tomb artefacts, such as halfas, although their young shoots were probably grazed by domestic animals. However, *Cyperus* achenes and tubers are found in the settlement samples, as cereal weeds, animal feed and human food.

Moreover, finished products like processed foods are available, whereas the trash deposits in domestic contexts only yield the raw materials and processing by-products. Such is the case of cereal products Whereas we have access to their threshing and dehusking by-products in the settlement mixed sediment, end products like "bread" can be found in the tombs. Thus, actual human food, though maybe biased by the funerary ritual choices, can be studied.

5.4 Ritual choices
The significance of offerings contained in ceramic vessels was discussed by Buchez (1998). She separates real from fictional and residual food offerings. Here, the deposit of whole fruit, cereal grains and "bread" could be real food offerings, because these products are either edible in that form or represent a storage of raw material (cereal grains). Barley processing remains, hearth residues and settlement soil deposits could be interpreted as symbolic or fictional food offerings, only the first being closely related to meals, the two others bearing a domestic signification. In the last cases (hearth and soil), the plant remains are residual, their presence as such is not intentional.

Funerary offerings in the Predynastic are not attested in texts before the second Dynasty, when lists of goods appear on funerary stelae (Saad, 1957). Stelae from the second, third and fourth Dynasties bear lists of goods to be taken by the deceased into the afterlife. Concerning the food part, they consist of ideal food preparations presented as a menu and a summary of the traditional funerary offering (Saad, 1957; Ziegler, 1990). From the late fifth Dynasty on, Pyramid Texts comprise lists of offerings to the deceased king; their function is to sustain his *Ka* (~soul) in the afterlife (*e.g.* Sethe, 1908-1922). These texts, written in the region of Memphis and Heliopolis, are in continuity with the earlier lists and probably with more ancient practices. A variety of animal and plant-based products are included. The identified plant products are frequently beer and several kinds of bread, but also wine, onions (Formula 125), figs (Formulas 152, 193), *Balanites aegyptiaca* fruit (Formula 160), carob (?) legumes (Formulas 168, 182), fruit and paste or "bread" made of *Ziziphus spina-christi* fruit (Formulas 166, 181 and 167 respectively), and cereal grains, including "roasted"/"grilled" grains (Formula 164). Some products have not yet been identified.

From this list, several items were found in the tombs in Adaïma: cereal grains, *Ziziphus spina-christi* fruit, maybe?? bread and beer. The *Faidherbia albida* legume can be identified as the "carob" (*Ceratonia siliqua*) mentioned in the lists. Indeed, the fruit of the two trees are similar morphologically, and bear the same name in Nubia, *kharub*, which shows the possible confusion between them, as may have happened in the Pharaonic period. They may have had interchangeable roles depending on the eco-geographical region considered – *Ceratonia siliqua* in the Mediterranean, *Faidherbia albida* further south in the Nile valley. Onions, figs and *Balanites aegyptiaca* fruit were not found in Adaïma tombs. However, *Cyperus esculentus* tubers found in a tomb are not mentioned in the texts. Since the function of the deposit of offerings is to ensure the sustenance of the *Ka*, one can presume that all available food products may contribute to it, including fresh vegetables and fruit, and processed foods.

There are differences in the lists of edible plants found in the settlement samples and in the tombs; emmer, *Balanites aegyptiaca* and Doam were identified in settlement samples but not in the tombs, whereas processed foods and the *Faidherbia albida* legume found in the tombs were not present in settlement samples. These differences highlight the importance of the archaeological context on the recovery and interpretation of plant remains. The fact that some products are present as funerary offerings bears witness to their status as food products, not only in the world of the dead, but in the world of the living. In some cases – *Ziziphus spina-christi* fruit, sedge tubers, cereals and in particular barley – it confirms the settlement data and enhances their cultural status, in others it complements it – *Faidherbia albida* fruit. The frequent presence of barley in the tombs and the absence of emmer wheat except for in hearth residues also implies the greater social status of barley, if not in the economy.

The presence of plant and animal food offerings in the Predynastic tombs of Adaïma indicates the existence of this type of ritual long before they were made official in texts from the Old kingdom in the Pyramid Texts, the Middle Kingdom in the Sarcophagus Texts and in the New Kingdom in the Book of the Dead. Moreover, the deposit of domestic hearth residues and perhaps settlement soil in tombs offers an insight into Predynastic and early use of cultural symbols relating to daily life, and testify to a practice that disappeared during the Old kingdom. Other testimonies of the deposit of "ash-jars" in Egypt come from the following cemeteries: Predynastic (N II) HK43 cemetery in Hierakonpolis (Friedman et al,. 1999: pp. 14-18), Diospolis parva cemetery H (Petrie, Mace, 1901: p. 35), Nagada (Petrie & Quibell, 1896: pp. 23, 24, 27, 29), 2nd Dynasty Elkab (Newton, 2002), early Dynastic Minshat Abu Omar (Thanheiser, 1992) and Tell Ibrahim Awad (mentioned in Thanheiser, 1992: pp. 169-170). Only the plant remains from Hierakonpolis, Minshat Abu Omar and Elkab were studied. The remains, although different in the details from those of Adaïma, are similar in their composition: a mixture of cereal remains including chaff and straw, seeds of wild/weed taxa, animal dung, bone remains and in particular fish bones, and charcoal. The proportion of emmer to barley varies geographically: emmer is more abundant only in Minshat Abu Omar, Lower Egypt. Their interpretation varies between domestic hearth refuse and remains of a funerary meal. It seems that all these ashy deposits come from domestic contexts. If they do come from ritual hearths, the ritual is not significantly different from daily practices, at least concerning the type of fuel used. The comparison therefore confirms our interpretation of the data from Adaïma and highlights a widely practised funerary ritual in Egypt in Predynastic and early Dynastic times.

In the Egyptian late Predynastic context, the results from this archaeobotanical study from domestic and funerary contexts at the Upper Egyptian site of Adaïma both confirm general observations on the plant economy of the time and throw light on the diversity of the landscape. Indeed, if the main crops were similar in all Upper Egypt, the relative importance of the different crops, in particular emmer and barley, varied from site to site and probably also through time. The morphological variability of crop remains and the abundance of wild plant remains also testify to changing/unstable growth conditions. The importance of products from non-cultivated plants is shown through the residues of domestic activities, the manufactured artefacts found in tombs and the funerary offerings. In the domestic and ritual spheres of activity, similarities with later ancient Egyptian practices are found.

6 Acknowledgements

This work was undertaken as part of a PhD project at the University Montpellier II, under the direction of Jean-Louis Vernet and George Willcox, whom I thank for their support. It was made possible by a PhD grant from the French Ministry of Education and Research (allocation de recherche), and by travel grants and technical support from the Institut Français d'Archéologie Orientale. René Cappers provided constructive discussions, and permitted the measurements of emmer grains at the Groningen Institute of Archaeology.

7 References

Arbonnier, M., 2000. *Arbres, arbustes et lianes des zones sèches d'Afrique de l'Ouest.* CIRAD, MNHN, IUCN, Paris.

Benchelah, A.-C., Bouziane, H., Maka, M. & C. Ouahes, 2000. *Fleurs du Sahara, Voyage ethnobotanique avec les Touaregs du Tassili.* Ibis Press, Paris.

Broun, A.F. & R.E. Massey, 1929. *Flora of the Sudan.* Thomas Murby & Co., London.

Buchez, N., 1999. Le mobilier céramique et les offrandes à caractère alimentaire au sein des dépôts funéraires prédynastiques: éléments de réflexion à partir de l'exemple d'Adaïma. *Archéo-Nil* 8, pp. 85-103.

Cappers, R.T.J., Thuyne, T. van & L. Sikking, 2003. Plant remains from predynastic El Abadiya-2 (Naqada area, Upper Egypt). In: Vermeersch, P.M., Hendrickx, S. & W. Van Neer (eds.), *El Abadiya 2, A Naqada I Occupation, Upper Egypt.* Leuven University Press, Leuven.

Crubézy, É., Janin, T. & B. Midant-Reynes (eds.), 2002. *Adaïma II : La nécropole prédynastique.* IFAO, Cairo.

Curtis, R., 2001. *Ancient food technology. Technology and change in history*, vol. 5. Brill, Leiden, Boston, Köln.

Dietrich, A., 2002. Les analyses de bois. In: Midant-Reynes, B. & N. Buchez (eds.), *Adaïma I: Économie et habitat.* IFAO, Cairo, pp. 517-519.

El-Hadidi, N., Fahmy, A.G. & & Willerding, 1996. The Palaeoethnobotany of locality 11C, Hierakonpolis (3800-3500 cal.BC); Egypt. 1. Cultivated crops and wild plants of potential value. *Taeckholmia* 16, pp. 45-60.

Fahmy, A.G., 2003. Palaeoethnobotanical studies of Egyptian Predynastic cemeteries: new dimensions and contributions. In: Neumann, K., Butler, A. & S. Kahlheber (eds.), *Food, Fuel and Fields. Progress in African Archaeobotany.* Heinrich-Barth-Institut, Köln, pp. 95-106.

Friedmann, R., Maish, A., Fahmy, A.G., Darnell, J. & E. Johnson, 1999. Preliminary report on field work at Hierakonpolis: 1996-1998. *Journal of the American Research Center in Egypt* 36, pp. 1-35.

Gast, M., 2000. *Moissons du désert. Utilisation des ressources naturelles au Sahara central.* Ibis Press, Paris.

Germer, R., 1988. *Katalog der altägyptischen Pflanzenreste der Berliner Museen.* Ägyptologische Abhandlungen 47. Otto Harrassowitz, Wiesbaden.

Greiss, E., 1957. *Anatomical identification of some Ancient Egyptian plant materials.* Mémoires de l'Institut d'Égypte tome 55. Institut d'Égypte, Cairo.

Helbaek, H., 1955. Ancient Egyptian wheats. *Proceedings of the Prehistoric Society* 21, pp. 93-95.

Henein, N., 1988. *Mari Girgis, Village de Haute-Égypte.* Bibliothèque d'Étude XCIV. IFAO, Le Caire.

Hepper, N., 1990. *Pharaoh's flowers. The botanical Treasures of Tutankhamun.* HMSO Publications, London.

Jauzein, P., 1995. *Flore des champs cultivés.* INRA, Paris.

Kamal, A. Bey, 1913. Le pain de nebaq des anciens Égyptiens. *Annales du Service des Antiquités de l'Égypte* 12, pp. 240-244.

Keimer, L., 1932. *Ceruana pratensis* Forsk.. dans l'Égypte ancienne et moderne. *Annales du Service des Antiquités de l'Égypte* 32, pp. 30-37

Kosinová, J., 1975. Weed communities of winter crop in Egypt. *Preslia* 47, pp. 58-74

Midant-Reynes, B. & N. Buchez (eds.), 2002. *Adaïma I: Économie et habitat.* IFAO, Cairo.

Neef, R., 2000. Umwelt und Landwirtschaft. In: H. Parzinger & R. Sanz (eds.), *Das Castro von Soto de Bureba. Archäologische und historische Forschungen zur Bureba in vorrömischer und römischer Zeit.* Marie Leidorf, Rahden, pp. 219-239.

Newton, C., 2001. Le Palmier Argoun *Medemia argun* (Mart.) Württemb. ex Wendl. In: Aufrère, S. (ed.), *Encyclopédie religieuse de l'univers végétal: croyances phytoreligieuses de l'Egypte ancienne,* vol. II. Orientalia Monspeliensa X. Université Paul Valéry, Montpellier III, Montpellier, pp. 141-153.

Newton, C., 2002. Environnement végétal et économie en Haute-Égypte à Adaïma au Prédynastique; Approches archéobotaniques comparatives de la IIe dynastie à l'Époque romaine. Unpublished PhD thesis, University Montpellier II.

Newton, C., 2004. Plant tempering of Predynastic pisé at Adaïma in Upper Egypt: Building material and taphonomy. *Vegetation History and Archaeobotany* 13(1), p. 55-64.

Newton, C., 2005. Upper Egypt: Vegetation at the beginning of the 3rd millennium BC inferred from charcoal analysis. *Journal of Archaeological Science* 32 (3), pp. 355-367.

Pernaud, J.-M., 2002. Anthracologie. In: Midant-Reynes, B., & N. Buchez (eds.), *Adaïma I: Économie et habitat.* IFAO, Cairo, pp. 502-506.

Petrie, F. & A. Mace, 1901. *Diospolis Parva, the cemeteries of Abadieh and Hu, 1898-9.* Oxford University Press, Oxford.

Petrie, F. & J.E. Quibell, 1896. *Naqada and Ballas.* British School of Archaeology in Egypt, Egyptian Research Accounts 23. Bernard Quaritch, London.

Saad, Z., 1957. *Ceiling Stelae in second dynasty tombs from the excavations at Helwan.* Supplément aux Annales du Service des Antiquités de l'Egypte, cahier No 21. IFAO, Cairo.

Samuel, D., 2000. Brewing and baking. In: Nicholson, P. & I. Shaw (eds.), *Ancient Egyptian materials and technology.* Cambridge University Press, Cambridge, pp. 537-576.

Sethe, K., 1908-1922. *Die altägyptischen Pyramidentexte.* Heinrichs, Leipzig.

Sidiyene, E.A., 1996. *Des arbres et des arbustes spontanés de l'Adrar des Iforas (Mali). Étude ethnolinguistique et ethnobotanique.* ORSTOM/CIRAD, Paris.

Thanheiser, U., 1992. Plant remains from Minshat Abu Omar: first impressions. In: Brink, E.C.M. van den (ed.), *Nile delta in transition: 4th-3rd milenium BC.* Proceedings of the seminar held in Cairo, 21-24 October 1990, at the

Netherlands Institute of Archaeology and Arabic studies. Tel Aviv-Jerusalem, pp. 167-170.

Van Zeist, W. & G.J. de Roller, 1993. Plant remains from Maadi, a Predynastic site in Lower Egypt. *Vegetation History and Archaeobotany* 2, pp. 1-14.

Vartavan, C. de, 1990. Contaminated plant foods from the tomb of Tutankhamun: a new interpretative system. *Journal of Archaeological Science* 17, pp. 473-494.

Vartavan, C. de, 1991. Rapport préliminaire sur les restes végétaux d'Adaïma. In: B. Midant-Reynes *et al.*, Le site prédynastique d'Adaïma. Rapport préliminaire de la deuxième campagne de fouille. *BIFAO* 91: pp. 244-246.

Vartavan, C. de, 2002. Carpologie, In: B. Midant-Reynes & N. Buchez (eds.), *Adaima I: Économie et habitat*. IFAO, Cairo, pp. 483-502.

Wendrich, W., 2000. Basketry. In: P. Nicholson & I. Shaw (eds.), *Ancient Egyptian materials and technology*. Cambridge University Press, Cambridge, pp. 254-267.

Wetterstrom, W., in press. Palaeoethnobotanical studies at Predynastic sites in the Nagada-Khattara Region. In: Hassan, F. (ed.), *Predynastic studies in the Nagada-Khattara Region of Upper Egypt*. New York: Academic Press.

Ziegler, C., 1990. *Catalogue des stèles, peintures et reliefs égyptiens de l'Ancien Empire et de la Première Période Intermédiaire vers 2686-2040 avant J.-C.* Paris: Réunion des Musées Nationaux, Ministère de la Culture, de la Communication, des Grands Travaux et du Bicentenaire.

Zohary, D. & M. Hopf, 2000. *Domestication of plants in the Old World. The origin and spread of cultivated plants in West Asia, Europe and the Nile Valley*. Third edition. Oxford: Oxford University Press.

New discoveries at Qasr Ibrim, Lower Nubia

A.J. CLAPHAM & P.A. ROWLEY-CONWY
Department of Archaeology, University of Durham, United Kingdom

The majority of the plant remains at Qasr Ibrim are preserved by desiccation. The remarkable preservation of the plant remains has led to various ancient biomolecular investigations. The first results on the study of the plant remains were published in 1989. Since then further work has been carried out and the evidence from these analyses is compared with those of 1989. This has shown that some of the main crop types were present at Qasr Ibrim before the date proposed by the 1989 study and that wild sorghum continues to be present after the Roman period. One new important crop plant was identified, lablab (*Lablab purpureus* (L.) Sweet), which first appears in the Late Meroitic period and becomes a common crop in the post-Meroitic period. The presence of lablab at Qasr Ibrim may be the earliest record of its cultivation in the Nile valley, if not Africa. Both studies agree that there is a change in agricultural practice in the late 4th century AD, with the introduction of more tropical summer crops. The introduction of these appears to coincide with the appearance of the *saqia*.

1 Introduction

Qasr Ibrim was a major settlement and cult centre in Lower Nubia from the early 1st Millennium BC (Rowley-Conwy, 1988) until the early 19th Century AD (see table 1). The site was located on a prominent hill top overlooking the river on the east bank of the Nile, about 70 km north of the modern Sudanese border and 40 km north-east of Abu Simbel. Since the construction of the High Dam at Aswan and the subsequent filling of Lake Nasser linked with the artificially maintained high water level to be exploited by the Toshka Project, Qasr Ibrim is now effectively an island. The results from the site are important as Qasr Ibrim is the only *in situ* site left in Lower Nubia since the flooding of the Nile valley.

A major programme of excavations has been carried out on the site since the early 1960s by the Egypt Exploration Society. Situated in an effectively rainless environment, conditions of preservation on the site are quite remarkable.

2 Chronology of occupation at Qasr Ibrim

Due to the length of occupation at Qasr Ibrim, it is not certain when the site was first occupied. The earliest date achieved so far is 2800±80 BP (OxA 1062) which gives a calibrated date (at two standard deviations) of 1250-810 BC. It is not known what the nature of occupation was at that time, but is associated with what seems to be a fortification.

Although excavations have been carried out for forty years, the depth of the deposits has prevented the production of a comprehensive chronology of the site for its earlier occupation levels considered here and therefore the dates given here are tentative. The earlier occupation levels of the site are now the main focus of current excavations and it is hoped that these will produce a more precise chronology.

3 History of previous archaeobotanical research

As mentioned above, the almost rainless conditions have led to the remarkable preservation of animal and plant material such as wooden artefacts, basketry, papyrus documents, leather, textiles, bones and seeds (e.g. Driskell *et al.*, 1989; Rowley-Conwy, 1991; Rose and Edwards, 1998).

This has led to the use of the unusually well-preserved plant remains in a variety of studies which would not normally be possible. These include the detection of ancient biomolecules such as lipids and DNA (O'Donoghue *et al.*, 1994) which have then aided identifications (O'Donoghue *et al.*, 1996) and helped to determine the history and origins of certain crops such as sorghum (Rowley-Conwy, 1991; Rowley-Conwy *et al.*, 1997; Deakin *et al.*, 1998; Rowley-Conwy *et al.*, 1999; Shaw *et al.*, 2001).

The well-preserved seeds have also provided evidence for the decomposition of biomolecules in desiccated environments (Evershed *et al.*, 1997a). The preservation of lipids within ceramics has provided the first evidence for the processing of palm fruits, dates (*Phoenix dactylifera* L.) and dom (*Hyphaene thebaica* (L.) Mart.) in the Nile valley (Copley *et al.*, 2001a & b).

Table 1: The archaeological periods and their approximate dates for Qasr Ibrim, Lower Nubia (based on ceramic evidence).

Archaeological Period	Approximate dates
Napatan	Early 7th century BC- ?3rd century BC
Roman	?25 BC-100AD
Meroitic (M)	AD100-350
Late Meroitic (LM + LM/X)	AD300-400
Post-Meroitic (X2)	AD500-550
Post-Meroitic/Christian (XC)	AD550-650
Early Christian (EC)	AD650-850
Christian (in general)	AD550-1500
Islamic	AD1500-1800

The methods developed in the preceding studies were also used to identify the presence of frankincense and a pine resin from the post-Meroitic occupation phase (Evershed et al., 1997b).

Apart from the biomolecular studies, the plant remains have provided data for site formation processes and sampling strategies (Rowley-Conwy, 1989; 1994) and the importance of Qasr Ibrim within its immediate hinterland (Rose & Rowley-Conwy, 1989; Rose, 1996).

Although these studies are important for understanding the history and origins of specific crops, and how the site developed, the preserved plant remains also provide an ideal opportunity to study the history of agricultural innovation in Lower Nubia. This work was started by Rowley-Conwy, who published a preliminary report covering the excavations between 1982 and 1986 (Rowley-Conwy, 1989). Rowley-Conwy also introduced systematic sampling to the site. Since 1998 this work has been continued by Clapham and is still on-going with the support of a research grant from NERC.

This paper will outline the results obtained by Rowley-Conwy and these will be compared to those achieved since 1998. Only the major crop species are discussed here.

3.1 Rowley-Conwy's results
Before discussing the results, it is important to indicate that, due to the unusual state of preservation of the plant remains at Qasr Ibrim, there are several problems involved in sampling and quantification.

Rowley-Conwy (1989) was well aware that if normal sampling procedures used on European and Middle Eastern sites, where the main method of preservation of plant remains is by charring, were employed at Qasr Ibrim, the preliminary sorting, identification and quantification of a single sample would be very time consuming and counter-productive. He also found that few primary activity areas survived intact, giving a limited 'snapshot' of past domestic activity. Therefore a compromise was established in which most on-floor deposits were sampled at the one litre level while other particular areas were sampled much more intensely. All observed remains of stored material and other concentrations have also been sampled (Rowley-Conwy, 1989; 1994). This sampling policy is still in operation.

Rowley-Conwy also noticed another problem associated with desiccated plant remains. On a site which has been occupied for nearly three millennia, where the plants tend not to decompose, it is likely that a lot of post-discard mixing will occur, with many types of rubbish mingling in the deposits which may hinder interpretation of the remains (Rowley-Conwy, 1989).

The processing of the samples also causes difficulties. On most sites archaeobotanical samples are processed by some method of water separation. It is not however possible to use flotation methods on the desiccated material of Qasr Ibrim, as damage to the remains will occur, as has been attested by observations at the site. Due to the high water levels and the porous nature of the bedrock (Nubian Sandstone), water seepage is now a major problem, and it has been noticed that areas affected by seepage produce very little in the way of plant remains. The water reduces the plant remains to an evil-smelling mass resembling decomposed peat. Therefore contact with water is best avoided, and the samples are processed by dry sieving through a series of geological screens.

Whilst identification of the remains is aided by the remarkable preservation, quantification is complicated. On sites where charring is the main method of preservation, in the majority of cases only the most robust plant remains are preserved, such as seeds. Thus quantification usually involves the counting of each individual seed. With desiccation the profusion of plant remains makes this difficult as very little is lost, thus giving a more complete but more complex assemblage. Therefore standard quantification and statistical

Table 2: First appearance of the main crops at Qasr Ibrim, Lower Nubia, according to Rowley-Conwy (1989) d = race durra.

Crops / Period	Napata	Roman	Meroitic	Post Meroitic	Christian & Islamic	Common name
Cereals						
Hordeum vulgare L.	★	★	★	★	★	Barley
Panicum miliaceum L.	★	★				Broomcorn millet
Pennisetum glaucum L.				★		Pearl millet
Sorghum bicolor (L.) Moench race bicolor			★	★	d★	Sorghum
Triticum aestivum ssp. vulgare (Vill.) MacKey			★			Bread wheat
Triticum turgidum L. ssp. dicoccon (Schrank) Thell	★	★	★	★		Emmer wheat
Triticum turgidum ssp. turgidum conv. durum (Desf.) MacKey			★			Durum wheat
Zea mays L.					★	Maize
Condiments/herbs/spices						
Allium cepa L.	★	★				Onion
Allium sativum L.	★	★				Garlic
Coriandrum sativum L.	★	★				Coriander
Pulses						
Lens culinaris Medik.	★	★				Lentil
Lupinus albus L.				★		Termis bean
Pisum sativum L.				★		Pea
Vigna unguiculata (L.) Walp.				★		Cowpea
Fruit & nuts						
Balanites aegyptiaca (L.) Delile	★	★				Desert date
Citrullus lanatus (Thunb.) Mats. & Nakai	★	★				Watermelon
Cucumis sativus L.	★	★				Cucumber
Ficus sp.	★	★				Fig
Phoenix dactylifera L.	★	★				Date
Fibre/oil crops						
Gossypium sp.		★	★			Cotton
Linum usitatissimum L.	★	★				Flax/linseed
Moringa peregrina (Forssk.) Fiori	★	★				Ben-oil
Ricinus communis L.	★	★				Castor Bean
Sesamum orientale L.				★		Sesame
Miscellaneous						
Sorghum bicolor ssp arundinaceum (Desv.) de Wet et Harlan	★	★				wild sorghum

analyses used on charred samples cannot be applied to those preserved by desiccation, and a more qualitative approach to the data is required.

Bearing these problems in mind it is still possible to interpret the activities at Qasr Ibrim, in terms of agricultural innovation and economic activity.

3.2 Results from the 1989 preliminary report

Rowley-Conwy's report set the framework for the further study of the agricultural history at Qasr Ibrim. Rowley-Conwy (1989) identified three main phases of agricultural activity. The earliest plant remains identified are those of the Napatan XXVth Dynasty (mid 8th–mid 7th century BC). The crop assemblage is very similar to those found further north in the Nile valley in Pharaonic Egypt (see table 2). The main crops identified are emmer wheat (*Triticum turgidum* L. ssp. *dicoccon* (Shrank) Thell.) and hulled barley (*Hordeum vulgare* L.), and small amounts of broomcorn millet (*Panicum miliaceum* L.). Flax (*Linum usitatissimum* L.) was the main fibre crop. The Roman period (c25 BC – mid 1st century AD) showed little difference from the Napatan occupation apart from an increase in the abundance of broomcorn millet and the first appearance of cotton seeds (*Gossypium* sp.). It is uncertain whether this fibre crop was cultivated locally or imported.

The major phase of change identified by Rowley-Conwy occurs after the Roman presence in the Meroitic period (cAD100-300) and the post-Meroitic period (cAD300-550). During the Meroitic period there is the first appearance of cultivated sorghum (*Sorghum bicolor* (L.) Moench. race *bicolor*), which dom-

Table 3: The recorded first appearance and subsequent presence of the main crops at Qasr Ibrim, Lower Nubia, since 1998 (Since the publication of Rowley-Conwy's work (1989) advances have been made in the differentiation of the periods of occupation in Nubia (see Edwards 1996). The abbreviations in the table are: N = Napatan, R = Roman, M = Meroitic; LM & LM/X = late Meroitic; X2 = Post Meroitic; XC = Post Meroitic/Christian and EC = Early Christian. (for dates see table 1).

Crops / Period	N	R	M	LM	LM/MX	X2	XC	EC	Common name
Cereals									
Avena sp.				★		★	★	★	Oats
Hordeum vulgare L.	★	★	★	★	★	★n	★	★	Barley
Panicum miliaceum L.		★		★	★	★	★		Broomcorn Millet
Pennisetum glaucum (L.) R. Br.		★	★	★	★	★	★	★	Pearl Millet
Setaria italica (L.) P. Beauv.	★	★	★	★	★	★	★	★	Foxtail Millet
Sorghum bicolor (L.) Moench.		★	★	★	★	★	★		Sorghum
Triticum aestivum ssp. *vulgare* (Vill.) MacKey				★		★	★	★	Bread wheat
Triticum turgidum L. ssp. *dicoccum* (Schrank) Thell. ★		★	★	★	★	★	★	★	Emmer wheat
Triticum turgidum conv. *durum* (Desf.) MacKey		★	★	★	★	★	★		Durum wheat
Condiments/herbs/spices									
Allium cepa L.	★	★	★		★	★	★	★	Onion
Anetheum graveolens L.	★	★	★	★		★	★	★	Dill
Ceratonia siliqua L.				★	★	★			Carob
Coriandrum sativum L.	★	★	★	★	★	★	★	★	Coriander
Juniperus phoenicea L.							★		Juniper
Lepidium sativum L.	★	★	★	★	★	★	★	★	Cress
Piper nigrum L.						★			Black Pepper
Senna italica Mill.★				★	★				Senna
Trachyspermum copticum (L.) Link		★							Ajowan
Trigonella foenum-graecum L.	★				★	★		★	Fenugreek
Pulses									
Cajanus cajan (L.) Millsp.				★					Pigeon pea
Cicer arientinum L.						★	★		Chickpea
Lablab purpureus (L.) Sweet		?		★	★	★	★		Lablab
Lens culinaris Medik.	★	★	★	★	★	★	★	★	Lentil
Lupinus albus L.			★	★	★	★	★	★	Termis
Pisum sativum L.		★		★	★	★	★	★	Pea
Vigna unguiculata (L.) Walp.								★	Cowpea
Fruits and nuts									
Amygdalus communis L.				★					Almond
Balanites aegyptiaca (L.) Delile★			★	★	★	★	★		Desert date
Citrullus lanatus (Thunb.) Mats. & Nakai	★	★	★	★	★	★	★	★	Watermelon
Cordia myxa L.★				★	★	★	★		Sebesten
Cordia sinensis Lam.★		★							
Corylus avellana L.						★	★		Hazel
Cucumis sativus L.	★	★	★	★		★	★	★	Cucumber
Ficus sp.	★	★	★	★	★	★	★	★	Fig
Mimusops schimperi Hochst.★	★		★	★					Persea
Olea europaea L.		★			★	★	★	★	Olive
Phoenix dactylifera L.	★	★	★	★	★	★	★	★	Date
Prunus persica (L.) Batsch						★			Peach
Vitis vinifera L.		★	★	★	★	★	★	★	Grape
Ziziphus spina-christi (L.) Desf.★		★	★	★	★	★	★		Christ's thorn
Fibre/oil crops									
Carthamus tinctoria L.	★	★	★	★	★	★	★	★	Safflower
Gossypium sp.		★	★	★	★	★	★	★	Cotton
Linum usitatissimum L.	★	★		★					Flax/Linseed

Crops / Period	N	R	M	LM	LM/MX	X2	XC	EC	Common name
Moringa peregrina (Forssk.) Fiori★	★	★		★		★	★		Ben-oil
Raphanus sp. L	★	★	★	★	★	★	★	★	Radish
Ricinus communis L.	★	★	★	★	★	★	★	★	Castor Bean
Sesamum orientale L.				★	★	★	★	★	Sesame
Miscellaneous									
Beta vulgaris L.									Beet
Lagenaria siceraria (Molina) Standley			★	★	★	★			Bottle Gourd
Luffa cylindrica Roem.						★			Loofah
Sorghum bicolor ssp. *arundinaceum* (Desv.) de Wet et Harlan	★	★		★	★	★			Wild Sorghum
Number of samples	2	4	7	9	9	28	17	4	

★ possibly collected from the wild
n = naked barley present
? - doubtful identification

inates the cereal remains. Barley remains common but emmer wheat becomes of minor importance. Two new species of wheat, both free-threshing are found in this period, durum wheat (*Triticum turgidum* ssp. *turgidum* conv. *durum* (Desf.) MacKey) and bread wheat (*Triticum aestivum* ssp. *vulgare* (Vill) MacKey).

In the post-Meroitic period, sorghum is still the predominant cereal. Emmer is present in small amounts and barley is still common. Cotton seeds and other remains such as capsules become more abundant, suggesting that it was cultivated close by Qasr Ibrim. Pulses other than lentil (*Lens culinaris* Medik.) are found including the termis bean (*Lupinus albus* L.) and peas (*Pisum sativum* L.). Pearl millet (*Pennisetum glaucum* L.) and the oil crop sesame (*Sesamum orientale* L.) are identified for the first time. Other oil crops present at the site include ben-oil (*Moringa peregrina* (Forssk.) Fiori) and castor oil (*Ricinus communis* L.) which are first found in the Napatan and Roman periods. Flax and cotton can also be used as sources of oil.

In the Christian and Ottoman periods (cAD550-1800) there appears to be very little change. The abundance of *Sorghum bicolor* race *bicolor* decreases at the start of the Ottoman period and is replaced by another race of sorghum, *Sorghum bicolor* race *durra*. Other new introductions include cowpea (*Vigna unguiculata* (L.) Walp.) and much later, maize (*Zea mays* L.).

Rowley-Conwy's work showed that there was a major shift in crops in the Meroitic and post-Meroitic periods, in which the crop assemblage associated with Pharaonic occupation of the Nile valley further north was replaced by a more African assemblage dominated by sorghum and other millets.

3.3 Work since 1998
81 Samples were used in this analysis covering 8 broad chronological periods. The results are presented in table 3, which should be considered as preliminary as work is still in progress.

In agreement with Rowley-Conwy, the earliest period from which samples are available, the Napatan occupation shows that the crop assemblage was very similar to that found further north in Pharaonic Egypt and supports Rowley-Conwy's conclusion that only wild sorghum (*Sorghum bicolor* ssp. *arundinaceum* (Desv.) de Wet et Harlan) is present, with no cultivated sorghum indicated. One major difference is that the broomcorn millet has been identified as foxtail millet (*Setaria italica* (L.) P. Beauv.) on re-examination.

In the Roman period, a number of crops make their first appearance. These include durum wheat which Rowley-Conwy identified as first appearing in the Meroitic period. *Sorghum bicolor* race *bicolor* was also identified from the Roman period, as was pearl millet, peas, olives (*Olea europaea* L.) and grape (*Vitis vinifera* L.). The appearance of these crops predates that proposed by Rowley-Conwy. It remains questionable whether the pulse lablab (*Lablab purpureus* (L.) Sweet) first appears in the Roman period, but as the number of finds is small it may be present as a contaminant from a later occupation. Cotton seeds were also identified from this period and as with Rowley-Conwy it is uncertain whether the crop was grown locally or imported.

In the Meroitic period, the first occurrence of the termis bean is noted. This was originally thought to have been introduced in the post-Meroitic period. Other early appearances include the bottle gourd (*Lagenaria siceraria* (Molina) Standley), *Senna italica* Mill. (possibly used as a medicine) and the desert date (*Balanites aegyptiaca* (L.) Delile), which were probably harvested from the wild.

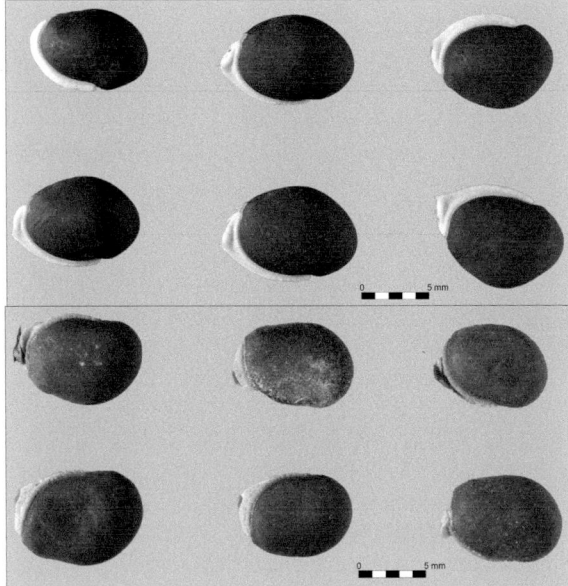

Figure 1: Beans of Lablab purpureus *(L.) Sweet. Top: Ancient (ssp.* uncinatus*) from Qasr Ibrim, Lower Nubia. Bottom: Modern (ssp.* purpureus*) collected from Aswan, Egypt. (Photograph AJC).*

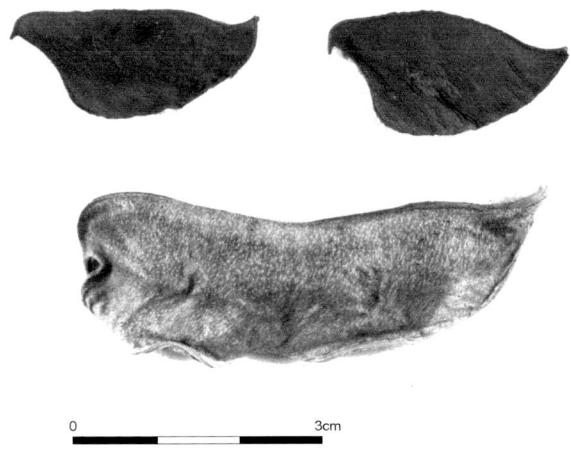

Figure 2: Pods of Lablab purpureus *(L.) Sweet. Top: Ancient (ssp.* uncinatus *) from Qasr Ibrim, Lower Nubia. Bottom: Modern (ssp.* purpureus*) collected from Aswan, Egypt. (Photograph AJC).*

In the late Meroitic period, crops introduced include oats (*Avena* sp.), although this may be present as a weed, pigeon pea (*Cajanus cajan* Millsp.) and carob (*Ceratonia siliqua* L.), almond (*Amygdalus communis* L.) and sesame. *Sorghum bicolor* race *bicolor* remains the dominant cereal, and barley is also common. The wheat species are of less importance, although bread wheat is introduced at this time. One major introduction in this period is a pulse, lablab. Both the distinctive seeds and the pods of this bean were found (see figures 1-2) and are found with increasing abundance in the post-Meroitic occupation of the site.

Lablab purpureus (hyacinth bean or bonavist) is an important legume crop in parts of the old world tropics. As a rainfed crop it requires approximately 600 mm annual rainfall or more, although it is also grown in the drier regions of the modern Sudan on flooded lands or river banks (Westphal, 1974). Lablab is one of the most important pulses in modern agriculture of the northern Sudan, where it is often intersown with sorghum (Hewison, 1948). It is very drought tolerant and can survive for some time into the dry season once it has become established. Various parts of the plant are used as food (including the green pods, leaves, flowers and dried seeds), as well as fodder. Perennial forms produce a root tuber which is edible (Westphal, 1974; Smartt, 1990).

The evidence from Qasr Ibrim in the form of pods and beans suggests that it was grown locally and consumed primarily as a pulse. The large quantities of dry pods suggest that it was brought on to the site after harvesting to be threshed and winnowed. The resulting pod fragments could have been used as fodder.

From the size and shape of the pods and beans, the lablab remains are those of ssp. *uncinatus*. *Lablab purpureus* ssp. *uncinatus* is considered to be wild in parts of East Africa, and sometimes locally brought under cultivation (Verdcourt, 1970; 1971; Maréchal *et al.*, 1978; Smartt, 1990).

Rowley-Conwy (1991) states that after the Roman period only cultivated races of sorghum are present. This is no longer correct as many heads and spikelets of wild sorghum were found in later periods (see figure 3). This suggests that wild sorghum was present throughout the occupation of the site.

In the post-Meroitic occupation of the site other crops, fruit and nuts are found for the first time. These include black pepper (*Piper nigrum* L.). Chickpea (*Cicer arientinum* L.), peach (*Prunus persica* (L.) Batsch.), beet (*Beta vulgaris* L.) and hazel nuts (*Corylus avellana* L.) are also found. Sorghum and barley are the dominant cereals. The wheats are of little importance but the small-grained millets, *Setaria italica* and *Pennisetum glaucum* remain signficant. A crypt in a post-Meroitic house was found to be full of beans and pods of lablab which suggests that this legume has become an important component of the crop assemblage.

4 Discussion and conclusion

The recent research is in general agreement with the finds reported by Rowley-Conwy (1989), with a few changes and additions. These include the re-identifi-

*Figure 3: Heads of wild sorghum (*Sorghum bicolor *ssp.* arundinaceum *(Desv.) de Wet et Harlan), from a late Meroitic deposit at Qasr Ibrim, Lower Nubia. (Photograph G. Owen).*

cation of broomcorn millet as foxtail millet and the extension of the occurrence of wild sorghum beyond the Roman period, which may have some bearing on the history of sorghum cultivation at Qasr Ibrim.

One of the important discoveries of Qasr Ibrim is the adoption of summer crops such as sorghum, which were added to the existing regime of winter cropping. This list has now been extended by the identification of lablab from the late Meroitic period onwards. Archaeobotanical records for lablab in Africa are scarce with only a single record from a pastoralist cave site at Geduld, northern Namibia, dating between 2300±50 BP (Pta-5872) and 1790±40 BP (Pta-4419) and the latest occupation occurring around 800±50 BP (Pta-4416). The lablab bean was found at level 8, dating to 1970± 40 BP (Pta-5875) (Smith & Jacobson, 1995). Although this is an earlier occurrence than at Qasr Ibrim, there is no evidence that it was cultivated, and the finds at Qasr Ibrim may represent the earliest record of cultivated lablab in Africa.

An important difference between the tropical crops and the already established winter crops of Near Eastern origin is the photoperiodicity which constrains the season of cultivation (Willcox, 1992). Winter crops such as barley and wheat are planted in the autumn with lengthening days after the winter solstice inducing flowering which subsequently leads to a spring harvest. This is suited to the middle and lower Nile where the flood season between August to October provides sufficient water.

Summer crops such as sorghum, cotton and lablab require shortening day lengths after the mid-summer solstice to bring on flowering. The cultivation of these summer crops in hyperarid Lower Nubia during the low Nile season would have required irrigation and it is likely that newly available irrigation technology allowed crops to be adopted from the south.

The *saqia* (an animal driven waterwheel) is generally thought to have been introduced into Nubia in the 1st Century AD, (Trigger, 1965; Adams, 1977; Welsby, 1996: 156) although increasing evidence places this introduction much later, probably as late as the end of the Meroitic period in 4th century AD (Rowley-Conwy, 1989; Horton, 1991; Edwards, 1996), which coincides with the introduction of many of the new crops. These additional crops would have increased productivity by adding an extra growing season, which may have in turn have allowed for some increase in population, a trend which has been argued for in the post-Meroitic period (Edwards, 1996). The adoption of these crops may also suggest that there is a change in the composition of the population towards the end of Meroitic period.

Future work on the plant remains from Qasr Ibrim will include radiocarbon dating in order to pinpoint the date of introduction of the new crops at Qasr Ibrim and also biomolecular studies of two of the main crops, lablab and cotton, which may elucidate their origins.

5 References

Adams, W.Y., 1977. *Nubia: Corridor to Africa*. Allen Lane, Penguin Books, London.

Copley, M.S., Rose, P.J., Clapham, A., Edwards, D.N., Horton, M.C. & R.P. Evershed, 2001a. Processing Palm Fruits in the Nile valley – biomolecular evidence from Qasr Ibrim. *Antiquity* 75, pp. 538-542.

Copley, M.S., Rose, P.J., Clapham, A., Edwards, D.N., Horton, M.C. & R.P. Evershed, 2001b. Detection of Palm Fruit lipids in archaeological pottery from Qasr Ibrim, Egyptian Nubia. *Proceedings of the Royal Society, London. Series B* 268, pp. 593-597.

Deakin, W.J., Rowley-Conwy, P. & C.H. Shaw, 1998. The Sorghum of Qasr Ibrim: Reconstructing DNA templates from Ancient Seeds. *Ancient Biomolecules* 2, pp. 117-124.

Driskell, B.N., Adams, N.K. & P.G. French, 1989. A Newly Discovered Temple at Qasr Ibrim, Preliminary Report with appendix by Rowley-Conwy, P., Bird Bones from the Temple at Qasr Ibrim. *Archaeologie du Nil Moyen* 3, pp. 11-54.

Edwards, D.N., 1996. *The archaeology of the Meroitic State: New Perspectives on its social and political organisation*. (BAR International Series 640. Cambridge Monographs on African Archaeology 38). Tempus Repartum, Oxford.

Evershed, R.P., Bland, H.A., van Bergen, P.F., Carter, J.F., Horton, M.C. & P.A. Rowley-Conwy, 1997a. Volatile compounds in Archaeological Plant remains and the Maillard Reaction during decay of Organic Matter. *Science* 278, pp. 432-433.

Evershed, R.P., Van Bergen, P.F., Peakman, T.M., Leigh-Firbank, E.C., Horton, M.C., Edwards, D., Biddle, M., Kjølbye-Biddle, B., & P.A. Rowley-Conwy, 1997b. Archaeological Frankincense. *Nature* 390, 667-668.

Hewison, J.W., 1948. Northern Province Agriculture. In: J.D. Tothill (ed.), *Agriculture in the Sudan*. Oxford University Press, London, pp739-760.

Horton, M., 1991. Africa in Egypt: New Evidence from Qasr Ibrim. In: W.V. Davies (ed.), *Egypt and Africa*. British Museum Press, London), pp 264-277.

Maréchal, R., Mascherpa, J-M & F. Stainer, 1978. Etude taxonomique d'un groupe complexe d'espéces des genres Phaseolus et Vigna (Papilionaceae) sur la base de données morphologiques et polliniques, traitées par l'analyse informatique. *Boissiera* 28, pp. 10-273

O'Donoghue, K., Brown, T.A., Carter, J.F. & R.P. Evershed, 1994. Detection of Nucleotide Bases in Ancient Seeds using Gas Chromotography/Mass Spectrometry & Gas Chromotography/Mass Spectrometry/Mass Spectrometry. *Rapid Communications in Mass Spectrometry* 8, pp. 503-508

O'Donoghue, K., Clapham, A., Evershed, R.P. & T.A. Brown, 1996. Remarkable Preservation of Biomolecules in Ancient Radish seeds. *Proceedings of the Royal Society, London. Series B* 263, pp. 541-547

Rose, P.J., 1996. *Qasr Ibrim: the hinterland survey* by Pamela Rose; with a contribution by Penelope Wilson; original fieldwork by P. Rose, P. Rowley-Conwy *et al*. Egypt Exploration Society, London.

Rose P., & D.N. Edwards, 1998. Excavations at Qasr Ibrim. Part I: Qasr Ibrim 1998 by Rose, P. Part II: 1998 excavation in the Trench 10/14 by D.N. Edwards. *Sudan and Nubia* Bulletin 2, pp. 61-65

Rose, P. & P. Rowley-Conwy, 1989. Qasr Ibrim Regional Survey. Preliminary results. *Archaeologie du Nil Moyen* 3, pp. 121-130

Rowley-Conwy, P., 1988. The Camel in the Nile Valley: new radiocarbon accelerator (AMS) dates from Qasr Ibrim. *Journal of Egyptian Archaeology* 78, pp. 245-248

Rowley-Conwy, P., 1989. Nubia AD0-550 and the "Islamic" Agricultural Revolution: Preliminary Botanical Evidence from Qasr Ibrim, Egyptian Nubia. *Archaeologie du Nil Moyen* 3, 121-130

Rowley-Conwy, P., 1991. Sorghum from Qasr Ibrim, Egyptian Nubia, c800BC-AD1811: A preliminary study. In: J.M. Renfrew, J.M. (ed.), *New Light on Early Farming. Recent developments in Palaeoethnobotany*. Edinburgh University Press. Edinburgh, pp. 191-211

Rowley-Conwy, P., 1994. Dung, dirt and Deposits: Site formation under conditions of near-perfect preservation at Qasr Ibrim, Egyptian Nubia. In: R. Luff, & P. Rowley-Conwy (eds), *Whither Environmental Archaeology?* (Oxbow Monograph 38). Oxbow Books, Oxford, pp. 25-32

Rowley-Conwy, P.A., Deakin, W.J. & C.H. Shaw, 1997. Ancient DNA from archaeological sorghum (Sorghum bicolor) from Qasr Ibrim, Nubia. Implications for domestication and evolution and a review of the archaeological evidence. *Sahara* 9, pp. 23-34

Rowley-Conwy, P.A., Deakin, W.J, & C.H. Shaw, 1999. Ancient DNA from Sorghum. The evidence from Qasr Ibrim, Egyptian Nubia. In: M. van der Veen (ed.) *The Exploitation of Plant Resources in Ancient Africa*. Kluwer Academic/Plenum Publishers, New York, pp. 55-61

Shaw, C.H., Deakin, W.J. & P. Rowley-Conwy, 2001. Ancient DNA from Archaeological Sorghum from Qasr Ibrim, Egyptian Nubia: Methods and Results. Millard, A., (ed.), *Archaeological Sciences '97. Proceedings of the Conference held at the University of Durham. 2nd-4th September 1997*. (BAR International Series 939). Oxbow Books, Oxford, pp. 96-99.

Smartt, J. 1990. *Grain Legumes: evolution and genetic resources*. Cambridge University Press, Cambridge.

Smith, A.B. & L. Jacobson, 1995. Excavations at Geduld and the appearance of early domestic stock in Namibia. *South African Archaeological Bulletin* 50, pp. 3-14

Trigger, B.G., 1965. *History and settlement in Lower Nubia*. Yale University Publications in Anthropology 69.

Verdcourt, B., 1970. Studies in the *Leguminosae-Papilionoideae* for the 'Flora of Tropical East Africa': III. *Kew Bulletin* 24(3), pp.379-443

Verdcourt, B., 1971. Phaeseoleae. In: E. Milne-Redhead, E. and R.M. Polhill (eds) *Flora of Tropical East Africa, Leguminosae (Part 4), Papilionidae (2)*. Crown Agents for Overseas Governments and Administrations, London.

Welsby, D., 1996. *The Kingdom of Kush-The Napatan and Meroitic Empires*. British Museum Press, London.

Westphal, E., 1974. *Pulses in Ethiopia, their taxonomy and agricultural significance*. (Agricultural Research Reports 815). Centre for Agricultural Publishing and Documentation, Wageningen.

Willcox, G. 1992. Some differences between crops of Near Eastern Origin and those from the tropics.. In: C. Jarrige (ed.), *South Asian Archaeology 1989*.(Monographs in World Archaeology 14). Prehistory Press, Madison, pp 291-299

Ancient Egyptian plant remains in the Agricultural Museum (Dokki, Cairo)

R.T.J. Cappers

Groningen Institute of Archaeology, University of Groningen, the Netherlands

R. Hamdy

Botany Department, Faculty of Science, Cairo University Herbarium, Egypt

A unique collection of subfossil plant remains is stored at the 'Graeco-Roman, the Coptic and the Islamic Museum' and the 'Ancient Egyptian Agriculture Museum'. With the exception of a few samples, these plant remains were obtained between 1932 and 1938 and represent 27 sites and 124 taxa. This publication presents a description of the botanical samples based on the labels accompanying the samples and the description in the catalogue and a new description based on a re-examination of these samples by microscopic analysis. In this description predominating plant species are distinguished from plant species which are only represented by a few plant remains.

1 Introduction

This publication presents a description of the botanical samples kept in the 'Graeco-Roman, the Coptic and the Islamic Museum' and the 'Ancient Egyptian Agriculture Museum', both of which belong to the Agricultural Museum in Dokki (Cairo). The Agricultural Museum has a unique collection of plant and animal remains from previous excavations. Associated objects, such as agricultural tools, are also included in the collection. Recently, a part of the Agricultural Museum has been re-organized and a special exhibition has been put on display which presents the development of ancient Egyptian agriculture.

The botanical samples have been previously studied by several persons, including G.A. Schweinfurth (German explorer and botanist), Hassan Khalifa (former director of the Agricultural Museum), E.A.M. Greiss (Faculty of Science, Cairo University), L. Keimer (German Egyptologist), E. Schiemann (Kaiser-Wilhelm-Institut für Kulturpflanzenforschung), E. Åberg and Purraisig (Uppsala University), V. Täckholm (Systematic botany, Cairo university), M. Drar (Systematic Botany, Cairo university), H. Helbaek (National Museum, Copenhagen), Dr. Abel el-Gazzar (Botany department, Cairo University) and M.M. Kislev (Tel Aviv University).

Previous identifications of the plant remains from the Agricultural Museum have found their way in many publications, but a complete and unabridged edition of the records has never been published. Comprehensive studies of ancient Egyptian plant materials have incorporated the records from the Agricultural Museum with those from other sources and moreover, most of the information has been summarized (*e.g.*: Täckholm & Täckholm, 1941; Täckholm & Drar, 1950, 1954 and 1969; Darby *et al.*, 1977, 1977; Germer, 1985; De Vartavan & Amorós, 1997).

The aim of this publication is twofold. First of all, as most of the identifications were based on floras dating back to the first part of the last century, an update of these identifications was necessary. This can be seen, for example, in the new cereal names, whose classification has recently been updated on the basis of cytogenetic affinities (Zohary & Hopf, 1994). Secondly, this catalogue is aimed at presenting the complete spectrum of plant remains in a particular sample. With regard to the economic plants listed, the enumerations can be used to study the possible relationship between plant species in specific archaeological contexts. The inclusion of the contamination of wild plants enables the study of, for example, ancient weed floras and the introduction of particular plant species.

2 Current study of the plant remains

The plant remains of the Agricultural Museum were studied during two visits: November 2004 and March 2005. Two additional short visits in October 2005 and June 2006 were necessary to check some identifications and descriptions.

All samples of archaeobotanical remains have been studied with the exception of (fragments) of garlands and bouquets. These decorations were previously studied by R. Hamdy for her PhD-thesis and will be summarized separately (Hamdy, 2004, this volume and forthcoming articles). Not included were samples of prepared food, such as the many well-preserved pieces of bread, and imitations of food items, such as those made from clay, wood and faience. Although all cases in the museum were processed systematically, it appeared that some earlier published samples had not been seen. Examples of such ommisions are a sample of persea fruits from Tut-Ankh-Amun's tomb (No. 4472), and fruits identified as sugar dates (*Balanites aegyptiaca*) from Saqqara (IIIe Dynasty; number on the fruit is illegible), both depicted by Darby *et al.*, 1977 (1977: figs. 18.20 and 19.9). The identification of the sugar dates is most probably incorrect: the shape of the fruits does not fit with that of sugar dates and even for the dôm palm, the shape is considered to be atypical. The fruits bear cuttings which have been interpreted as incisions for extracting the fat-containing seeds. This seems unlikely judging the shape and hardness of the sugar date fruits and the size of the seeds (for a discussion on oil extraction see also: Cappers, 2006).

Large samples were sieved with a stack of sieves with mesh sizes 2.0 mm, 1.0 mm and 0.5 mm to aid the sorting out of the samples. The samples or sieve fractions were studied under a dissecting microscope with 5-56 X magnification and a Euromex glass fibre cold light illuminator. The dominant plant species have been distinguished from those which are present as contaminants. Small samples were completely investigated but in the case of some very large samples only a subsample could be checked and minor contaminations might have been missed. In such cases, it has been stated in the catalogue that only a subsample has been thoroughly checked.

As no reference material was at our disposal while working in the Museum, the identification as presented in the catalogue and on the label could not always be confirmed. This is especially true for vegetative plant remains, such as tubers and stem fragments. On occasion it was possible to identify such specimens afterwards with the aid of drawings and notes that were made with the herbarium of the Cairo University and the reference collection of the Groningen Institute of Archaeology. Most samples of Juniper (*Juniperus*), for example, have, for this reason, not been identified beyond the level of the genus. A positive identification is still possible, however, as the fruits are diagnostic in their size, number of scales and seeds. More research in the future is necessary to improve the catalogue.

From many samples the number of specimens were determined and measurements were taken. When dealing with small samples, all specimens were counted and measured. From large samples an estimate of the number was made and only measurements were taken from the smallest and biggest specimens. The measurements are based on drawings of the seeds and fruits. In addition, special features were recorded such as artificial cut marks, damage by insects and the presence of fungi.

3 Presentation of data

The information for each record is presented in a standardized format. First, the record number is presented. Samples are presented in numerical order; those without a number are placed at the end. For a limited number of samples only (stored in the laboratory of the museum) no number and additional information is available.

Next the information on the botanical samples is presented and the specific catalogue number is given. This information was obtained from the four-volume catalogue, kept at the Agricultural Museum, and from the labels which go with the plant remains. To reduce redundancy, it has been decided to integrate the information from the catalogue with that of the labels. The labels include the ones exposed with the samples as well as the wrapped papers that are present inside the boxes. It is assumed that in most cases the labels are based on the descriptions in the catalogue, but that also original labels are still present with the samples. For some samples the information in the catalogue does not match that presented on the labels. When appropriate, both the description from the catalogue and that of the label is presented. A few numbered samples are not documented in the catalogue. For these samples the description is based solely on the labels.

To optimize the catalogue references, it was decided to translate the French and Arabic notes into English. As a result the description is not always an accurate reproduction of the original texts. When necessary, the English descriptions was upgraded and if the handwriting was difficult to read, the concerning text was put in brackets.

The description of the samples has also been standardized with respect to the sequence of the information. First the botanical name(s) of the plant remain(s) is presented together with the archaeological context and measurements (if available). Latin names are only included when mentioned in the catalogue or on the labels. These Latin names have not been updated to facilitate the comparison with previous publications in which they are recorded. Next, the location and period are mentioned, followed by the excavator or the institute responsible for the excavation and the year of excavation. Unfortunately, this information is not always documented. Then information is presented dealing with the previous identification of the sample, in most cases the name of the person and the date of the identification is given. Most descriptions end with an (alpha)numerical code which refers to the archaeological context, such as the pottery in which the plant remains were found. This code consists either of three units (originally presented vertically) or of four units (originally presented in a cross, reading clockwise from upper right). Very often these codes are crossed out, indicating that they are no longer valid.

A distinction has been made between plant species that predominate in the sample and plant species that are only present in low quantities. Plant names follow Boulos (1999; 2000; 2003 and 2005) for the wild plant species and Wiersema & León (1999) for the cultivated plants. When a particular identification is reliable, but could not be confirmed because of the lack of reference material, the person who made the original identification is mentioned again.

In most cases both the plant part as well as the number of fragments (sometimes only an estimate) are mentioned in addition to the Latin plant names. The preservation condition is desiccated unless otherwise stated. Additional information such as the presence of cut marks or damage by other organisms is also mentioned, if applicable.

Special attention has been paid to the presence of diploid (viz.: einkorn [*Triticum monococcum*]), tetraploid (viz.: emmer [*Triticum turgidum* ssp. *dicoccon*] hard wheat [*Triticum turgidum* ssp. *durum*]) and hexaploid wheats (viz.: *Triticum aestivum*). Because 2-seeded varieties of einkorn wheat occur, the distinction between einkorn and emmer might be problematic. Both primitive wheat species can be easily distinguished, however, by their upper spikelet. In the diploid einkorn the rachis internode of the apacial spikelet has the same orientation as that of the other spikelets present in the ear. In tetraploid and hexaploid wheats, the orientation of the rachis internode of the apacial spikelet is different: the internode is 'twisted' as a result of which the glumes are at a 90° angle to the rachis (see Cappers 2004 for illustrations). If upper spikelets with a twisted rachis internode are present, it is mentioned in the description of the sample. The distinction between free-threshing tetraploid wheat (viz.: [*Triticum turgidum* ssp. *durum*]) and free-threshing hexaploid wheat (viz.: [*Triticum eastivum*]) is based on diagnostic features of the rachis fragments (Hillman, 2001; Maier, 1995; Cappers & Jans, this volume). With the exception of some samples with small grain kernels, which are strikingly charred (see the discussion below), samples with only grain kernels have been identified as free-threshing wheat (viz.: traditionally labelled as '*Triticum aestivum/durum*').

Some samples have an additional comment, including references to photographs depicted in the second volume of 'Food: The Gift of Osiris' (Darby *et al.*, 1977).

4 Some concluding remarks

The catalogue comprises 442 samples representing 27 sites and 124 taxa (Tables 1 and 2). For over 80% of the samples, the year is documented in which the sample was donated to the Agricultural Museum. Almost all these documented samples were obtained by the museum between 1932 and 1938. Only a few of the documented samples were obtained in a later period: 1942 (N=4), 1946 (N=2) and 1953 (N=3).

Most samples originate from Deir El Medineh (New Kingdom; N=89), Karanis (Roman period; N=62), Saqqara (Old Kingdom; N=47), Thebes (New Kingdom; N=37) and Gebelein (Old Kingdom and Pharaonic period; N=32). Unfortunately, a considerable number of samples (ca. 25%) were obtained from antique dealers, for the most part in Luxor. Most probably the majority of these samples were illegally taken from the tombs of the Nobles on the west bank of modern Luxor (ancient Thebes) by the inhabitants of Qurna. It may not be excluded, however, that some of these samples are of modern date. Because botanical samples were in demand, it is quite likely that also recent plant remains were offered for sale as antiquities. In this respect, samples No. 4461 and No. 5003 may serve as an example, as they concern plant species which were not likely to be found in ancient Egypt. Sample No. 4461 consists of four complete fruits of the custard-apple (*Annona squamosa* L.) and although it is currently widely cultivated in the tropics, its probable origin is the West Indies. Nine blackish flower buds of cloves (*Syzygium aromaticum* [L.] Merr. & M. Perry) together with a brownish spikelet of emmer wheat (*Triticum turgidum* ssp. *dicoccon*) were obtained from Mohamed Mansour in Luxor. It is possible that an ancient emmer spikelet was added to a sample of modern cloves. Cloves have also been recorded from Karanis (Kôm Aushim; sample No. 0293), but the concerning fragments are in fact stem fragments with bud-like structures consisting of many imbricate scales. Similar plant fragments were frequently found in a trash deposit in Karanis that was sampled during the RuG/UCLA excavation (started in 2005). Although these fragments have not yet been identified, they are, on no account, *Syzygium aromaticum*. Other suspicious samples are, for example, No. 1355 (a possible fruit of pistachio [*Pistacia*]), No. 3357 (predominately consisting of hard wheat [*Triticum turgidum* ssp. *durum*], but with contaminations of emmer [*Triticum turgidum* ssp. *dicoccon*], which was not grown in Egypt anymore at the time the sample was bought, and sorghum [*Sorghum bicolor*], which was most probably taken into cultivation in the Graeco-Roman period) and No. 4200 (consisting of a few fruits of citron [*Citrus medica*]).

It has been suggested that einkorn was not a crop of its own right in ancient Egypt (Murray, 2000). The absence of einkorn in the samples, at least as a dominant plant, supports this view. Only four samples contained wheat that has been identified as bread wheat (viz.: No. 0135, No. 1358, No. 2054 and No. 3379). Although no threshing remains are present, the grain kernels have a typical compact shape and are similar to the variety *compactum*, as has been suggested by H. Helbaek for the sample Nos 1359 and 3379. Only sample No. 2054 is well documented (Karanis, Roman period), the other three sample are bought from antique dealers without information on provenance and period. Remarkably, all four samples of bread wheat contain charred grain kernels whereas most of the wheat samples in the Agricultural Museum have been preserved by desiccation. It may not be excluded that charring has deformed the original shape and that we are dealing with emmer wheat (Braadbaart, 2004). On the other hand, samples with typically charred emmer grains are also present in the collection and are dated to the Predynastic period (viz.: No. 4088, No. 4089 and No. 4295).

The present catalogue supports, for the time being, the late import of termis (*Lupinus albus*) and the fungal disease covered smut (*Ustilago hordei* (Pers.) Lagerh.) as well as the shift to free-threshing wheats in the Graeco-Roman period. With the exception of sample No. 5234 (Thebes; Late period), all documented samples point to the Roman period. In two samples originating from Roman Karanis (viz. No. 0274 and No. 2057) some rachis fragments of hulled 6-row barley (*Hordeum vulgare* ssp. *vulgare* (hulled) are infected with the fungus covered smut. This kind of infection was also found in barley from Roman Berenike and Shenshef (Cappers, 2006). So far, no such infection has been reported from earlier periods. Samples with free-threshing wheat are conspicuously originating from the Graeco-Roman period and support the hypothesis that in this period emmer wheat was replaced on a large scale by hard wheat (*Triticum turgidum* ssp. *durum*).

5 Acknowledgements

We would like to express our thanks to Mr. Mohamed El Hossainy El Akaad (General Supervisor of the Agricultural Museum and Exhibitions) for the permission to study the plant remains. We are also indebted to Mr. Hassan Abd El-Rahman Khattab (former General Director of the Agricultural Museum, concultant of the Agricultural Museum and expert in ancient agriculture) for his hospitality and discussions on the archaeobotanical collection of the Agricultural Museum. The staff members Michel Wasfi Abd el-Motagaly, Raef Ali Ezzat, Keryakes Lous Keryakes, Omar Besheer, Dalia Mohamed Salem, Manal Ali Ahmed and Ahmed Khalil Ibrahim were very helpful in their assistance while working on the samples.

6 References

Boulos, L., 1999. *Flora of Egypt: 1.* Cairo, Al Hadara.

Boulos, L., 2000. *Flora of Egypt: 2.* Cairo, Al Hadara.

Boulos, L., 2003. *Flora of Egypt: 3.* Cairo, Al Hadara.

Boulos, L., 2005. *Flora of Egypt: 4.* Cairo, Al Hadara.

Braadbaart, F., 2004. *Carbonization of peas and wheat – a window into the past. A laboratory study*. PhD-thesis Leiden University. Heemstede, Gravé.

Cappers, R. T. J., T. Van Thuyne & L. Sikking, 2004. Plant Remains from Predynastic El Abadiya-2 (Naqada Area, Upper Egypt). In: S. Hendrickx, R.F. Friedman, K.M. Cialowicz & M. Chlodnicki (eds), *Egypt at Its Origins. Studies in Memory of Barbara Adams. Proceedings of the International Conference "Origin of the State. Predynastic and Early Dynastic Egypt," Krakow, 28th August–1st September 2002.* Leuven, Peeters, pp. 277-293.

Cappers, R.T.J., 2006. *Roman foodprints at Berenike. Archaeobotanical evidence of trade and subsistence in the Eastern Desert of Egypt*. Monograph Series 55, Los Angeles, Cotsen Institute of Archaeology.

Darby, W. J., P. Ghaliounghi & L. Grivetti, 1977. *Food: The Gift of Osiris*. Vol. 2. London, Academic Press.

Germer, R., 1985. *Flora des pharaonischen Ägypten*. Mainz, Verlag Philipp von Zabern.

Hamdy, R., 2004. *Documentary and ethnobotanical study of floral bouquets and garlands in Egypt since the 18th Dynasty (± 1700 BC)*. PhD-thesis Cairo University.

Hamdy, R. Plant remains from the intact garlands present at the Egyptian Museum in Cairo. In: R.T.J. Cappers (ed.), *Fields of change. Progress in African Archaeobotany. Proceedings of the 4th International Workshop of African Archaeobotany*. Groningen Archaeological Studies No. 5. Groningen, Barkhuis.

Hillman, G.C., 2001. Archaeology, Percival and the problems of identifying wheat remains. In: P.D.S. Caligari & P.E.

Brandham (eds), Wheat taxonomy: the legacy of John Percival. *The Linnean*, special issue No. 3: pp. 27-36.

Maier, U., 1995. Morphological studies of free-threshing wheat ears from a neolithic site in southwest Germany, and the history of the naked wheats. *Vegetation History and Archaeobotany* 5: pp. 39–55.

Murray, M. A., 2000. Cereal production and processing. In: P.T.Nicholson & I. Shaw (eds), *Ancient Egyptian Materials and Technology*. Cambridge University Press, Cambridge, pp. 505-536..

Täckholm, V. & G. Täckholm, 1941. *Flora of Egypt*. Vol. 1. Cairo, Fouad I University Press.

Täckholm, V. & M. Drar, 1950. *Flora of Egypt*. Vol 2. Cairo, Fouad I University Press.

Täckholm, V. & M. Drar, 1954. *Flora of Egypt*. Vol 3. Cairo, Cairo University Press.

Täckholm, V. & M. Drar, 1969. *Flora of Egypt*. Vol 2. Cairo, Cairo University Press.

Vartavan De, C. & V. A. Amorós, 1997. *Codex of Ancient Egyptian Plant Remains*. London, Triade Exploration.

Wiersema, J. H. & B. León, 1999. *World Economic Plants: A Standard Reference*. London, CRC Press.

Zohary, D. & M. Hopf, 2000. *Domestication of Plants in the Old World*. 3rd edition. Oxford, University Press.

7 Catalogue

No. 0081
Cat. ? *Triticum vulgare*; Deir El Medineh; French Institute of Oriental Archaeology D.M.; [4|3|27|29].
Dominant ● *Triticum aestivum/durum*
50-100 grain kernels.

No. 0130
Cat. 1/13 Bottle gourland seed (diameter: 0.065 cm); donated by the Egyptian Museum (29-09-1934); [5|5|27|2].
Dominant ● *Lagenaria siceraria*
Ca. 8.7 (width) x 5.1 (height) x 0.3 cm (thickness). Small opening at top of fruit (ca. 1 cm in diameter).

No. 0135
Cat. ? -
Dominant ● *Triticum aestivum*
Rather short grain kernels (slightly charred).
Contam. ● *Triticum turgidum* ssp. *durum*
Grain kernels, slightly burned, and desiccated rachis fragment.
● *Hordeum vulgare* ssp. *vulgare* (hulled)
Spikelets, partly charred and partly desiccated.

No. 0136
Cat. 1/14 Seeds of lucerne (*Medicago sativa* L.); treated by Hossam Abd el-Hamid and Kamal Tantawi, Ministry of Archaeology and Hussein el-Monayer, Agricultural Museum (14-04-1970); part given from a sample present in the Egyptian Museum (hall 53, case P) (1934); [4|8|27|22].
Dominant ● *Medicago sativa*
Large number (ca. 50 ml) of seeds with only a few fragments of fruits.

No. 0271
Cat. 1/32 Quantity of *Ricinus communis*; Karanis (Kôm Aushim); Roman period; donated by the Antiquity Department (11-1935); [28|BS 170|A (crossed out)].
Dominant ● *Ricinus communis*
Ca. 200 seeds, only few with insect holes.

No. 0272-A
Cat. 1/33 Quantity of grains of *Carthamus tinctorius*; Karanis (Kôm Aushim); Roman period; donated by the Antiquity Department (11-1935); [29|B132B★|Q].
Dominant ● *Carthamus tinctorius*
Several thousands of fruits. Preservation condition different from 0272-B. Only small part of the sample has been examined for possible contamination.

No. 0272-B
Cat. 1/33 Quantity of grains of *Carthamus tinctorius*; Karanis (Kôm Aushim); Roman period; donated by the Antiquity Department (11-1935).
Dominant ● *Carthamus tinctorius*
Several thousands of fruits. Preservation condition differs from sample No. 0272-A.
Contam. ● *Triticum turgidum* ssp. *durum*
Few rachis fragments.

No. 0273
Cat. 1/33 Quantity of dates (*Phoenix dactylifera*); Karanis (Kôm Aushim); Roman period;; donated by the Antiquity Department (11-1935); [H6|B40|A].
Dominant ● *Phoenix dactylifera*
28 fruits and 1 seed.

No. 0274
Cat. 1/33 Quantity of barley; Karanis (Kôm Aushim); Roman period; donated by the Antiquity Department (11-1935).
Dominant ● *Hordeum vulgare* ssp. *vulgare* (hulled)
Spikelets and a few rachis fragments. Few

	rachis fragments are infected with the fungus covered smut (*Ustilago hordei* (Pers.) Lagerh.).
Contam.	● *Triticum turgidum* ssp. *durum* Single grain kernel and rachis fragment ● *Raphanus raphanistrum* Fruit fragment. ● *Phalaris paradoxa* Spikelets. ● *Sinapis* cf. *arvensis* Fruit fragments (valves, 6-nerved) and seeds. ● *Galium aparine* Fruit. ● *Malva nicaeensis/sylvestris* Fruit with reticulate dorsal surface. ● *Lotus* Seed. ● Unknown fruit fragment Possibly of Fabaceae. ● Unknown seeds Shape varies from round to oval, ca. 4 mm in diameter.
No.	0275
Cat. 1/33	Mixture of wheat and barley; Karanis (Kôm Aushim); Graeco-Roman period; donated by the Antiquity Department (11-1935).
Dominant	● *Triticum turgidum* ssp. *durum* Grain kernels and some rachis fragments. ● *Hordeum vulgare* ssp. *vulgare* (hulled) Spikelets (ca. 10-15%) and some rachis fragments.
Contam.	● *Medicago* cf. *polymorpha* Fruit.
.	● Some pieces of charcoal
No.	0276
Cat. 1/33	Quantity of garlic bulbs (*Allium sativum*); Karanis (Kôm Aushim); Roman period; donated by the Antiquity Department (11-1935); [29	C137A	VII (crossed out)].
Dominant	● *Allium sativum* Bundle of garlic bulbs, bulbs partly fallen off. Length of a complete bulb: 3 cm.
No.	0277
Cat. 1/33	Mixture of wheat and barley; Karanis (Kôm Aushim); Roman period; donated by the Antiquity Department (11-1935).
Dominant	● *Triticum turgidum* ssp. *durum* For the most part grain kernels, only a few spikelets present.

Contam.	● *Hordeum vulgare* (hulled) 1 spikelet.
No.	0279
Cat. 1/33	Fruit (*Juglans regia*); Karanis (Kôm Aushim); Roman period; donated by the Antiquity Department (11-1935).
Dominant	● *Juglans regia* Whole fruit (40 x 35 mm).
Comment	See figure 18.28 in Darby *et al.*, 1977.
No.	0280
Cat. 1/33	2 fruits (*Corylus avellana*); Karanis (Kôm Aushim); Roman period; donated by the Antiquity Department (11-1935).
Dominant	● *Corylus avellana* 2 fruits, both gnawed.
No.	0281
Cat. 1/33	Quantity of *Carthamus tinctorius*; Karanis (Kôm Aushim); Roman period; donated by the Antiquity Department (11-1935); [28	132*	K].
Dominant	● *Carthamus tinctorius* Complete fruits.
Contam.	● *Triticum turgidum* ssp. *durum* Few rachis fragments. ● Gramineae tribe Triticeae Few chaff (lemma and palea) fragments. ● *Trifolium* Dozens of seeds, 1 also with fruit with ca. 12 veins.
No.	0282
Cat. 1/33	Some seeds of *Lupinus termis*; Karanis (Kôm Aushim); Roman period; donated by the Antiquity Department (11-1935); [28	136★	P (crossed out)].
Dominant	● *Lupinus albus* 8 seeds.
No.	0283
Cat. 1/33	Quantity of lettuce seeds (*Lactuca sativa*); Karanis (Kôm Aushim); Roman period; two boxes; donated by the Antiquity Department (11-1935); [29	137 A^2	D (crossed out)].
Dominant	● Cruciferae Large number of small seeds with reticulate pattern (diameter: ca. 1 mm).
No.	0284
Cat. 1/34	Quantity of lentils (*Lens esculenta*); Karanis (Kôm Aushim); Roman period; donated by the Antiquity Department (11-1935); [29	C 162 B	C (crossed out)].

Dominant	• *Lens culinaris* Diameter: ca. 4 mm.		Dominant	• *Trifolium* cf. *alexandrinum* Large number of seeds.
Contam.	• *Vicia* cf. *ervilia* Seed (4.5 mm long). • *Hordeum vulgare* cf. ssp. *vulgare* (hulled) Spikelet.		Contam.	• *Hordeum vulgare* (hulled) Spikelet. • *Malva sylvestris* Fruit. • *Anthemis* Fruit.

No.   0285
Cat. 1/34   Grains of *Carthamus tinctorius*; Karanis (Kôm Aushim); Roman period; donated by the Antiquity Department (11-1935); [30|C 123 CZ|A (crossed out: 29|B 191 K|X)].

Dominant • *Carthamus tinctorius*
Hundreds of fruits, all split into halves.

Contam. • *Olea europaea*
2 endocarps.
• *Triticum turgidum* ssp. *durum*
Ear.
• *Hordeum vulgare* cf. ssp. *vulgare* (hulled)
Rachis fragment.
• Gramineae tribe Triticeae
Chaff (lemma and palea).
• *Phoenix dactylifera*
Pedicel.

No.   0286
Cat. 1/34   Quantity of *Cocculus pendulus* (label: seeds of *Galium tricorne* Stocks); Karanis (Kôm Aushim); Roman period; donated by the Antiquity Department (11-1935).

Dominant • *Galium tricornutum*
Many fruits (present on 2.0 mm sieve).
• cf. *Vaccaria hispanica*
Many seeds (present on 1.0 mm sieve). Black, round and small papillae present on some of the specimens.

Contam. • *Maerua crassifolia*
Fruit.
• *Triticum aestivum/durum*
1 charred grain kernel.
• *Hordeum vulgare* (hulled)
Spikelets.
• *Beta vulgaris*
Fruit clusters.
• *Raphanus raphanistrum*
Seeds.
• cf. *Sinapis*
Fruit stalk.

No.   0287
Cat. 1/34   Black-mustard seeds (*Brassica nigra*); Karanis (Kôm Aushim); Roman period; donated by the Antiquity Department (11-1935); [26|A4|V (crossed out)].

No.   0288
Cat. 1/34   Pine cone; Karanis (Kôm Aushim); Roman period; donated by the Antiquity Department (11-1935); [31|I 11 2L*|B (crossed out)].

Dominant • *Pinus*
Complete cone, ca. 7 cm long and 3 cm wide, thick stalk (15 x 5 mm). Scales closed.

No.   0289
Cat. 1/34   Flax seeds; Karanis (Kôm Aushim); Roman period; donated by the Antiquity Department (11-1935); [28|B 136 A|G].

Dominant • *Linum usitatissimum*
Ca. 200-300 seeds.

No.   0290
Cat. 1/34   Fruit of Cucurbitaceae. A bottle of gourland seeds (against worms), *Lagenaria vulgaris*; donated by the Antiquity Department (11-1935); [31|I-111N| D].

Dominant • *Lagenaria siceraria*
Fruit (5.7 x 5.1 cm) and 4 relatively short seeds.

No.   0292
Cat. 1/34   Turnip root (*Brassica rapa*); Karanis (Kôm Aushim); Roman period; donated by the Antiquity Department (11-1935); [31|II 201 H|D].

Dominant • ?
Swollen root (8.5 x 5.5 cm) with other roots still attached.

No.   0293
Cat. 1/34   Part (stem) of clove-tree (*Eugenia aromatica* Baill); Karanis (Kôm Aushim); Roman period; donated by the Antiquity Department (11-1935); [26|B36A|G (crossed out)].

Dominant • ?
Stem fragments with bud-like structures. Similar plant fragments are also frequently found in a trash deposit in Karanis that was sampled during the RuG/UCLA excavation (started in 2005). On no account *Syzygium aromaticum*.

No.	0294
Cat. 1/35	Flowerbuds of pomegranate; Karanis (Kôm Aushim); Roman period; donated by the Antiquity Department (11-1935); [27	203A	C (crossed out)].
Dominant	● *Punica granatum*
	Flower buds and seeds.

No.	0295
Cat. 1/35	Branches of *Laggera aurita* Sch.Bip.; Karanis (Kôm Aushim); probably Graeco-Roman period; identified by V. Täckholm as *Conyza aurita* L. (28-01-1950); donated by the Antiquity Department (11-1935).
Dominant	● *Pseudoconyza viscosa*
	Woolly branches without generative parts. Identified by V. Täckholm, only Latin name has been updated.

No.	0351
Cat. ?	Animal remains; [4	32	?	22].
Dominant	● *Lens culinaris*
	10 seeds (blackish, but not charred).

No.	0371
Cat. 1/50	Pottery (diameter: 14 cm) containing 8 unincised sycamore fruits (*Ficus sycomorus*) and 2 fruits of *Medemia argun* (cf. No. 0372); probably from Gebelein; G.C.C. Maspero (label: probably from Deir El Medineh; ca. 1400 BC); donated by the Antiquity Department (1935); [4	2	27	22 (pottery)].
Dominant	● *Ficus sycomorus*
	8 whole accessory fruits (syconia).

No.	0372
Cat. 1/50	2 fruits of *Medemia argun*; Gebelein; G.C.C. Maspero; donated by the Antiquity Department (1935); [4	2	27	22 (pottery)].
Dominant	● *Medemia argun*
	Whole fruit (43 x 32 mm) and endocarp (40 x 27 mm).

No.	0373
Cat. 1/50	Small alabaster dish (diameter: 18 cm) containing barley as well as 3 fruits of *Medemia argun* (cf. No. 0374) and modelled bread made of papyrus (cf. No. 0375, not included in this catalogue); Saqqara; Ve Dynasty; donated by the Antiquity Department; [4	3	27	22 (pottery) (crossed out)].
Dominant	● *Triticum turgidum* ssp. *dicoccon*
	Thousands of spikelets. Upper spikelets with twisted internode. Sample not completely checked.

No.	0374
Cat. 1/50	3 fruits of *Medemia argun* (cf. No. 0373); Saqqara; Ve Dynasty [copied from No. 0373]; donated by the Antiquity Department; [4	3	27	22 (pottery)].
Dominant	● *Medemia argun*
	2 fruits, still connected to each other. Size of 1 specimen: 45 x 34 mm.
Comment	See figure 18.11 in Darby *et al.*, 1977.

No.	0377
Cat. 1/50	Small pottery containing seeds of *Juniperus*; hidden place at Deir El Bahari; XXe or XXIe Dynasty; donated by the Antiquity Department (1935); [4	9	27	22 (pottery)].
Dominant	● *Juniperus*
	Hundreds of seeds.

No.	0378
Cat. 1/51	Small pottery containing grapes and 3 seeds of a pine (cf. No. 0379); tomb at Gebelein; G.C.C. Maspero (1885); donated by the Antiquity Department (1935); [4	11	27	22 (pottery)].
Dominant	● *Vitis vinifera*
	Ca. 50 fruits.

No.	0379
Cat. 1/51	3 seeds of pine (cf. No. 0378); tomb at Gebelein; Maspero (1885); donated by the Antiquity Department (1935); [4	11	27	22 (pottery)].
Dominant	● *Pinus pinea*
	3 complete seeds (18-20 mm long).

No.	0380
Cat. 1/51	Small pottery (diameter: 12.5 cm) containing 8 sacrificed fruits of *Ficus sycomorus* rolled with dates (*Phoenix dactylifera*); tomb at Gebelein; G.C.C. Maspero (1885); donated by the Antiquity Department (1935); [4	12	27	22 (pottery)] and [4	13	27	22 (dates)[2]].
Dominant	● *Ficus sycomorus*
	8 accessory fruit (syconia), some with fragments of textile still attached to the outside. It seems as if the accessory fruits were cut open.

No.	0381
Cat. 1/51	Small alabaster pottery (diameter: 6 cm) containing fruits of dates (*Phoenix dactylifera* L) wrapped up with some sycamore figs; tomb at

Gebelein; Pharaonic period; G.C.C. Maspero (1885); donated by the Antiquity Department (1935); [4|13|27|22 (crossed out)] and [4|13|27|22 (sycamore).

Dominant
- *Phoenix dactylifera*
  7 small fruits.

No. 0382
Cat. 1/51 A small pottery dish (diameter: 15.5 cm) containing squeezed barley, used in beer making; tomb at Gebelein; G.C.C. Maspero (1885); identified as *Hordeum* sp. by H. Helbaek (22-6-1955); donated by the Antiquity Department; [4|14|27|22].

Dominant
- *Hordeum vulgare* (hulled)
  Threshing remains (lemma, palea and rachis fragments).

Contam.
- *Triticum turgidum* ssp. *dicoccon*
  Spikelets and spikelet forks.

No. 0383
Cat. 1/51 A small pottery dish (diameter: 13.9 cm) containing fruits of *Ziziphus spina-christi*; tomb at Gebelein; G.C.C. Maspero (1885); donated by the Antiquity Department (1935); [4|15|27|22].

Dominant
- *Ziziphus spina-christi*
  Dozens of whole fruits.

Contam.
- *Cyperus esculentus*
  Small tubers.
- *Pinus pinea*
  Seed.

No. 0384
Cat. 1/52 Pine cone (length: 10 cm); tomb at Gebelein; G.C.C. Maspero (1885); donated by the Antiquity Department (1935); [4|20|27|22].

Dominant
- *Pinus pinea*
  Complete cone, ca. 10 cm long, a few seeds still present.

No. 0385
Cat. 1/52 A small pottery dish (diameter: 14 cm) containing fruits of *Maerua crassifolia* Forssk.; Gebelein (Qena); G.C.C. Maspero (1885); donated by the Antiquity Department (1935); [4|21|27|22 (crossed out)].

Dominant
- cf. *Lathyrus sativus*
  Ca. 60-80 seeds.

Contam.
- *Maerua crassifolia*
  2 fragments of fruits.
- *Hordeum vulgare* ssp. *vulgare* (hulled)
  1 spikelet.

No. 0386
Cat. 1/52 A small pottery dish (diameter: 13 cm) [or: halfa baskets?] containing grains of *Hordeum hexastichum*; Gebelein (1886); XIe Dynasty?; see publication of Schweinfurth in which he described the plants found in a tomb of a certain Ani; donated by the Antiquity Department (1935); [4|23|27|22 (pottery)].

Dominant
- *Hordeum vulgare* ssp. *vulgare* (hulled)
  Hundreds of conspicuously small and long-awned spikelets.

No. 0389
Cat. 1/53 Basket of halfa (9 x 8 cm) containing fruits of *Ziziphus spina-christi*; tomb at Gebelein (1886); donated by the Antiquity Department (1935); [4|28|27|22].

Dominant
- *Ziziphus spina-christi*
  17 complete fruits, some with insect damage.

No. 0392
Cat. 1/53 A small pottery dish (diameter: 5.2 cm) containing fruits of *Juniperus* sp.; Deir El Bahari (Thebes) (1881/2); XXIe Dynasty; donated by the Antiquity Department (1935); [4|34|27|22].

Dominant
- *Juniperus*
  2 fruits (diameter: 9 mm).

No. 0393
Cat. 1/53 Alabaster vase (diameter: 6 cm) containing fruits of *Juniperus* sp.; Cheikh Abd El Qurnah (Thebes); XXVIe Dynasty; H. Helbaek added (25-06-1955): horse bean (*Vicia faba*), raisin (*Vitis vinifera*) and wheat (*Triticum dicoccum*); donated by the Antiquity Department (1935); [4|38|27|22 or 28].

Dominant
- *Vicia faba* var. *minor*
  9 seeds (length: ca. 10 mm).

Contam.
- *Vitis vinifera*
  5 fruits.
- *Triticum aestivum/durum*
  5 grain kernels.

No. 0394
Cat. 1/53 Pottery dish (diameter: 8.3 cm) found in a little basket containing barley grains; Gebelein, probably in the tomb of Ani; XIe Dynasty?; H. Helbaek added (22-06-1955): spikelets of wheat (*Triticum dicoccum*); donated by the Antiquity Department (1935); [4|39|27|22 (pottery)].

Dominant
- *Triticum turgidum* ssp. *dicoccon*

	Ca. 50 spikelets, including top spikelets with twisted internode.
Contam.	● *Cyperus esculentus* 2 tubers.

No.	0395
Cat. 1/54	Small pottery dish (diameter: 9 cm) containing a number of fruits of *Lathyrus sativus*; tomb at Gebelein (1885); donated by the Antiquity Department (1935); [4|40|27|22 – CS$_2$|214|70].
Dominant	● *Cupressus* Ca. 30 fruits. Diameter: 7-9 mm.

No.	0396
Cat. 1/54	Small basket of halfa (9 x 14 cm) filled with fragments of wheat and leaves of sycamore; tomb at Gebelein (1886); according to Schweinfurth (1936), the wheat is no longer present in the sample; donated by the Antiquity Department (1935); [4|43|27|22].
Dominant	● *Vitis vinifera* Branches with leaves, fragmented.
Contam.	● *Triticum turgidum* ssp. *dicoccon* Spike.

No.	0397
Cat. 1/54	Pottery dish (diameter: 7.4 cm) containing fruits of *Cocculus pendulus* (syn.: *Cocculus leaeba*) found in a small basket; the tomb of Ani at Gebelein; XIe Dynasty; part of the fruits are described in the catalogue of the Egyptian Museum, Journal d'entrée (JE), No. 0137; donated by the Antiquity Department (1935); [4|44|27|22 (pottery)].
Dominant	● *Cocculus* cf. *pendulus* 51 endocarps.

No.	0398
Cat. 1/54	Pottery dish (diameter: 5.1 cm) containing fruits of *Cocculus leaeba*; part of the fruits are described in the catalogue of the Egyptian Museum, Journal d'entrée (JE) No. 0137; donated by the Antiquity Department (1935); [4|47|27|22 (pottery)].
Dominant	● *Maerua crassifolia* Fruit fragment.
Contam.	● Unknown fruit Fibrous mesocarp (27 x 21 mm).

No.	0399
Cat. 1/54	Pottery dish containing fruits of *Ficus sycomorus*; tomb of Ani at Gebelein (1886); XIe Dynasty?; donated by the Antiquity Department (1935); [4|49|27|22].
Dominant	● *Ficus sycomorus* 9 accessory fruits (syconia).

No.	0400
Cat. 1/55	Pottery dish (diameter: 6 cm) containing a seed of *Sinapis arvensis*; Dra Abu El Naga (Thebes), probably New Kingdom; identified by Schweinfurth; donated by the Antiquity Department (1935); [4|50|27|22 (crossed out) (pottery)].
Dominant	● ? On no account *Sinapis arvensis*. Size of seed ca. 8 x 7 mm.

No.	0401
Cat. 1/55	Small alabaster vase (diameter: 5 cm) containing *Cyperus esculentus*; Dra Abu El Naga (Thebes); label of G.C.C. Maspero: No. 4489; (on paper that belongs to the sample: Deir El Medinah (Luxor); New Kingdom (ca. 1300 BC); donated by the Antiquity Department (1935); [4|51|27|22].
Dominant	● *Cyperus esculentus* 71 tubers in small alabaster dish.
Contam.	● *Vitis vinifera* 1 seed. ● *Hordeum vulgare* (hulled) 1 complete spikelet and 2 halves. ● *Triticum turgidum* ssp. *dicoccon* 1 glume base.

No.	0403
Cat. 1/55	Barley ears; tomb at Gebelein; Pharaonic period, since 3300 years; G.C.C. Maspero (1885); [An illegible word is written on a small paper]; donated by the Antiquity Department (1935); [4|53|27|22] and [4|6|27|29].
Dominant	● *Cicer arietinum* Stem with 4 fruits and 7 seeds. ● *Hordeum vulgare* ssp. *vulgare* (hulled) 2 ears.

No.	0404
Cat. 1/55	Saucepan (diameter: 16.5 cm) containing seeds and some fragments of roasted pine cones; excavations at Saqqara (1905-06); J.E. Quibell?; picture taken by Quibell of the original 3 pieces before restoration (Cairo 1907, pl. XXXIII, p. 30); donated by the Antiquity Department (1935); [4|7|27|29].
Dominant	● *Pinus pinea* Hundreds of charred cones scales and seeds (only a few desiccated specimens).

Contam.	• *Olea europaea* Desiccated endocarp. • *Phoenix dactylifera* charred seed Sample partly checked.	No. Cat. 1/56	0408 3 endocarps and some fruits of *Prunus amygdalus* (*Amygdalus communis*); Hawara (Fayum); Ptolemaic period; donated by the Antiquity Department (1935); [4|11|27|29].
Comment	See figure 18.27 in Darby *et al.*, 1977; possibly only the few desiccated specimens are depicted.	Dominant	• *Amygdalus communis* 3 endocarps. • ? Ca. 20 fruits (4.3 x 4.1 mm) with wrinkled surface.

No. 0405
Cat. 1/55 Fragments to be studied: *Rumex dentatus*; Saqqara; donated by the Antiquity Department (1935); [4|8|27|29].
Dominant • *Rumex* cf. *dentatus*
Part of inflorescence (5.0 x 1.5 cm). Each valve distinctly toothed and with a well-developed tubercle.

No. 0406
Cat. 1/56 Plant fragments to be studied: jujube, lentil and leek (on the label together with the sample: barley and jujube); Thebes; G.J. Chester; donated by the Antiquity Department (1935); [4|9|27|29].
Dominant • *Ziziphus spina-christi*
4 complete fruits.
Contam. • *Triticum turgidum* ssp. *dicoccon*
Glume base.
• *Nigella sativa*
Many seeds.
• *Ambrosia maritima*
1 fruit.
• *Lens culinaris*
1 seed.
• *Cuminum cyminum*
Several fruits.
• *Phalaris*
Spikelet.
• *Lolium* cf. *temulentum*
Spikelet.
• cf. *Trigonella foenum-graecum*
• *Brassica*
• Leguminosae
Seed.

No. 0407
Cat. 1/56 Plant fragments to be studied: *Lychnis coeli-rosa* fruits; period uncertain; identified by V. Täckholm as *Eudianthe coeli-rosa* Endl. (28-01-1950); donated by the Antiquity Department (1935).
Dominant • *Silene coeli-rosa*
2 flowers and 9 fruits.

No. 0409
Cat. 1/56 Grain from sample at foundation of Montouhotep temple; Deir El Bahari (Thebes) (1922); XIe Dynasty; donated by the Antiquity Department (1935); [6|2|22|29].
Dominant • *Hordeum vulgare* ssp. *vulgare* (hulled)
Dozens of spikelets.

No. 0460
Cat. 1/63 Large quantity of cake disks of the remains of pressed olives (length: 58 cm); Karanis (Kôm Aushim); Roman period; identified by Dr. Ayadi (1949); probably used as fodder for cattle; excavated by the American excavations and donated by the Antiquity Department (1936).
Dominant • *Olea europaea*
Large pile of pressed olives (ca. 55-60 cm high).

No. 0939
Cat. 1/160 2 fruits of *Balanites aegyptiaca* (max. length: 2.5 cm) found at the crocodile gallery (northwest basement) of the step pyramid; Saqqara; IIIe Dynasty, reign of Zoser; donated by J-Ph. Lauer, architect of the Antiquity Service at Saqqara (16-11-1936).
Dominant • *Balanites aegyptiaca*
2 endocarps (1 gnawed).

No. 0940
Cat. 1/160 Fruits of *Ziziphus spina-christi* found in the cellar of the step pyramid (north-west basement); Saqqara (25-05-1933); IIIe Dynasty, reign of Zoser; donated by J-Ph. Lauer, architect of the Antiquity Service at Saqqara (16-11-1936).
Dominant • *Ziziphus spina-christi*
13 endocarps.

No. 1047
Cat. 1/173 Wheat with some barley and some fruits of flax (*Linum usitatissimum*); found in the West part of Karanis (Kôm Aushim) (02-1903); Graeco-Roman period; donated by the Antiquity

	Department (26-11-1936), Egyptian Museum.
Dominant	● *Triticum turgidum* ssp. *durum* No chaff present.
Contam.	● *Hordeum vulgare* ssp. *vulgare* (hulled) Few spikelets.

No.	1213
Cat. 2/2	Tiger-nuts (*Cyperus esculentus*); Badari (Upper Egypt); Prehistoric period; found by G. Brunton.
Dominant	● *Cyperus esculentus* Several hundreds of tubers.

No.	1214
Cat. 2/3	Tiger-nuts (*Cyperus esculentus*) found in a basket in the tomb of Ani; Gebelein (Upper Egypt); XIe or XIIe Dynasty (ca. 2000 BC); part given from a sample in the Egyptian Museum (hall No. 53, case O) (1936); [4	24	27	22].
Dominant	● *Cyperus esculentus* Ca. 100-200 tubers.
Contam.	● *Triticum turgidum* cf. ssp. *dicoccon* 1 grain kernel.

No.	1215
Cat. 2/3	Tiger-nuts (*Cyperus esculentus*); period uncertain; part given from a sample in the Egyptian Museum (hall No. 53, case P) (1933); [4	6	27	22].
Dominant	● *Cyperus esculentus* Large quantity of tubers.
Contam.	● *Hordeum vulgare* ssp. *vulgare* (hulled) Ca. 15-20 spikelets.

No.	1216
Cat. 2/3	Tiger-nuts (*Cyperus esculentus*); Gebelein; XIe or XIIe Dynasty (ca. 2000 BC); found by G.C.C. Maspero (1885); part given from a sample in the Egyptian Museum (1933).
Dominant	● *Cyperus esculentus* Dozens of tubers.
Contam.	● *Triticum turgidum* ssp. *dicoccon* Rachis fragments (glume base and spikelet fork). ● *Daucus carota* Fruit.

No.	1217
Cat. 2/3	Tiger-nuts (*Cyperus esculentus*); period uncertain; part given from a sample in the Egyptian Museum (hall No. 53, case P) (1933); [4	14 or 17	27	22].
Dominant	● *Cyperus esculentus* Large quantity of tubers.

Contam.	● *Hordeum vulgare* ssp. *vulgare* (hulled) 1 spikelet. ● cf. *Lathyrus sativus* Only part of the hilum is preserved.

No.	1218
Cat. 2/3	Tiger-nuts (*Cyperus esculentus*); Deir El Medineh (Luxor); New Kingdom, XVIIIe Dynasty (ca. 1400 BC); excavations of the French Institute of Oriental Archaeology; donated by the Egyptian Museum (1933).
Dominant	● *Cyperus esculentus* 47 tubers.

No.	1219
Cat. 2/3	Tiger-nuts (*Cyperus esculentus*); Saqqara; date uncertain, but probably Late period; donated by the Egyptian Museum (1933).
Dominant	● *Cyperus esculentus* Large quantity of tubers. Horizontal compartimentation visible in part of the specimens.

No.	1312
Cat. 2/17	*Cyperus longus* (galangal) found in a toilet box; Gebelein; Old Kingdom; excavations of the Italian Mission (1935-37); according to V. Täckholm: 5 fruits and part of the stem of *Cyperus articulatus* (not *rotundus*); donated by the Egyptian Museum (04-1937).
Dominant	● *Cyperus articulatus* 5 tubers. Identification by V. Täckholm.

No.	1313
Cat. 2/18	Inflorescence of Egyptian millet (*Sorghum vulgare*) (length: 0.13 cm); Gebelein; Old Kingdom?; excavations of the Italian Mission (season 1935-37), Journal d'entrée (JE) No. 66854; donated by the Egyptian Museum (04-1937).
Dominant	● *Sorghum* Inflorescence.

No.	1314
Cat. 2/18	Fragments of *Calotropis procera*; Gebelein; Old Kingdom; excavations of the Italian Mission (1935-37), Journal d'entrée (JE) No. 66854; identified by E.A.M. Greiss (1939); donated by the Egyptian Museum (04-1937).
Dominant	● ? Thin fruitscale, with fibrous tissue inside Probably not *Calotropis procera*.

No.	1316
Cat. 2/18	Leaves of Sycamore (*Ficus sycomorus*); Gebelein;

	Old Kingdom; excavations of the Italian Mission (1935-37), Journal d'entrée (JE) No. 66854; donated by the Egyptian Museum (04-1937).
Dominant	● *Ficus sycomorus* Ca. 5 leaves.
No. Cat. ?	1337 Seeds of *Mimusops laurifolia*.
Dominant	● *Mimusops laurifolia* 10 seeds.
No. Cat. ?	1342 -
Dominant	● *Hyphaene thebaica* 1 endocarp and 1 endocarp with remains of mesocarp and exocarp.
No. Cat. 2/26	1350 *Cyperus esculentus*; Deir El Medineh; XVIIIᵉ Dynasty; donated by the French Institute of Oriental Archaeology (1933).
Dominant	● *Cyperus esculentus* Ca. 60-80 tubers.
Comment	See figure 16.15 in Darby *et al.*, 1977.
No. Cat. 2/27	1355 A quantity of red gum; period uncertain; bought in Luxor by Mm. Mohamed Zulfikar Bey, ex. director of the museum, and donated to the museum (1937).
Dominant	● cf. *Pistacia* Fruit. ● Piece of resin
No. Cat. 2/27	1356 Remains of wheat spikelets (*T. diccocum*); uncertain provenance, probably Thebes; possibly Late or Graeco-Roman period; bought in Luxor by Mm. Mohamed Zulfikar Bey, ex. director of the museum, and donated to the museum (1937).
Dominant	● *Triticum turgidum* ssp. *dicoccon* Spikelets, most specimens empty as a result of insect damage.
No. Cat. 2/28	1357 Seeds of persea (*Mimusops schimperi*); uncertain provenance and date; bought in Luxor by Mm. Mohamed Zulfikar Bey, ex. director of the museum, and donated to the museum (1937).
Dominant	● *Mimusops laurifolia* 10 seeds.
No. Cat. 2/28	1358 Carbonised wheat grains; date uncertain; identified by H. Helbaek as *Triticum compacteum* (21-04-1955); bought in Luxor by Mm. Mohamed Zulfikar Bey, ex. director of the museum, and donated to the museum (1937).
Dominant	● *Triticum aestivum* Rather short grain kernels (charred).
Contam.	● *Hordeum vulgare* ssp. *vulgare* (hulled) ● Gramineae tribe Triticeae Culm nodes. ● Pieces of charcoal
No. Cat. 2/28	1359 Remains of flax-fibre (*Linum usitatissimum*); period uncertain; bought in Luxor by Mm. Mohamed Zulfikar Bey, ex. director of the museum, and donated to the museum (1937).
Dominant	● *Linum usitatissimum* Fragments of stem.
No. Cat. 2/28	1361 A paste of sant fruits and seeds of *Acacia nilotica*; period uncertain; probably used for tanning or for medical purposes; bought from Luxor by Mm. Mohamed Zulfikar Bey, ex. director of the museum, and donated to the museum (1937).
Dominant	● *Acacia* Pressed cake, in which several whole seeds of *Acacia* are present.
No. Cat. 2/29	1363 2 seeds of dates (max. length: 2 cm); bought in Luxor by Mm. Mohamed Zulfikar Bey, ex. director of the museum, and donated to the museum (1937).
Dominant	● *Phoenix dactylifera* 2 seeds.
No. Cat. 2/29	1365 Wheat grains; possibly Roman period; bought in Luxor by Mm. Mohamed Zulfikar Bey, ex. director of the museum, and donated to the museum (1937).
Dominant	● *Triticum turgidum* ssp. *durum* Grain kernels.
Contam.	● *Sorghum halepense* 1 spikelet. ● Gramineae tribe Triticeae Rachis fragment.
No. Cat. 2/29	1366 Head of nape (*Brassica napus*) (max. length:

	3.8 cm); period uncertain; bought in Luxor by Mm. Mohamed Zulfikar Bey, ex. director of the museum, and donated to the museum (1937).
Dominant	● ?
	Small bulblike structure. Differs from storage organs (taproot and lower part of the stem), typical for some members of the genus *Brassica*.
No. Cat. 2/30	1367 Seeds of *Acacia nilotica*; period uncertain; bought in Luxor by Mm. Mohamed Zulfikar Bey, ex. director of the museum, and donated to the museum (1937).
Dominant	● *Acacia nilotica* Several hundreds of seeds and a few fruit fragments.
No. Cat. 2/30	1368 Unincised *Ficus sycomorus*; bought in Luxor by Mm. Mohamed Zulfikar Bey, ex. director of the museum, and donated to the museum (1937).
Dominant	● *Ficus sycomorus* 9 accessory fruits (syconia) (18-38 x 22-43 mm).
Comment	See figure 18.25 in Darby *et al.*, 1977.
No. Cat. 2/30	1370 Masses of wine lees containing seeds of grape; period uncertain; bought in Luxor by Mm. Mohamed Zulfikar Bey, ex. director of the museum, and donated to the museum (1937).
Dominant	● *Vitis vinifera* Few pieces of pressed grapes (pericarp, seeds and pedicels).
No. Cat. 2/30	1372 2 dôm nuts (max. diameter: 5-6 cm); uncertain provenance and period; bought in Luxor by Mm. Mohamed Zulfikar Bey, ex. director of the museum, and donated to the museum (1937).
Dominant	● *Hyphaene thebaica* 1 endocarp and 1 endocarp with remains of mesocarp and exocarp.
No. Cat. 2/31	1373 Fragments including 2 dôm fruits (*Hyphaene thebaica*); Luxor; date uncertain but probably New Kingdom; bought in Luxor by Mm. Mohamed Zulfikar Bey, ex. director of the museum, and donated to the museum (1937).
Dominant	● *Hyphaene thebaica* 1 fragment with exocarp present (no cut mark); 1 fragment of endocarp (possible cut mark); 3 whole, immature fruits; 1 seed (2 parts); 1 fragment of pedicel.
No. Cat. 2/31	1374 Fragments of carbonised dôm (*Hyphaene thebaica*); period uncertain; bought in Luxor by Mm. Mohamed Zulfikar Bey, ex. director of the museum, and donated to the museum (1937).
Dominant	● ? Sun-burnt plant fragment. On no account *Hyphaene thebaica*. Possible part of a rhizome.
No. Cat. 2/31	1375 Seeds of thorn tree (*Balanites aegyptiaca*); period uncertain; bought in Luxor by Mm. Mohamed Zulfikar Bey, ex. director of the museum, and donated to the museum (1937).
Dominant	● *Balanites aegyptiaca* 14 endocarps, 4 with insect hole and 1 gnawed.
No. Cat. 2/36	1401 Leaves of Persea (*Mimusops laurifolia*) and 2 fruit offsprings; Deir El Medineh; XVIII[e] Dynasty; remains from a funerary bouquet; donated by the French Institute of Archaeology, with permission of the Antiquity Department (1933).
Dominant	● *Mimusops laurifolia* Unripe fruits and leaves. Probably from a bouquet.
No. Cat. 2/37	1402 Fruits of Persea (*Mimusops schimperi*); Saqqara; unknown period; donated by the Antiquity Department (04-07-1933).
Dominant	● *Mimusops laurifolia* 8 fruits, partly fragmented.
No. Cat. 2/37	1404 Vase of pottery (height : 8.2 cm; diameter: 12 cm) containing seeds of Persea (*Mimusops schimperi*); Luxor; New Kingdom; bought from an antique dealer in Luxor (26-06-1934).
Dominant	● *Mimusops laurifolia* Ca. 80-100 fruits, partly broken and seeds exposed.
No. Cat. 2/38	1407 Plate of pottery (diameter: 17 cm) full of dried

	grapes; Deir El Medineh; XVIIIe Dynasty; excavations of the French Institute of Oriental Archaeology (1933).
Dominant	● *Vitis vinifera* Hundreds of fruits, some of them with pedicel still attached, including some large fruits.
Comment	See figure 18.6 in Darby *et al.*, 1977.
No. Cat. 2/37	1411 Unincised fruits of *Ficus sycomorus*; Luxor; probably Graeco-Roman period; bought from an antique dealer in Luxor (15-02-1933).
Dominant	● *Ficus sycomorus* Dozens of accessory fruits (syconia) (biggest specimen: 47 x38 mm).
No. Cat. 2/38	1412 Incised fruits of *Ficus sycomorus*; Pyramid of Zoser at Saqqara; IIIe Dynasty; part of a considerable quantity kept in the Scientific Section; donated by the Antiquity Department (1934).
Dominant	● *Ficus sycomorus* Ca. 50 flattened accessory fruits (syconia), outside brown, inside blackish (obviously not charred).
No. Cat. 2/39	1414 4 figs (*Ficus carica*); Deir El Medineh; XVIIIe Dynasty; excavations of the French Institute of Oriental Archaeology (1933).
Dominant	● *Ficus carica* 4 accessory fruits (syconia).
No. Cat. 2/39	1416 2 fruits of *Hyphaene thebaica* (max. length: 6.8 cm); Deir El Medineh; XVIIIe Dynasty; excavations of the French Institute of Oriental Archaeology (1933).
Dominant	● *Hyphaene thebaica* 2 whole fruits.
No. Cat. 2/39	1417 Fruits of dôm (*Hyphaene thebaica*), a piece of bread and some incense, all enveloped in a fragment of linen cloth; Deir El Medineh; XVIIIe Dynasty; excavations of the French Institute of Oriental Archaeology (1933).
Dominant	● *Hyphaene thebaica* Whole fruit.
Contam.	● Piece of bread in textile bag
No. Cat. 2/40	1421 Small fruits of Pomegranate in a pottery plate (diameter: 18 cm); Deir El Medineh; XVIIIe Dynasty; excavations of the French Institute of Oriental Archaeology (1933).
Dominant	● *Punica granatum* 10 whole fruits (38-56 x 26-48 mm).
No. Cat. 2/40	1423 Fruits of *Juniperus* and *Ziziphus* in addition to resin; uncertain provenance and period; bought from an antique dealer in Luxor (M. Tadros).
Dominant	● *Ziziphus spina-christi* Endocarp. ● *Juniperus* cf. *oxycedrus* Fruits. ● Resin
No. Cat. 2/41	1425 Olive seeds (*Olea europaea*); Karanis (Kôm Aushim); Graeco-Roman period; the University of Michigan's excavations; [29 | GB 172 | A-D30 | G123bZ | A].
Dominant	● *Olea europaea* Hundreds of endocarps, some specimens gnawed.
No. Cat. 2/41	1426 Fruits and kernels of *Balanites aegyptiaca* in a pottery plate (diameter: 10.9 cm); Deir El Medineh; XVIIIe Dynasty; excavations of the French Institute of Oriental Archaeology (1933).
Dominant	● *Balanites aegyptiaca* 28 endocarps, 7 endocarps with remains of mesocarp and exocarp and 1 endocarp with small hole of uncertain origin.
No. Cat. 2/41	1427 Pottery dish (diameter: 24 cm) containing wheat grains (*Triticum dicoccum*); Deir El Medineh; XVIIIe Dynasty; excavations of the French Institute of Oriental Archaeology (1933).
Dominant	● *Triticum turgidum* ssp. *dicoccon* Thousands of spikelets, upper spikelets with twisted internode not found, but spikelets rather large and 2-seeded. Few lower rachis fragments are non-brittle and a few fragments disarticulated in a similar way as in spelt wheat.
No. Cat. 2/41	1429 Ears of wheat; Karanis (Kôm Aushim); Roman period; the University of Michigan's excavations; [29 | C 132 V' | A].

Dominant	• *Triticum turgidum* ssp. *durum* Complete ears. 4 specimens with hairy glumes and 6 specimens with smooth glumes.	No. Cat. 2/42	1433 Broad bean seeds (*Vicia faba*); Karanis (Kôm Aushim); Roman period; the University of Michigan's excavations.
No. Cat. 2/41	1430 Fragments of ears of wheat (*Triticum dicoccum*), part of sample ; storehouses north of the Step Pyramid of Zoser at Saqqara; IIIe Dynasty; donated by J-Ph. Lauer (1933).	Dominant Comment	• *Vicia faba* ssp. *minor* 17 seeds (0.8-0.9 mm long). See figure 17.10 in Darby *et al.*, 1977.
Dominant	• *Triticum turgidum* ssp. *dicoccon* Several hundreds of ears, top spikelets missing, however.	No. Cat. 2/42	1434 Pod of Chick pea (*Cicer arietinum*) (length: 2.1 cm); uncertain provenance and period; bought from an antique dealer in Luxor (1934).
Contam.	• *Hordeum vulgare* ssp. *vulgare* (hulled) Rachis fragments and spikelets.	Dominant	• *Cicer arietinum* 1 fruit with 2 seeds inside.
Comment	See figure 11.13 in Darby *et al.*, 1977 (presented as *Triticum*).	No. Cat. 2/42	1436 Seeds of *Lupinus termis*, part of a quantity in a Roman vase; Karanis (Kôm Aushim); Roman period; bought from an antique dealer in Luxor (Moharrib) (26-06-1934).
No. Cat. 2/41	1431 Carbonised wheat grains (*Triticum dicoccum*); Merimde (west of Bani Salama) (1933); Prehistoric period.	Dominant	• *Lupinus alba* 156 seeds (8-12 mm).
Dominant	• *Triticum turgidum* ssp. *dicoccon* Hundreds of charred spikelets, a few loose grain kernels.	No. Cat. 2/43	1437 Safflower (*Carthamus tinctorius* var. *inermis* Schweinf.); Saqqara; Late period (ca. 7th century BC); from J.E. Quibell (1898) and donated by the Antiquity Department (04-07-1933).
No. Cat. 2/42	1432 Pottery plate containing seeds of Chick pea, mixed with other seeds; Deir El Medineh, XVIIIe Dynasty; information added by H. Helbaek (23-06-1955): *Lathyrus sativus*, *Triticum dicoccum*, *Hordeum tetrastichum*, *Coriandrum sativum*, *Vitis vinifera* and *Carum carvi*; excavations of the French Institute of Oriental Archaeology (1933).	Dominant	• *Carthamus tinctorius* Ca. 20 inflorescences with ripe fruits and partly with stem remains.
		No. Cat. 2/43	1438 Capsules of flax (*Linum usitatissimum*); Thebes; New Kingdom; donated by the Antiquity Department (04-07-1933); another quantity of flax kept in a bronze plate is still present in the Egyptian Museum (hall 53, case P); [4	1	27	22].
Dominant	• *Lathyrus sativus* Hundreds of seeds, damage by insects.	Dominant	• *Linum usitatissimum* Hundreds of whole fruits, also with upper part of culm.
Contam.	• *Coriandrum sativum* Fruit. • *Vitis vinifera* Fruit. • *Cuminum cyminum* Fruit. • *Triticum turgidum* ssp. *dicoccon* Several spikelets. • *Hordeum vulgare* (hulled) Charred spikelet. • *Lathyrus hirsutus* Seed. • *Citrullus colocynthis* Half seed. • *Echium rauwolfii* Fruit.	Contam.	• *Hordeum vulgare* (hulled) 1 spikelet. • Cruciferae Unknown fruit.
		No. Cat. 2/43	1439 Watermelon seeds; Deir El Medineh; New Kingdom; excavations of the French Institute of Oriental Archaeology (1933).
		Dominant	• *Citrullus lanatus* Ca. 50 seeds.
		Contam.	• *Hordeum vulgare* ssp. *vulgare* (hulled) Spikelet.

No.	1440
Cat. 2/43	Fragments of curved serpent Cucumber (*Cucumis flexuosus* L.) in a pottery plate (max. diameter: 17 cm); Deir El Medineh; New Kingdom; excavations of the French Institute of Oriental Archaeology (1933).
Dominant	● *Cucumis sativus* Fruit fragment with seeds in rows (morphology of the seeds has been checked).
No.	1441
Cat. 2/43	Pottery dish (diameter: 11.7 cm) containing watermelon seeds; Deir El Medineh; New Kingdom; excavations of the French Institute of Oriental Archaeology (1933).
Dominant	● *Citrullus lanatus* Ca. 100 seeds.
Contam.	● dung particle of sheep/goat
No.	1442
Cat. 2/44	Onions (*Allium cepa*); El Rizeikat near Luxor; period uncertain; bought from Said Mulatham in Luxor (15-02-1933).
Dominant	● *Allium cepa* Dozens of bulbs (ca. 50 x 20 mm).
No.	1444
Cat. 2/44	Garlic specimen (*Allium sativum*); Karanis (Kôm Aushim); Roman period; the University of Michigan's excavations (1933); [29|C 137A|VII].
Dominant	● *Allium sativum* Ca. 5 specimens, partly disintegrated. Whole specimens 3.7 cm in diameter.
No.	1445
Cat. 2/44	Garlic stem tied with halfa; found in El Rizeikat near Luxor; period uncertain; bought from Said Mulatham in Luxor (15-02-1933).
Dominant	● *Allium sativum* 4 bundles, including 3 bulb bases, tied together with a rope. ● ? Bulb-like specimen, ca. 6.5 cm long and 1.5 cm width, with both long and short stalk. Looks similar to unknown fruit of No. 4201.
No.	1541
Cat. 2/60	*Cordia myxa* seeds; period uncertain; bought in Luxor (1935).
Dominant	● *Cordia myxa* 26 endocarps.
Contam.	● *Olea europaea* 1.5 endocarp, badly preserved.
Comment	See figure 18.3 in Darby *et al.*, 1977.
No.	1542
Cat. 2/60	*Cordia myxa* seed; uncertain provenance and period; bought from an antique dealer in Luxor (1935).
Dominant	● *Cordia myxa* Well-preserved complete endocarp (recent?).
No.	1543
Cat. 2/61	Almond fruit (height: 1.8 cm), pierced at the conic end; Deir El Medineh; possibly used for whistling by children; donated by P.H.H. Bovier-Lapierre from the French Institute of Oriental Archaeology (1933).
Dominant	● ? Unknown pierced fruit.
No.	1544
Cat. 2/61	One Cordia seed; uncertain provenance and period; bought from an antique dealer in Luxor (1932).
Dominant	● *Cordia myxa* 1 endocarp.
No.	1545
Cat. 2/61	*Cordia myxa* seeds found with wheat and barley; Deir El Medineh; XVIIIe Dynasty; excavated by the French Institute of Oriental Archaeology and donated to the museum (1933).
Dominant	● *Cordia myxa* Endocarp and part of mesocarp. Rather angular.
No.	1546
Cat. 2/61	Seeds of *Cordia myxa*; Saqqara; period uncertain; excavated by J.E. Quibell and donated to the museum (1933).
Dominant	● *Cordia myxa* 33 endocarps, badly preserved.
No.	1576
Cat. 2/70	Small pottery dish (diameter: 12 cm; height: 5 cm) containing: fine linen threads, a few fruits of *Ficus sycomorus*, a few seeds of *Acacia nilotica*, one almond fruit, one rhizome of *Cyperus rotundus*; Hermopolis Tuna El Gebel, Saitic period; excavated by Samy Gabra from the Egyptian University and donated by the Faculty of Arts.
Dominant	● *Ficus sycomorus*

	8 small accessory fruits (syconia) and 2 leaves.
Contam.	• *Amygdalus communis* 1 endocarp. • *Acacia nilotica* Fruit fragment. • cf. *Cyperus* Tuber.

No.	1966
Cat. 2/108	Seeds of dates; uncertain provenance (probably Luxor); Coptic period; bought from an antique dealer in Luxor (1933).
Dominant	• *Phoenix dactylifera* Ca. 50 seeds.

No.	1968
Cat. 2/108	Grains and straw of barley found in box; Karanis (Kôm Aushim); Roman Period; donated to the museum (1935); [28	B 136 B	C (box)].
Dominant	• *Hordeum vulgare* ssp. *vulgare* (hulled) Spikelets and a few rachis fragments.
Contam.	• *Triticum turgidum* ssp. *durum* Rachis fragment. • Gramineae tribe Triticeae Few culm nodes. • *Galium aparine* Fruit. • *Beta vulgaris* 2-germed fruit cluster. • *Convolvulus arvensis* Seed. • *Sinapis* cf. *arvensis* • *Raphanus raphanistrum* Fruit fragment. • *Malva nicaeensis/sylvestris* Fruit with reticulate dorsal surface. • Charcoal pieces • Bone fragment

No.	2054
Cat. 2/117	Carbonised wheat; Karanis (Kôm Aushim); Roman period; donated by the Antiquity Department (1935).
Dominant	• *Triticum aestivum* Rather short grain kernels (charred).
Contam.	• *Triticum turgidum* ssp. *durum* Few grain kernels and rachis fragments. • *Hordeum vulgare* ssp. *vulgare* (hulled) Spikelets and rachis fragments. • Charcoal fragments

No.	2055
Cat. 2/118	Barley; Karanis (Kôm Aushim); Roman period; donated by the Antiquity Department (1935); [29	C 137 A'	X].
Dominant	• *Hordeum vulgare* ssp. *vulgare* (hulled) Spikelets, most of them empty due to insect damage, and a few rachis fragments.
Contam.	• *Tamarix* Stem fragment. • ? Unknown plant remains. Possibly unripe fruit of a member of the Fabaceae.

No.	2056
Cat. 2/118	Grains and straw of barley and wheat; Karanis (Kôm Aushim); Roman period; donated by the Antiquity Department (1935); [27	CS 60	MII].
Dominant	• *Triticum turgidum* ssp. *durum* Grain kernels, badly affected by insects, and a few rachis fragments. • *Hordeum vulgare* ssp. *vulgare* (hulled) Spikelets (ca. 10-15%).
Contam.	• *Sinapis* cf. *arvensis* • *Raphanus raphanistrum* Seed. • *Convolvulus arvensis* Seed. • *Galium aparine* Fruit. • Gramineae tribe Triticeae Culm node (charred).

No.	2057
Cat. 2/118	Grains and straw of barley; Karanis (Kôm Aushim); Roman period; donated by the Antiquity Department (1935); [28	237★	P].
Dominant	• *Hordeum vulgare* ssp. *vulgare* (hulled) Spikelets and a few rachis fragments. Few rachis fragments are infected with the fungus covered smut (*Ustilago hordei* (Pers.) Lagerh.).
Contam.	• *Zygophyllum* cf. *coccineum* Fruit. • *Convolvulus arvensis* Seed. • *Galium aparine* Fruit. • Charcoal pieces

No.	2103
Cat. 2/122	Bunch of *Hyphaene thebaica* (length: 11 cm); uncertain provenance (possibly Thebes); bought from an antique dealer in Luxor (1933).
Dominant	• *Hyphaene thebaica* Part of fruiting stalk (103 x 18 mm).

No.	2104		No.	2252
Cat. 2/122	Fragment of burnt *Hyphaene thebaica* (length: 11 cm); uncertain provenance (possibly Thebes); bought from an antique dealer in Luxor (1933).		Cat. 2/133	Fragments of bottle-gourd (*Lagenaria vulgaris*); period uncertain; bought from an antique dealer in Luxor (1934).
Dominant	● *Hyphaene thebaica* Part of fruiting stalk (52 x 19 mm).		Dominant	● *Citrullus lanatus* Large part of fruit, some seeds have fallen out.

No. 2104
Cat. 2/122 Fragment of burnt *Hyphaene thebaica* (length: 11 cm); uncertain provenance (possibly Thebes); bought from an antique dealer in Luxor (1933).
Dominant ● *Hyphaene thebaica*
Part of fruiting stalk (52 x 19 mm).

No. 2105
Cat. 2/122 Fragment of *Hyphaene thebaica* (length: 8 cm); Elephantine; probably Roman or Coptic period; excavations of P.H.H. Bovier-Lapierre from the French Institute of Oriental Archaeology (1933).
Dominant ● *Hyphaene thebaica*
Part of fruiting stalk, slightly curved.

No. 2106
Cat. 2/122 Fragments of dôm-kernels; Saqqara; late Period; excavated by J.E. Quibell at Saqqara and donated by the Egyptian Museum (1935).
Dominant ● *Hyphaene thebaica*
2 seeds, fragmented.

No. 2249
Cat. 2/133 Tiger nut seeds (*Cyperus esculentus*); Meir; excavated by Ahmed Bey Kamal (09-1910) and donated by the Egyptian Museum (11-1937); [4|5|27|29].
Dominant ● *Cyperus esculentus*
Ca. 100 tubers.

No. 2250
Cat. 2/133 *Cordia myxa*; period uncertain; bought from an antique dealer in Luxor (Molatham) (1934).
Dominant ● *Cordia myxa*
63 whole fruits, some specimens without exocarp.
Contam. ● *Ficus* cf. *sycomorus*
1 whole accessory fruit (syconium) and some fragments.

No. 2251
Cat. 2/133 Moderate quantity of olive seeds; period uncertain; bought from an antique dealer in Luxor (1934).
Dominant ● *Olea europaea*
Hundreds of endocarps.
Contam. ● *Cicer arietinum*
1 seed
● *Vicia ervilia*
Several seeds.
● Unknown Leguminosae
1 seed (diameter: 3.7 mm; glossy, dark brown).

No. 2252
Cat. 2/133 Fragments of bottle-gourd (*Lagenaria vulgaris*); period uncertain; bought from an antique dealer in Luxor (1934).
Dominant ● *Citrullus lanatus*
Large part of fruit, some seeds have fallen out.

No. 2273
Cat. 2/135 A pottery dish (height: 29 cm) containing black-mustard seeds (*Brassica nigra*), mixed with Medick seeds (probably *Medicago sativa*); period uncertain (Roman period?); Hassan Khalifa, former director of the Agricultural Museum, identified the sample as *Brassica nigra*; bought in Luxor (14-05-1935).
Dominant ● *Brassica nigra*
Huge quantity of seeds and fruits. Valves (ca. 11 mm long) with vague vein pattern, beak without seeds.
Contam. ● *Trifolium*
Fruits (upper part leathery, lower part membranous) and seeds.
● cf. *Vaccaria hispanica*
● *Triticum*
Rachis fragment.

No. 2662
Cat. 2/188 Dish of polished red pottery (diameter: 13 cm) containing Persea (*Mimusops schimperi*) seeds; Deir El Medineh; New Kingdom; donated by the French Institute of Oriental Archaeology (1933)
Dominant ● *Mimusops laurifolia*
Ca. 50 seeds.

No. 2663
Cat. 2/188 Fruits of *Mimusops schimperi*; Deir El Medineh; New Kingdom; donated by the French Institute of Oriental Archaeology (1933).
Dominant ● *Mimusops laurifolia*
11 whole fruits (33-19 x 22-14 mm).

No. 2664
Cat. 2/188 Persea seeds; Deir El Medineh; New Kingdom; XVIIIe Dynasty; donated by the French Institute of Oriental Archaeology (1933).
Dominant ● *Mimusops laurifolia*
1 fruit and 80 seeds.

No. 2665
Cat. 2/188 Persea fruits; Saqqara; period uncertain; donated by the Antiquity Department (1934).
Dominant ● *Mimusops laurifolia*
11 seeds.

No. 2666
Cat. 2/188 Persea seeds; period uncertain; bought in Luxor (1933/34).
Dominant • *Mimusops laurifolia*
50-100 seeds.

No. 2667
Cat. 2/188 One fruit and one seed of persea (*Mimusops schimperi*); period uncertain; bought in Luxor (1933/34).
Dominant • *Mimusops laurifolia*
2 fruits (29 x 19 and 23 x 14 mm) and 32 seeds.
Contam. • *Hordeum vulgare* ssp. *vulgare* (hulled)
3 spikelets.

No. 2686
Cat. 2/191 Small onion bulb (*Allium cepa*) with short leaves; Deir El Medineh; New Kingdom; donated by B.C.M.J. Bruyère from the French Institute of Oriental Archaeology (1932).
Dominant • cf. *Allium cepa*
Small bulb (8 x 30 mm).

No. 2687
Cat. 2/192 Garlic leaves (*Allium sativum*) in 4 small bundles tied with halfa and linen strings (label: bundles of onion and part of the outer skin (scale) [*Allium cepa*]); bought in Luxor (27-02-1933).
Dominant • *Allium cepa*
Bulbs are tied together with ropes made of cf. *Phoenix dactylifera*.

No. 2688
Cat. 2/192 Dish of red polished pottery (diameter: 13 cm) containing unincised sycamore fruits (*Ficus sycomorus*); Deir El Medineh; New Kingdom; donated by the French Institute of Oriental Archaeology, excavation No. J.2 (1933).
Dominant • *Ficus sycomorus*
29 accessory fruits (syconia) (12-28 x 15-38 mm), some with artificial incisions.

No. 2689
Cat. 2/192 Small plate of red pottery (diameter: 10.8 cm) containing incised sycamore fruits (*Ficus sycomorus*); Deir El Medineh; New Kingdom; donated by the French Institute of Oriental Archaeology, excavation No. J.4 (1933).
Dominant • *Ficus sycomorus*
15 accessory fruits (syconia), some with artificial incisions.

No. 2693
Cat. 2/193 Safflower seeds (*Carthamus tinctorius*); Karanis (Kôm Aushim), Roman period; donated by the Antiquity Department (1935); [29|B 191 K|X].
Dominant • *Carthamus tinctorius*
Hundreds of fruits, most of them split into halves.
Contam. • *Triticum turgidum* ssp. *durum*
Rachis fragment.
• *Hordeum vulgare* (hulled)
Rachis fragment and chaff (lemma and palea).
• *Olea europaea*
Small fragment of endocarp.

No. 2694
Cat. 2/193 Safflower seeds (*Carthamus tinctorius*); Karanis (Kôm Aushim); Roman period; donated by the Antiquity Department (1935) [28||132★|K] and [26].
Dominant • *Carthamus tinctorius*
Hundreds of complete fruits. Top of fruit is partly ribbed.
Contam. • Gramineae tribe Triticeae
Small amount of threshing remains (lemma and palea).
• *Trifolium*
2 seeds.

No. 2695
Cat. 2/193 Carthamus seeds bearing flower hairs, found in a Roman vase containing a large quantity of the same seeds; Roman period; bought in Luxor (1933).
Dominant • *Carthamus tinctorius*
Hundreds of complete fruits with pappus still present.

No. 2696
Cat. 2/193 Carthamus flowers; Karanis (Kôm Aushim) (label: also called Tebtynis); Roman period; donated by the Antiquity Department (1935); [29|C 132V'|A].
Dominant • *Carthamus tinctorius*
Part of inflorescence and some fruits.
Contam. • Gramineae
Culm fragment, possibly of cereal.

No. 2697
Cat. 2/193 2 Carthamus flowers (*Carthamus tinctorius*); Tebtynis (Fayum); Roman period; Italian Excavation and donated by the Antiquity Department (1935); [29|C 132V'|A].

Dominant	• *Carthamus tinctorius* Part of inflorescence, no fruits could be observed.
No. Cat. 2/193	2698 Castor-oil seeds (*Ricinus communis*); Roman period; bought in Luxor (1932).
Dominant	• *Ricinus communis* Ca. 100 seeds, damaged by animals.
No. Cat. 2/193	2699 *Lupinus termis* seeds; Karanis (Kôm Aushim); Roman period; donated by the Antiquity Department (1935); [28	C 136★	P].
Dominant	• *Lupinus alba* 44 seeds (8-10 mm).
No. Cat. 2/193	2700 Lettuce seeds (*Lactuca sativa*) found in a Roman pottery vase; Roman period; bought in Luxor (1933).
Dominant	• Cruciferae Thousands of small round seeds (1 mm in diameter). Single valve present with vague vein pattern and small pedicel (same shape as *Brassica* and *Sinapis*).
Comment	See figure 17.9 in Darby *et al.*, 1977 (erroneously presented as *Lactuca sativa*).
No. Cat. 3/6	2785 3 Fruits of *Ficus sycomorus* and 1 seed of *Balanites aegyptiaca*; Deir El Medineh; New Kingdom; [Journal d'entrée (JE) No. 68821]; donated by the Antiquity Department, Egyptian Museum (05-04-1938).
Dominant	• *Ficus sycomorus* 3 whole accessory fruits (syconia).
No. Cat. 3/6	2786 One seed of *Balanites aegyptiaca* (diameter: 3.7 mm); Deir El Medineh; New Kingdom; excavations of the French Institute of Oriental Archaeology and donated by the Antiquity Department (05-04-1938).
Dominant	• *Balanites aegyptiaca* 1 endocarp.
No. Cat. 3/7	2791 Few fruits of tiger-nuts (*Cyperus esculentus*); El Sheik Ibada (Antinoöpolis); Coptic period; excavated by the Florence University expedition and donated by the Egyptian Museum (05-04-1938); [6	4	22	29].
Dominant	• *Cyperus esculentus* Ca. 35 tubers.
Contam.	• *Lupinus digitatus* 2 Seeds, top of hilum clearly exposed.
No. Cat. 3/7	2792 Few fruits of thorn-christ (*Ziziphus spina-christi*); El Sheik Ibada (Antinoöpolis); Coptic period; excavated by the Florence University expedition and donated by the Egyptian Museum (05-04-1938); [6	3	22	29].
Dominant	• *Ziziphus spina-christi* 5 complete fruits and 2 endocarps.
Contam.	• *Ficus* cf. *sycomorus* Accessory fruit (syconium), with insect holes (16 x 19 mm).
No. Cat. 3/72	3205 Small quantity of rush-nuts (*Cyperus esculentus*); tomb of Ra-hotep at Gebelein; XIIe Dynasty; Meir (1910); donated by the Egyptian Museum (1937/38); [4	5	27	29].
Dominant	• *Cyperus esculentus* Ca. 30 tubers.
No. Cat. 3/72	3206 Rush-nut (*Cyperus esculentus*); Gebelein tomb; G.C.C. Maspero (1885); donated by the Egyptian Museum (1937/38); [4	37	27	22].
Dominant	• *Cyperus esculentus* Several hundreds of tubers.
Contam.	• *Triticum turgidum* ssp. *dicoccon* Rachis fragments (glume base and spikelet fork). • *Daucus carota* Fruits. • cf. *Vicia* Seed (diameter: 2.8 mm, hilum ca. two-third of the diameter). • ? Unknown incomplete fruit (at least 4 mm long).
No. Cat. 3/72	3207 Fruits of thorn-christ; Saqqara; probably Late period; donated by the Antiquity Department (No. 2322, Tank F) (1933).
Dominant	• *Ziziphus spina-christi* Ca. 100 fruits, probably all complete when secured.
No. Cat. 3/72	3208 Fruits and kernels of *Ziziphus spina-christi*; Medinet Habu; New Kingdom; donated by Antiquity Department (No. Eye 33b) (1933).

Dominant ● *Ziziphus spina-christi*
12 endocarps, some fragmented.

No. 3209
Cat. 3/72 Fruits of *Ziziphus spina-christi*; Medinet Habu; New Kingdom; donated by the Antiquity Department (No. Eye 55e) (1933).
Dominant ● *Ziziphus spina-christi*
9 mesocarps.

No. 3210
Cat. 3/72 Kernels of thorn-christ, some are incised and others split into sections; Saqqara; Late period; donated by the Antiquity Department (exc. No. 2459) (1933).
Dominant ● *Ziziphus spina-christi*
Ca. 50 endocarps.
Contam. ● *Olea europaea*
1 endocarp.

No. 3211
Cat. 3/73 Fruit and a kernel of thorn-christ; Saqqara; Late period; donated by the Antiquity Department (1933).
Dominant ● *Ziziphus spina-christi*
1 complete fruit and 1.5 endocarp.

No. 3212
Cat. 3/73 Bored kernel of thorn-christ; Elephantine; Late period; P.H.H. Bovier-Lapierre; donated by the Antiquity Department (1933).
Dominant ● *Ziziphus spina-christi*
8 endocarps, all gnawed.
Contam. ● *Olea europaea*
1 endocarp, gnawed.

No. 3214
Cat. 3/73 Dish of red pottery containing rush-nuts; Gebelein; XIe Dynasty; donated by the Egyptian Museum (1937/38); [4|37|27|22 (pottery)].
Dominant ● *Cyperus esculentus*
Hundreds of tubers.
Contam. ● *Hordeum vulgare* (hulled)
Spikelets and chaff.
● *Triticum turgidum* ssp. *dicoccon*
Spikelets and rachis nodes (spikelet forks).
● cf. *Daucus carota*
Fruit.
● *Linum usitatissimum*
Fruit fragments.
● *Lolium temulentum*
Spikelets.

No. 3215
Cat. 3/73 Dish of red pottery containing rush-nuts; tomb of Ani at Gebelein; XIe Dynasty; donated by the Egyptian Museum.
Dominant ● *Cyperus esculentus*
Ca. 200-300 tubers in pottery.
Contam. ● *Vitis vinifera*
Leaf fragments.
● *Linum usitatissimum*
Fruit fragment.
● *Hordeum vulgare* (hulled)
1.5 spikelet.

No. 3216
Cat. 3/74 Fruits of *Balanites aegyptiaca*; Deir El Medineh (Luxor); XVIIIe Dynasty (1400 BC); donated by the French Institute of Oriental Archaeology (1933/34).
Dominant ● *Balanites aegyptiaca*
28 fruits, most of them still whole specimens. Not infected.

No. 3217
Cat. 3/74 Kernels of thorn tree *(Balanites aegyptiaca)*, most of which are bored; period uncertain; donated by L. Keimer (1935).
Dominant ● *Balanites aegyptiaca*
4 complete fruits (dark coloured) and 8 endocarps, all gnawed and light coloured. Most probably from different archaeological contexts.

No. 3218
Cat. 3/74 Bored kernels of *Balanites aegyptiaca*; probably Thebes; period uncertain; on separate paper: 'Deir El Medineh (Luxor); from 3300 years'; bought in Luxor (1934).
Dominant ● *Balanites aegyptiaca*
Large quantity of endocarps. Only subsample checked: all specimens are gnawed.

No. 3219
Cat. 3/74 Juniperus fruits; Deir El Medineh (Luxor); New Kingdom; donated by the French Institute of Oriental Archaeology (1933/34).
Dominant ● *Juniperus*
24 fruits (9-15 mm long).

No. 3220
Cat. 3/74 Fruits of *Balanites aegyptiaca*; Deir El Medineh (Luxor); New Kingdom, XVIIIe Dynasty (1400 BC); donated by the French Institute of Oriental Archaeology (1933/34).

Dominant
- *Balanites aegyptiaca*
  5 complete fruits, not infected.

No. 3222
Cat. 3/75 Incised sycamore fruits; period uncertain; bought in Luxor (1935).
Dominant
- *Ficus sycomorus*
  Ca. 30 whole accessory fruits (syconia), some specimens incised.

No. 3223
Cat. 3/75 Red pottery dish (diameter: 13.5 cm) containing sycamore fruits; probably New Kingdom or late; bought in Luxor (1932).
Dominant
- *Ficus sycomorus*
  Ca. 50 immature accessory fruits (syconia).

No. 3224
Cat. 3/75 Red pottery dish (diameter: 12.5 cm) containing sycamore fruits; period uncertain, probably late; bought in Luxor (1934).
Dominant
- *Ficus* cf. *sycomorus*
  Accessory fruits (syconium): 1 whole specimen, 1 immature specimen and 1 fragment.

No. 3225
Cat. 3/75 Incised sycamore fruits; period uncertain; bought in Luxor (1934).
Dominant
- *Ficus sycomorus*
  Ca. 200 whole, small accessory fruits (syconia), long-stalked.

No. 3226
Cat. 3/75 Unincised sycamore fruits; period uncertain; bought in Luxor (1934).
Dominant
- *Ficus sycomorus*
  Ca. 100 accessory fruits (syconia).

No. 3227
Cat. 3/75 Small basket (margounah, diameter: 12 cm, height: 8 cm) of halfa and date palm leaves containing a few leaves and twigs of *Ficus sycomorus*; Deir El Medineh; XVIII-XIXe Dynasty; donated by the French Institute of Oriental Archaeology.
Dominant
- *Vitis vinifera*
  3 fruits.
Contam.
- *Ziziphus spina-christi*
  Whole specimen (exocarp removed for identification).

No. 3228
Cat. 3/76 Sycamore leaves (*Ficus sycomorus*); Assiut; period uncertain; probably excavated by J.E. Quibell from the French Institute of Oriental Archaeology and donated by the Egyptian Museum (1934); Quibell in a letter to G.C.C. Maspero: 'botanical specimens from Assiut, shall I send them to Schweinfurth. He has left to Berlin some days ago. Please keep them until next year. They are too brittle to travel safely. Quibell'.
Dominant
- *Ficus sycomorus*
  Leaves.

No. 3229
Cat. 3/76 Unincised sycamore fruits; Deir El Medineh; XVIII-XIXe Dynasty; donated by the French Institute of Oriental Archaeology.
Dominant
- *Ficus sycomorus*
  Ca. 50 accessory fruits (syconia).

No. 3235
Cat. 3/77 Persea kernels (*Mimusops laurifolia*); donated by the Egyptian Museum (1937-38); [4|52|27|22].
Dominant
- *Mimusops laurifolia*
  10 seeds.

No. 3265
Cat. 3/82 Red pottery vase (diameter: 8.3 cm) containing raisins; Deir El Medineh; New Kingdom; donated by the French Institute of Oriental Archaeology.
Dominant
- *Vitis vinifera*
  Ca. 100 fruits.
Contam.
- *Lathyrus sativus*
  Dozens of seeds.
- *Triticum turgidum* ssp. *dicoccon*
  Spikelets.
- *Phoenix dactylifera*
  Fruit.

No. 3267
Cat. 3/82 Pottery plate and linen wrappers containing incense and fruits of raisins, dates, *Balanites*, sycamore, figs and dôm; Deir El Medineh (Luxor); New Kingdom, XVIII-XIXe Dynasty; donated by the French Institute of Oriental Archaeology (1933).
Dominant Large box containing a pottery dish and several linen wrappings. Plant parts most probably originate from the pottery dish and the linen wrappings, some of which are still connected to each other.
Content of linen wrappings (identified through openings in the wrappings):

- *Ficus sycomorus*
  Accessory fruits (syconia).
- *Balanites aegyptiaca*
  Fruit.
- *Hyphaene thebaica*
  Fruit.
- *Ficus carica*
  Accessory fruit (syconium)
- Small stone-like particles (resin?)
- ?

Content of pottery disk:
- *Balanites aegyptiaca*
  3 whole fruits.
- *Phoenix dactylifera*
  3 whole fruits.
- *Vitis vinifera*
  6 fruits.
- *Hordeum vulgare* (hulled)
  Few spikelets.
- *Triticum turgidum* ssp. *dicoccon*
  Few spikelets.
- Gramineae tribe Triticeae
  String of beads made from folded culm fragments.
- *Lolium temulentum*
  Few spikelets.
- Prepared food

Plant remains out of context:
- *Balanites aegyptiaca*
  6 whole fruits.
- *Phoenix dactylifera*
  6 whole fruits.
- *Ficus sycomorus*
  2 accessory fruits (syconia).
- *Ficus carica*
  2 accessory fruits (syconia).
- *Vitis vinifera*
  2 fruits.
- Black resin
  2 pieces.

No.	3268
Cat. 3/82	Dried raisins (*Vitis vinifera*); Deir El Medineh; period uncertain (label: New Kingdom); donated by the Egyptian Museum (1933).
Dominant	• *Vitis vinifera* Ca. 30 desiccated fruits and 8 charred.

No.	3269
Cat. 3/82	Basket (margounah) of halfa and date palm leaves containing raisins in a red polished dish, husks of barley in some small pottery and mud vases (probably remnants of beer), natron in two small pottery vases; Deir El Medineh; XVIII-XIXe Dynasty; donated by the French Institute of Oriental Archaeology (17-12-1932).
Dominant	• *Vitis vinifera* Hundreds of fruits, some of them with pedicel still attached.

No.	3270
Cat. 3/82	2 dôm fruits; tomb of Tut-Ankh-Amun at Thebes; New Kingdom; donated by the Egyptian Museum (1937).
Dominant	• *Hyphaene thebaica* 2 whole fruits.
Comment	Possibly depicted by Darby et al., 1977 in figure 18.17 (see also No. 3375).

No.	3271
Cat. 3/82	Dôm fruits and kernels; Deir El Medineh; New Kingdom; donated by B.C.M.J. Bruyère from the French Institute of Oriental Archaeology (1933/34).
Dominant	• *Hyphaene thebaica* 11 whole fruits (3 immature), 4 exocarps (1 specimen cut) and 4 seeds.

No.	3272
Cat. 3/83	In catalogue: 1 dôm fruit and 5 kernels (diameter: 5-7 cm); Karanis (Kôm Aushim); Roman period. On label: fruits of *Hyphaene thebaica* taken from different tombs; New Kingdom; [30 | C147★ | T II-30 | C123C^{+5} | N-30 | C125CF^3 | D-28 | BS160 | U-30 | C141★ | BI-28 | 156★ | LI].
Dominant	• *Hyphaene thebaica* 1 whole fruit and 5 endocarps.

No.	3273
Cat. 3/83	4 dôm fruits, linen wrapper containing natron, halfa and vine twigs; period uncertain (probably New Kingdom or later); bought in Luxor (1935).
Dominant	• *Hyphaene thebaica* 4 whole fruits
Contam.	• *Vitis vinifera* Several stem fragments. • cf. Gramineae tribe Triticeae Culm fragment. • *Phoenix dactylifera* Small fragment of seed.

No.	3274
Cat. 3/84	Dôm fruits and kernels; period uncertain; bought in Luxor (Molatham) (1934).
Dominant	• *Hyphaene thebaica* Dozens of endocarps, only 2 specimens with remains of mesocarp, and ca. 50 seeds.

	2 endocarps have been cut: 1 transverse and 1 lengthwise.
No.	3275
Cat. 3/84	Dôm kernels; bought from an antique dealer in Luxor (Molatham) (1934).
Dominant	• *Hyphaene thebaica* 1 whole fruit (with large, gnawed hole); 1 fragment of exocarp (cut mark), 1 immature fruit and 8 seeds (partly fragmented).
Contam.	• cf. *Medemia argun* Oval (28 x 19 mm) and with small hole on top (same as No. 4398).
No.	3276
Cat. 3/84	Dôm kernels; Deir El Medineh; New Kingdom; donated by the French Institute of Oriental Archaeology (1933/34).
Dominant	• *Hyphaene thebaica* Several hundreds of fragments of exocarps, almost all with crosswise cut marks.
No.	3278
Cat. 3/84	Vase of white pottery containing dates; probably Roman period; bought from an antique dealer in Luxor (26-06-1934) and donated by the Egyptian Museum (11-1937); [S 56291]; [D 43251[;[C$_1$].
Dominant	• *Phoenix dactylifera* Pottery containing several hundreds of fruits, large specimens (ca. 45 x 20 mm).
No.	3279
Cat. 3/85	Red pottery (diameter: 13 cm; height: 12 cm) containing dates; Fayum; probably Graeco-Roman; donated by the Egyptian Museum (1937/38) (Journal d'entrée (JE) No. 30269).
Dominant	• *Phoenix dactylifera* Ca. 100 fruits (ca. 40 x 10 mm).
No.	3280
Cat. 3/85	Dates and seeds of *Phoenix dactylifera*; Deir El Medineh; New Kingdom donated by the French Institute of Oriental Archaeology (1933).
Dominant	• *Phoenix dactylifera* Ca. 30 fruits, partly disintegrated (ca. 30 x 12 mm).
No.	3281
Cat. 3/85	Vase of white pottery (diameter: 6.5 cm; height: 7 cm) with suspended rope and cover of linen containing dates (*Phoenix dactylifera*), a few grains of *Lupinus termis*, a few seeds of *Olea europaea*; Deir El Medineh; New Kingdom; bought from an antique dealer (Said Molatham) in Luxor (24-12-1934).
Dominant	• *Phoenix dactylifera* 16 fruits: 3 specimens partly disintegrated (longest specimen: 42 mm long; smallest specimen: 23 mm long). • *Balanites aegyptiaca* 1 endocarp with large gnawed hole. • *Lupinus alba* 5 whole seeds. • *Olea europaea* 4 endocarps. • ? Looks like flax fibres surrounded by pieces of textile.
No.	3282
Cat. 3/85	Seeds of dates (*Phoenix dactylifera*); period uncertain; bought from antique dealer in Luxor (1933).
Dominant	• *Phoenix dactylifera* Several hundreds of seeds, including some whole fruits.
No.	3284
Cat. 3/86	Dates and seeds of *Phoenix dactylifera*; Saqqara; probably Late; donated by the Antiquity Department (1933).
Dominant	• *Phoenix dactylifera* Ca. 100 fruits of date, relatively small specimens (ca. 24 x 12 mm), and 2 <u>peduncles.</u>
No.	3287
Cat. 3/86	Pine cones (*Pinus pinea*) (a: 9cm, b: 8cm, c: 7.5cm, d: 6cm); uncertain provenance; probably Graeco-Roman period; donated by the Antiquity Department (1936).
Dominant	• *Pinus pinea* Scales from *P. Pinea* cones (a, b and c) partly removed, no seeds present and no clear imprints of seeds on scales. • *Pinus* Unknown cone (d) is smaller.
No.	3288
Cat. 3/86	Pine cones (*Pinus pinea*) (height: a: 9cm, b: 7 cm); Saqqara; probably Graeco-Roman period; from excavations of J.E. Quibell and donated by the Egyptian Museum (1936).
Dominant	• *Pinus pinea* 2 cones, only lower part, no seeds present.

No. 3289
Cat. 3/87 Pine cones (*Pinus pinea*) (height: 7cm); Karanis (Kôm Aushim); probably Graeco-Roman period; donated by the Antiquity Department; [28 | C 88 M | B].
Dominant • *Pinus pinea*
Lower part of cone, no seeds present.

No. 3290
Cat. 3/87 Pine cone (*Pinus halepensis*); Saqqara; probably Graeco-Roman period; from excavations of J.E. Quibell and donated by the Egyptian Museum (1936).
Dominant • *Pinus*
Cone, most (small) scales removed. Partly burnt.

No. 3291
Cat. 3/87 Pine cone (*Pinus pinea*) (height: 10 cm); Graeco-Roman period; donated (1937/38).
Dominant • *Pinus pinea*
Whole, closed cone.

No. 3292
Cat. 3/87 Pine cone (*Pinus pinea*) (height: 8.5 cm); Hermopolis Tuna El Gebel; Graeco-Roman period; secured by L. Keimer; excavations of the Faculty of Arts by Dr. Samy Gabra (02/03-1931) and donated by Fouad 1st University (1936).
Dominant • *Pinus pinea*
Cone (75 x 90 mm) partly open, full-grown seeds present.

No. 3293
Cat. 3/87 *Juniperus* fruits, in the underground storage space of galleries within the enclosure of the step pyramid; Saqqara; IIIe Dynasty, reign of Zoser; found and donated by J-Ph. Lauer (25-11-1935); Reference: 'Lauer, Le pyramide a degres 1936, AI, page 184'.
Dominant • *Juniperus*
Several fruits, most of them fallen into dust. Seeds present, with resin.

No. 3294
Cat. 3/88 Almond fruit (height: 1.5 cm); Karanis (Kôm Aushim); Graeco-Roman period; donated by the Antiquity Department.
Dominant • *Amygdalus communis*
1 endocarp (18 x 13 mm).

No. 3295
Cat. 3/88 Peach stone (*Amygdalus persica*) (diameter: 2.1 cm); Karanis (Kôm Aushim); Graeco-Roman period; donated by the Antiquity Department; [28 | 156★ | K].
Dominant • *Prunus persica*
Complete endocarp (2.4 cm long).

No. 3296
Cat. 3/88 Peach stone (*Amygdalus persica*) (diameter: 3 cm); Karanis (Kôm Aushim); Graeco-Roman period; donated by the Antiquity Department; [28 | 156★ | MI].
Dominant • *Prunus persica*
Complete endocarp (3.0 cm long).

No. 3297
Cat. 3/88 Peach stone (*Amygdalus persica*) (diameter: 3 cm); Karanis (Kôm Aushim)/ Tebtynis; Graeco-Roman period; Italian excavations and donated by the Antiquity Department (1933).
Dominant • *Prunus persica*
Complete endocarp (3.2 cm long).

No. 3298
Cat. 3/88 Peach stone (*Amygdalus persica*) (diameter: 2.3 cm); Karanis (Kôm Aushim); Graeco-Roman period; donated by the Antiquity Department; [28 | B 136 A | O].
Dominant • *Prunus persica*
Complete endocarp (2.6 cm long).

No. 3299
Cat. 3/89 Peach stone (*Amygdalus persica*) (diameter: 2.5 cm); Karanis (Kôm Aushim)/ Saqqara; probably Late or Graeco-Roman period; excavated by J.E. Quibell and donated by the Egyptian Museum.
Dominant • *Prunus persica*
Complete endocarp (27 mm long).

No. 3300
Cat. 3/89 Pomegranate rind; Saqqara; Late period; probably excavated by J.E. Quibell and donated by the Egyptian Museum.
Dominant • *Punica granatum*
2 fruits, fragmented.

No. 3301
Cat. 3/89 Pomegranate rind and seeds; Saqqara; Late period; probably excavated by J.E. Quibell and donated by the Egyptian museum.
Dominant • *Punica granatum*
2 fruits, both open (possibly rather recent).

No. 3302
Cat. 3/89 Pomegranate rind and seeds;, provenance

	and date unknown; donated by the Egyptian Museum; [5|8|15|31].
Dominant	● *Punica granatum* 2 fruits, fragmented.
No. Cat. 3/100	3349 Fruit of red lotus (*Nelumbium speciosum*) (diameter: 8 cm); uncertain provenance; probably Graeco-Roman period; donated by the Egyptian Museum (11-07-1946).
Dominant	● *Nelumbo nucifera* Receptacles (torus) with some seeds still present (10 cm in diameter).
No. Cat. 3/100	3350 Fruit of red lotus (*Nelumbium speciosum*) (diameter: 8.7 cm); uncertain provenance; probably Graeco-Roman period; donated by the Egyptian Museum (11-07-1946).
Dominant	● *Nelumbo nucifera* Receptacles (torus) with some seeds still present (8 cm in diameter).
Comment	See figure 16.10 in Darby *et al.*, 1977.
No. Cat. 3/101	3356 Pottery dish (diameter: 16 cm) containing grains of wheat (*T. dicoccon*); Cheikh Abd El Qurnah (Thebes; 1888); New Kingdom (1580-1080 BC); identified by E. Schiemann (1939); donated by the Egyptian Museum (1937/38); [4|29|27|22].
Dominant	● *Triticum turgidum* ssp. *durum* Rachis fragments. ● *Triticum turgidum* ssp. *dicoccon* Few spikelets.
Contam.	● *Sorghum halepense* Few spikelets.
No. Cat. 3/101	3357 Pottery dish (diameter: 17.5 cm) containing naked wheat; probably Qurnah (Thebes); probably New Kingdom; bought in Luxor (1932).
Dominant	● *Triticum turgidum* ssp. *durum* Grain kernels, some partly gnawed by insects.
Contam.	● *Triticum turgidum* ssp. *dicoccon* ● *Sorghum bicolor* Spikelets. ● Gramineae tribe Triticeae Few threshing remains (chaff, rachis fragments and culm nodes).
No. Cat. 3/101	3358 Pottery dish (diameter: 15.5 cm) containing barley and a few wheat grains; uncertain provenance; probably New Kingdom; donated by the Egyptian Museum (1937/38); [4|16|27|22].
Dominant.	● *Hordeum vulgare* ssp. *vulgare* (hulled) Spikelets.
Contam.	● *Triticum turgidum* ssp. *durum* Few grain kernels. ● *Triticum* Badly preserved grain (charred).
In tube	● *Vitis vinifera* 6 seeds, 1 with fruit remains. ● *Vicia faba* Seed-coat with hilum. ● Cyperaceae Unidentified rhizome. ● Gramineae Unidentified spikelet.
No. Cat. 3/102	3359 Pot containing wheat of double-rowed ears; uncertain provenance and period; identified as *Triticum dicoccum* by E. Schiemann; donated by the Egyptian Museum (1937/38); [4|22|27|22 (pottery)].
Dominant	● *Triticum turgidum* ssp. *dicoccon* Grains and a few rachis fragments (slightly charred).
Contam.	● *Hordeum vulgare* ssp. *vulgare* (hulled) Spikelets (slightly charred).
No. Cat. 3/102	3360 Pot (diameter: 8.3 cm) containing carbonised wheat, barley, lentil and grass-pea, *Lathyrus sativus*, *Lens esculentus*; uncertain provenance, Ptolemaic period; *Triticum dicoccum* identified by E. Schiemann (1939), *Vicia* sp and *Lolium temulentum* identified by H. Helbaek (21-04-1955); donated by the Egyptian Museum; [4|36|27|22].
Dominant	● *Triticum turgidum* ssp. *dicoccon* Grains, spikelet forks and a few complete spikelets (charred).
Contam.	● *Hordeum vulgare* ssp. *vulgare* (hulled) Spikelets (charred) and chaff fragment (desiccated). ● *Pisum sativum* 1 seed (charred). ● *Lens culinaris* 1 seed (charred). ● *Lathyrus sativus* Ca. 5 Seeds (charred). ● Gramineae tribe Triticeae Culm node (charred).

No.	3361	No.	3366
Cat. 3/102	Barley (*Hordeum vulgare*), tomb of Tut-Ankh-Amun at Thebes; New Kingdom; identified by H. Helbaek (22-04-1955) as: barley with a few spikelets of *Triticum dicoccum*; donated by the Egyptian Museum (1937/38).	Cat. 3/103	Ear of wheat (max. length: 17 cm); Tebtynis (Fayum); Roman period; Italian excavations and donated by the Antiquity Department (1935).
Dominant	● *Hordeum vulgare* ssp. *distichum* (hulled) Large sample of spikelets, small grain kernels. Few rachis fragments.	Dominant	● *Triticum turgidum* ssp. *durum* Ear with 9 cm long culm. Glumes smooth and with awn.
Contam.	● *Triticum turgidum* ssp. *dicoccon* Spikelets.	No.	3367
Comment	See figure 11.12 in Darby *et al.*, 1977 (presented as *Hordeum vulgare*).	Cat. 3/104	Wheat (*Triticum dicoccum*), Abu-Sir (1903); Middle Kingdom (2000 BC); donated by the Egyptian Museum (1937/38); [4	4	27	29].
No.	3362	Dominant	● *Triticum turgidum* ssp. *dicoccon* Thousands of spikelets, including upper spikelets with twisted internode. Sample checked through glass cover.
Cat. 3/103	Barley (*Hordeum tetrastichum*) and wheat (*Triticum dicoccum*); tomb of Ra-hotep at Kuseih (near Meir); XIIe Dynasty; excavated by Ahmed Bey Kamal (09-1910); barley and wheat identified by E. Schiemann; donated by the Egyptian Museum (1937/38); [4	1	27	29].	No.	3368
		Cat. 3/104	Pottery dish (diameter:16.5 cm) containing rush-nuts (*Cyperus esculentus*); uncertain provenance and period; donated by the Egyptian Museum (1937/38); [4	17	27	22 (pottery)].
Dominant	● *Hordeum vulgare* ssp. *vulgare* (hulled) Spikelets, predominant. ● *Triticum turgidum* ssp. *dicoccon* Spikelets.	Dominant	● *Cyperus esculentus* Large quantity of tubers.
Contam.	● *Lotus* Fruit. Not present in separate sample (without number, but description of sample is similar). ● *Cyperus esculentus* Tubers. Only present in separate sample.	Contam.	● *Triticum turgidum* ssp. *dicoccon* 1 spikelet (maybe more specimens present).
		No.	3369
		Cat. 3/104	Pottery dish (diameter:12.5 cm) containing rush-nuts (*Cyperus esculentus*); uncertain provenance and period; donated by the Egyptian Museum (1937/38); [4	6	27	22 (pottery)].
No.	3364	Dominant	● *Cyperus esculentus* Hundreds of tubers.
Cat. 3/103	Pottery vase (diameter: 13 cm) containing lentil seeds, a pomegranate and a sycamore fruit; uncertain provenance; G.C.C. Maspero label No. 4628; Graeco-Roman period; H. Helbaek (22-06-1955): no millet observed, but many empty lentil seed-coats, a few *T. dicoccum* and a few *Hordeum*, 1 carbonised grain of *Triticum compacteum*; donated by the Egyptian museum (1937/38); [4	7	27	22].	Contam.	● *Hordeum vulgare* ssp. *vulgare* (hulled) Ca. 7 spikelets. ● *Triticum turgidum* ssp. *durum* Rachis fragment. ● *Mimusops laurifolia* 2 seed fragments. ● ? Fragment of unknown fruit (16 x 9 mm).
		No.	3370
Dominant	● *Lens culinaris* Many seeds.	Cat. 3/104	Fruits of juniper; probably Deir El Medineh; bought in Luxor (1933/34).
Contam.	● *Punica granatum* Small fruit (28 mm + calyx: 7 mm). ● *Ficus sycomorus* Accessory fruit (syconium). ● *Hordeum vulgare* ssp. *vulgare* (hulled) Few spikelets. ● *Cyperus esculentus* Few tubers. ● *Triticum turgidum* ssp. *dicoccon* 1 spikelet.	Dominant	● *Juniperus* Dozens of fruits (3 seeds are present in 1 fruit; only 1 specimen checked). ● *Ziziphus spina-christi* Dozens of endocarps.
		No.	3371
		Cat. 3/104	Pottery dish (diameter:10.5 cm) containing

	4 fruits of thorn tree; Deir El Medineh; New Kingdom; donated by the French Institute of Oriental Archaeology with permission of the Antiquity Department (1933).
Dominant	● *Balanites aegyptiaca* 4 Complete fruits in pottery dish.
No. Cat. 3/105	3372 Pottery dish (diameter:14 cm; height: 5 cm) containing fruits of *Balanites aegyptiaca*; tomb at Gebelein; probably Middle Kingdom; G.C.C. Maspero (1885); donated by the Egyptian Museum (1937/38); [4|18|27|22].
Dominant	● *Balanites aegyptiaca* 21 fruits, most of them still whole specimens. In some endocarps recent spikelets of *Panicum miliaceum* present.
Contam.	● *Pinus pinea* Seed. ● Resin A few pieces, one of them embedding a seed-like structure.
No. Cat. 3/105	3373 Pottery dish (diameter:14 cm) containing kernels of *Balanites aegyptiaca*; donated by the Egyptian Museum (1937/38); [4|10|27|22].
Dominant	● *Balanites aegyptiaca* 4 fruits with part of mesocarp (1 with insect damage) and 9 endocarps (3 gnawed).
No. Cat. 3/105	3374 4 dôm fruits (*Hyphaene thebaica*); Thebes; New Kingdom; donated by the Egyptian Museum (1933/34).
Dominant	● *Hyphaene thebaica* 4 whole fruits.
No. Cat. 3/105	3375 2 dôm fruits (*Hyphaene thebaica*); tomb of Tut-Ankh-Amun at Thebes; New Kingdom; donated by the Egyptian Museum (1937).
Dominant	● *Hyphaene thebaica* 2 whole fruits.
Comment	Possibly depicted by Darby et al., 1977 in figure 18.17 (see also No. 3270).
No. Cat. 3/105	3376 Pottery vase (diameter: 14 cm) containing 2 dôm fruits and 5 kernels of *Hyphaene thebaica*; Qurna (Thebes) (1888); probably New Kingdom; donated by the Egyptian Museum (1937/38); [4|30|27|22].
Dominant	● *Hyphaene thebaica*

	2 whole fruits and 1 endocarp.
	● cf. *Medemia argun* 4 fruits (exocarp, for the most part, missing).
No. Cat. 3/106	3378 Quantity of wheat (*Triticum dicoccum*) with empty ears found in the adjacent big pottery vase of Roman type (height: 40 cm); uncertain provenance; probably Graeco-Roman period; bought in Luxor (1934).
Dominant	● *Triticum turgidum* ssp. *dicoccon* Large vessel filled with spikelets. Rachis is semi-brittle: fragments frequently consist of several internodes, especially the basal part of the ear. In the box are only a few rachis fragments.
No. Cat. 3/106	3379 Pottery vase (height: 21 cm) containing carbonised grains of wheat and a few grains of barley; uncertain provenance and period; H. Helbaek (22-06-1955): *Triticum compacteum* Host. and a few grains of hulled barley. No certain *T. dicoccum* observed; bought in Luxor (1934).
Dominant	● *Triticum aestivum* Rather short grain kernels (charred).
Contam.	● *Hordeum vulgare* (hulled) Few spikelets (charred). ● Charcoal pieces ● Dung of sheep/goat Desiccated.
No. Cat. 3/107	3384 Two small pottery dishes (diameter: 4.5 cm and 5.2 cm) containing one grain of *Lupinus termis*; donated by the Egyptian Museum (1937/38); [4|48|27|22].
Dominant	● *Lupinus albus* 1 seed, no damage by insects.
No. Cat. 3/121	4059 Red pottery vase (diameter: 19.5 cm) in which was kept refuse of flax (straw and fibres); uncertain provenance; probably Graeco-Roman period; bought in Luxor (15-02-1933).
Dominant	● *Linum usitatissimum* Large box with stem fragments and pieces of rope.
No. Cat. 3/121	4060 Whitish red pottery vase (diameter: 21.5 cm) in which was kept refuse of flax (straw and fi-

bres); uncertain provenance; probably Graeco-Roman period; bought in Luxor (15-02-1933).

Dominant
- *Linum usitatissimum*
Large box with stem fragments and pieces of rope. Also small textile bag with unknown content.

No. 4079
Cat. 3/123 Pottery dish (diameter: 11.8 cm; height: 6 cm) containing *Medicago sativa* seeds; temple of Isis at Denderah; probably Graeco-Roman period; identified by H. Helbaek (25-06-1955); donated by the Egyptian Museum (1937/38); [4|8|27|22].

Dominant
- *Medicago sativa*
Many seeds and a few fruit fragments.

No. 4081
Cat. 3/123 Small carton box containing fragments of resinous substance (calafonia); uncertain provenance and period; donated by the Egyptian Museum (1937/38).

Dominant
- cf. resin

No. 4082
Cat. 3/123 Small alabaster dish (diameter: 5 cm) containing fragments of resinous substance (calafonia); uncertain provenance and period; G.C.C, Maspero label No. 4626; donated by the Egyptian Museum (1937/38).

Dominant
- cf. resin

No. 4083
Cat. 3/123 Pottery dish (diameter: 15 cm) containing capsules of *Linum usitatissimum*; Cheikh Abd El Qurna (Thebes) (1885); XXe Dynasty; part of 8 ardabs found in the tomb used as a store; donated by the Egyptian Museum (1937/38); [4|25|27|22 (pottery)].

Dominant
- *Linum usitatissimum*
Ca. 100 whole fruits.

No. 4084
Cat. 3/123 Small pottery dish (diameter: 5.2 cm) containing *Lupinus termis* seeds; uncertain provenance and period; donated by the Egyptian Museum (1937/38); [4|33|27|22].

Dominant
- *Lupinus albus*
33 seeds, no damage by insects.

No. 4085
Cat. 3/124 Sycamore figs; stores of step pyramid of Zoser at Saqqara; IIIe Dynasty; donated by the Egyptian Museum (1937/38) (Journal d'entrée (JE) No. 65913).

Dominant
- *Ficus sycomorus*
Several hundreds of accessory fruits (syconia).

No. 4086
Cat. 3/124 Small carton box containing wheat grains (*Triticum dicoccum*); tomb of Ra-hotep at Meir (Kuseih); XIIe Dynasty; excavated by Ahmed Kamal (1910); donated by the Egyptian Museum (1937/38); [4|3|27|29].

Dominant
- *Triticum turgidum* ssp. *dicoccon*
Ca. 40 spikelets, no top spikelet with twisted internode present.

Contam.
- *Hordeum vulgare* ssp. *vulgare* (hulled)
1 spikelet.

No. 4087
Cat. 3/124 Small carton box containing barley grains (*Hordeum tetrastichum*); tomb of Ra-hotep at Meir (Kuseih); XIIe Dynasty; excavated by Ahmed Kamal (1910); donated by the Egyptian Museum (1937/38); [4|2|27|29].

Dominant
- *Hordeum vulgare* ssp. *vulgare* (hulled)
Hundreds of spikelets and a few rachis fragments.

Contam.
- *Triticum turgidum* ssp. *dicoccon*
Rachis fragments.

No. 4088
Cat. 3/124 Carbonised wheat grains (*Triticum dicoccum*); Merimde (west of Bani Salama), Prehistoric period; excavated by H.J.B. Junker from the German Institute of Archaeology and donated by the Antiquity Department (1933).

Dominant
- *Triticum turgidum* ssp. *dicoccon*
Several hundreds of charred grain kernels, with clear impression of chaff.

No. 4089
Cat. 3/124 Carbonised wheat grains (*Triticum dicoccum*); Merimde (west of Bani Salama), Prehistoric period; excavated by H.J.B. Junker from the German Institute of Archaeology and donated by the Antiquity Department (1933).

Dominant
- *Hordeum vulgare* (hulled)
Charred spikelets/grain kernels (not mentioned in notebook).
- *Triticum turgidum* ssp. *dicoccon*
Charred grain kernels, with clear impression of chaff.
- *Lens culinaris*
1 charred seed.

• *Lolium temulentum*
  Charred spikelets.

No. 4095
Cat. 3/126 Bronze vase (diameter: 16 cm) containing linen capsules (*Linum humile*); period uncertain; G.C.C. Maspero, label No. 4620; donated by the Egyptian Museum (1937/38); [4 | 1 | 27 | 22].

Dominant
• *Linum usitatissimum*
  Large number of fruits, mostly complete and with part of the culm (up to 2 cm).

Contam.
• *Hordeum vulgare* ssp. *vulgare* (hulled)
  Few spikelets.
• *Triticum turgidum* ssp. *durum*
  Grain kernel.
• *Triticum turgidum* ssp. *dicoccon*
  1 spikelet.
• *Cyperus esculentus*
  Few tubers.
• Cruciferae
  Fruit fragment.
• *Zilla spinosa*
  2 fruits.

No. 4096
Cat. 3/126 Small pottery dish (diameter: 5.7 cm) containing 4 linen seeds; Gebelein; period uncertain; donated by the Egyptian Museum (1937/38); [4 | 45 | 27 | 22 (pottery)].

Dominant
• *Linum usitatissimum*
  4 complete fruits, 2 with part of culm still present.

No. 4097
Cat. 3/126 Seeds of watermelon in a small pottery dish (diameter: 5.6 cm); uncertain provenance and period (label: Late period); identified by Dr. Abel el-Gazzar of the Botany department (29-11-1980): *Citrullus vulgaris* var. *colocynthoides*; donated by the Egyptian Museum (1937/38); [4 | 46 | 27 | 22 hall 59, case O].

Dominant
• *Citrullus lanatus*
  112 Seeds and a few small fragments of the fruit.

No. 4167
Cat. 3/137 Big pottery vase (height: 56 cm), its cover (diameter: 10.3 cm) and quantity of lupin grains (*Lupinus termis*) found in the adjacent pottery vase of Roman type; uncertain provenance; bought in Luxor (Mansour) (26-06-1934).

Dominant
• *Lupinus alba*
  Large number of seeds.

No. 4168
Cat. 3/137 Small quantity of bitter vetch (*Lathyrus sativum*) (label: seeds of grass-pea); period uncertain; treated by Hossam Abd el-Hamid and Kamal Mantawi, Ministry of Archaeology and Hussein el-Monayere, Agricultural Museum (01-04-1970); bought in Luxor (Molatham) (15-02-1933).

Dominant
• *Vicia ervilia*
  Hundreds of seed.

Contam.
• *Linum usitatissimum*
  Half fruit.

No. 4169
Cat. 3/137 Quantity of chick-pea (*Cicer arietinum* L.) found in the adjacent red pottery vase of Roman type (height: 19 cm); probably Graeco-Roman period; treated by Hossam Abd el-Hamid and Kamal Mantawi, Ministry of Archaeology and Hussein el-Monayer, Agricultural Museum, 14-04-1970; bought in Luxor (1934).

Dominant
• *Cicer arietinum*
  Hundreds of seeds.

No. 4170
Cat. 3/137 Red pottery vase (height: 10 cm) containing a small quantity of *Lupinus termis*; uncertain provenance (maybe Deir El Medineh); probably New Kingdom or Graeco-Roman period; bought in Luxor (1934); (02-04-70).

Dominant
• *Lupinus albus*
  18 seeds, no damage.

No. 4171
Cat. 3/137 Few grains of grass pea (*Lathyrus sativus*); Deir El Medineh; New Kingdom; donated by the French Institute of Oriental Archaeology (1933); (02-04-70).

Dominant
• *Lathyrus sativus*
  Ca. 40-50 seeds, damage by insects.

No. 4172
Cat. 3/137 Quantity of grass pea (*Lathyrus sativus*); Deir El Medineh; New Kingdom; identified by H. Helbaek (25-06-1955); donated by the French Institute of Oriental Archaeology (1933).

Dominant
• *Lathyrus sativus*
  Thousands of seeds (from most of the specimens only the testa is preserved).

Contam.
• *Lathyrus hirsutus*
  Seeds.
  Sample not completely checked.

No. 4173
Cat. 3/138 Red pottery dish (diameter: 11.8 cm) containing grass pea (*Lathyrus sativus*); Deir El Medineh; New Kingdom; identified by H. Helbaek (25-06-1955); treated by Hossam Abd el-Hamid, Kamal Tantawi, Ministry of Archaeology and Hussein el-Monayer, Museum of Agriculture (01-04-1970); donated by the French Institute of Oriental Archaeology (1933); [CS$_2$].
Dominant • *Lathyrus sativus*
Hundreds of seeds.
Contam. • *Pisum sativum*
Dozens of seeds, slightly flattened and with small papillae (similar to No. 4176).
Large sample, not completely checked.
Comment See figure 17.11 in Darby et al., 1977 (erroneously presented as *Cicer arietinum*).

No. 4174
Cat. 3/138 Few grains of beans (*Vicia faba*); Saqqara; probably Late period; excavations of J.E. Quibell (1933).
Dominant • *Vicia faba*
13 seeds (ca. 14 x 10 mm).

No. 4175
Cat. 3/138 Few grains of beans (*Vicia faba*); Karanis (Kôm Aushim); Graeco-Roman period; donated by the Antiquity Department (1935); [29|B 166★|G] and [10].
Dominant • *Vicia faba* ssp. *minor*
Ca. 100 seeds, seed-coat mostly still present.

No. 4176
Cat. 3/138 Small quantity of grass pea (*Lathyrus sativus*); Deir El Medineh; New Kingdom; identified by H. Helbaek (25-06-1955); donated by the French Institute of Oriental Archaeology with permission of the Antiquity Department (1933).
Dominant • *Lathyrus sativus*
Several hundreds of seeds.
Contam. • *Pisum sativum*
Dozens of seeds, slightly flattened and with small papillae (similar to No. 4173).
• *Citrullus colocynthis*
1 seed.

No. 4177
Cat. 3/138 Big basket (diameter: 57 cm; height: 37 cm) of halfa and palm leaves containing big quantity of safflower; Karanis (Kôm Aushim); Roman period; donated by the Antiquity Department (16-10-1934).
Dominant • *Carthamus tinctorius*
Large number of seeds, small percentage affected by insect damage.

No. 4178
Cat. 3/138 Big pottery vase (height: 49 cm) of Roman type containing quantity of safflower; uncertain provenance; Roman period; bought in Luxor (1933).
Dominant • *Carthamus tinctorius*
Large number of seeds, minor part with insect damage.

No. 4179
Cat. 3/139 Big pottery vase (height: 55 cm) of Roman type containing olive seeds (*Olea europaea*); Roman period; bought in Luxor (Mansour) (26-06-1934).
Dominant • *Olea europaea*
Large number of endocarps, minor part with insect damage in amphora, together with linen cloth.
Contam. • *Cicer arietinum*
Half seed.

No. 4180
Cat. 3/139 3 olive fruits; Deir El Medineh; XVIIIe Dynasty; donated by the French Institute of Oriental Archaeology with permission of the Antiquity Department (1933).
Dominant • *Olea europaea*
3 whole fruits.

No. 4181
Cat. 3/139 Quantities of onions (*Allium cepa*) and garlic (*Allium sativum*) remains; probably Thebes; bought in Luxor (Tadros) (27-02-1933).
Dominant • *Allium cepa*
Threshing remains (bulb bases, scales and lower parts of stems). Bulb bases with ca. 3 impressions of bulbils.
Contam. • *Allium sativum*

No. 4182
Cat. 3/139 Few remains of onions; uncertain provenance and period; bought by L. Keimer, (09-04-1933).
Dominant • *Allium cepa* var. *aggregatum*
Several bulbs. Size of 1 specimen: 6 x 3.7 cm. A few bulbs bunched together at the root (shallots).

No.	4183
Cat. 3/139	Small quantity of *Allium sativum*; Deir El Medineh; XVIIIe Dynasty; donated by the French Institute of Oriental Archaeology with permission of the Antiquity Department (1933).
Dominant	● *Allium sativum* Ca. 10 bulbs, partly fallen apart.
Comment	See figure 17.2 in Darby *et al.*, 1977

No.	4184
Cat. 3/139	Small quantity of onions (*Allium cepa*) found among Royal garlands; Thebes; New Kingdom; donated by the Antiquity Department (26-11-1937); [4|13|27|29].
Dominant	● *Allium sativum* 4 cloves.

No.	4185
Cat. 3/140	Few bulbs of onions (label: samples of garlic and onions); uncertain provenance and period; bought in Luxor (Mansour) (14-04-1932).
Dominant	● *Allium cepa* 14 bulbs, disintegrated specimens. ● *Allium sativa* 1 bulb.

No.	4187
Cat. 3/140	Small red pottery vase (height: 9 cm) containing a few watermelon seeds (*Citrullus vulgaris*); uncertain provenance; Graeco-Roman period; bought from an antique dealer (Molatham) in Luxor (27-06-1934).
Dominant	● *Citrullus lanatus* 18 Seeds.
Contam.	● *Lagenaria siceraria* 0.5 seed. ● *Acacia* 1 Seed.

No.	4188
Cat. 3/140	Nape (*Brassica napus*); probably Thebes; period uncertain; bought from an antique dealer (Molatham) in Luxor (1933).
Dominant	● cf. *Brassica napus* Storage organs (taproot and lower part of the stem).

No.	4189
Cat. 3/140	Nape (*Brassica napus*); probably Thebes; period uncertain; bought from an antique dealer (Molatham) in Luxor (1933).
Dominant	● cf. *Brassica napus* Storage organs (taproot and lower part of the stem).
Comment	See figure 17.5 in Darby *et al.*, 1977

No.	4190
Cat. 3/140	Fruit parts of *Citrullus colocynthis*; probably Thebes; period uncertain; bought from an antique dealer (Mansour) in Luxor (1933).
Dominant	● *Citrullus colocynthis* 1 fruit, fragmented, no seeds present.

No.	4191
Cat. 3/141	Few seeds of Jew's-mallow (*Corchorus olitorius*); Karanis (Kôm Aushim); donated by the Antiquity Department (1934).
Dominant	● *Trifolium* Hundreds of seeds, blackish (2-2.5 mm long).
Contam.	● *Rumex dentatus* Inflorescence.
Comment	See figure 17.6 in Darby *et al.*, 1977 (erroneously presented as *Corchorus olitorius*).

No.	4192
Cat. 3/141	Few bulbs of garlic (*Allium sativum*); uncertain provenance and period; bought from an antique dealer in Luxor (1933).
Dominant	● *Allium sativum* Ca. 5 specimens: 1 whole specimen, the others disintegrated.

No.	4193
Cat. 3/141	A small piece of watermelon (*Citrullus vulgaris*); probably Thebes (Luxor); probably Graeco-Roman period; bought from an antique dealer in Luxor (1933).
Dominant	● *Citrullus lanatus* 3 fragments of a fruit. Seeds in rows, partly undeveloped.

No.	4194
Cat. 3/141	2 pieces of bottle-gourd (*Lagenaria vulgaris*); Karanis (Kôm Aushim); Roman period; donated by the Antiquity Department (1934); [28|BS 160|BI].
Dominant	● *Lagenaria siceraria* 2 fragments, 1 with a stalk (ca. 2 and 3 mm thick).

No.	4195
Cat. 3/141	3 pieces of bottle-gourd (*Lagenaria vulgaris*); Elephantine; Roman period; donated by the Antiquity Department (P.H.H. Bovier-Lapierre) (1934).
Dominant	● *Lagenaria siceraria*

	3 fragments, 1 with a stalk (ca. 3 mm thick).
No.	4196
Cat. 3/142	Piece of *Lagenaria vulgaris*; Deir El Medineh; New Kingdom; the excavations of the French Institute of Oriental Archaeology and donated by the Antiquity Department (1933).
Dominant	● *Lagenaria siceraria* Lower part of fruit with cut marks.
No.	4197
Cat. 3/141	A piece of vegetable marrow (*Cucurbita pepo* L); Karanis (Kôm Aushim); Roman period; donated by the Antiquity Department (1934); [29 | C 55 E.D. | E].
Dominant	● cf. *Cucumis sativus* No seeds present. Upper part of fruit (diameter: 3 cm) with thick stalk present (1.2 x 0.7 cm).
No.	4198
Cat. 3/142	Quantity of watermelon seeds (*Citrullus vulgaris*); tomb of Tut-Ankh-Amun at Thebes; XVIIIe Dynasty; donated by the Egyptian Museum (1937).
Dominant	● *Citrullus lanatus* Hundreds of seeds.
Contam.	● *Sorghum halepense* Spikelet. ● *Lens culinaris* Seed. ● *Triticum turgidum* ssp. *dicoccon* Spikelet. ● *Hordeum vulgare* ssp. *vulgare* (hulled) Spikelets and rachis fragments.
No.	4199
Cat. 3/142	One fruit of *Citrus medica* (diameter: 10 cm); Medinet Habu; XVIIIe Dynasty; the American excavations and donated by the Antiquity Department (1936).
Dominant	● cf. *Citrus medica* Whole fruit (95 x 43 mm), similar to No. 4200.
No.	4200
Cat. 3/142	Remainsof fruits of *Citrus medica*; uncertain provenance and period, but probably New Kingdom; bought in Luxor (1934).
Dominant	● cf. *Citrus medica* Few fruits, 1 fruit with 10 (or 11) partitions and large seeds. Size of whole fruit: 90 x 57 mm). Shape and morphology of seeds is similar to seeds of *Citrus medica*, although they are quite large.
Comment	See figure 18.2 in Darby *et al.*, 1977
No.	4201
Cat. 3/143	2 small fruits of *Citrus medica* L.; uncertain provenance and period (New Kingdom?); 'Ságit-il-nci de *Citrus medica* Risso var. *cedrata* Risso, d'après Ahmed Risso Bey p. 51; d'après Schftle [Hnam] p. 85; turung ([nilt] C)'; bought in Luxor (1933).
Dominant	● Cucurbitaceae Small fruit (4.0 x 1.2 cm). ● ? Large fruit (6.8 x 1.7 cm), looks similar to un known fruit of No. 1445. On no account *C. medica*.
No.	4203
Cat. 3/143	Germinated barley; Deir El Medineh; VXIIIe Dynasty; donated by the French Institute of Oriental Archaeology (1933).
Dominant	● *Hordeum vulgare* (hulled) A large sample of germinated barley kernels.
No.	4204
Cat. 3/143	Parts of stem and flowers of *Hyoscyamus muticus* L.; Saqqara; period uncertain; found by J.E. Quibell; identified by V. Täckholm (28-01-1950); donated by the Egyptian Museum (1933).
Dominant	● *Hyoscyamus muticus* 2 fruiting branches. Identification by R.T.J. Cappers only to genus level.
No.	4205
Cat. 3/143	Parts of *Calotropis procera* fruits; Assiut; period uncertain, identified by V. Täckholm (1951); associated letter: 'Sir G. Maspero, Botanical specimens from associated, I shall send them to Schweinfurth. He has left to Berlin some days ago. Please keep them until next year. They are too brittle to travel safely. Quibell'; donated by the Egyptian Museum (1934).
Dominant	● *Calotropis procera* Fruit fragments.
No.	4206
Cat. 3/144	Quantity of coriander (*Coriandrum sativum*) and *Cyperus esculentus*; tomb of Tut-Ankh-Amun at Thebes; XVIIIe Dynasty; identified by H. Helbaek (25-06-1955); donated by the Egyptian Museum (1937/38).

Dominant	● *Coriandrum sativum* Thousands of fruits
Contam.	● *Ziziphus spina-christi* Whole fruit and endocarp. ● *Vitis vinifera* 1 Seed. ● *Grewia* Endocarp. ● *Lathyrus hirsutus* 1 Seed. ● *Emex spinosa* 1 Fruit. ● *Sesamum indicum* 1 Seed. Only one-fifth of sample has been checked.

No.	4207
Cat. 3/144	*Cocculus hirsutus* and *Grewia tenax*; tomb of Tut-Ankh-Amun at Thebes; XVIIIe Dynasty; identified and published by E. Schiemann (1941); donated by the Egyptian Museum (1937/38).
Dominant	● *Grewia* Thousands of fruits, exocarp partly missing.
Contam.	● *Juniperus* 1 fruit ● *Ziziphus spina-christi* 1 endocarp. ● *Phoenix dactylifera* 1 seed. ● cf. *Sorghum halepense* 1 spikelet. ● Gramineae tribe Paniceae 1 spikelet.

No.	4210
Cat. 3/144	Jericho-rose fruit (*Anastatica hierochuntica*); uncertain provenance and period; identified by V. Täckholm; bought from an antique dealer (Mansour) in Luxor (1935).
Dominant	● *Anastatica hierochuntica* Whole plant, including taproot.

No.	4211
Cat. 3/145	Sheba, probably *Parmelia furfuracea*; Deir El Medineh; New Kingdom; B.C.M.J. Bruyère; identified by V. Täckholm; donated by the French Institute of Oriental Archaeology with permission of the Antiquity Department (1933).
Dominant	● lichen

No.	4212
Cat. 3/145	Basket (height: 22 cm) of date palm leaves containing henna leaves (*Lawsonia inermis*); Fayum; probably Graeco-Roman period; bought at Cairo (Nohman).
Dominant	● *Myrtus communis* Small leaves, many glands visible at the lower surface (in the reference collection of the Groningen Institute of Archaeology, some samples of leaves are present which have glands on both sides, leaf margin slightly recurved.

No.	4213
Cat. 3/145	Rootstocks of narrow-leaved Asphodel (*Asphodelus tenuifolius*); probably Thebes; period uncertain; identified by V. Täckholm (20-02-1951); eaten boiled as a vegetable and a good fodder [additional information on label is difficult to read]; bought in Luxor (Mansour) (14-04-1935).
Dominant	● *Asphodelus tenuifolius* At least 10 different specimens of rootstocks. Leaves flattened, ca. 4 mm broad. Identified by V. Täckholm.

No.	4216
Cat. 3/146	Pottery vase (diameter: 23.5 cm) containing carob fruits (*Ceratonia siliqua*); uncertain provenance; Coptic period; identified by V. Täckholm (1950); bought from an antique dealer (Tadros) in Luxor (1933).
Dominant	● *Ceratonia siliqua* Pods with seeds.
Comment	See figure 18.1 in Darby *et al.*, 1977

No.	4219
Cat. 3/146	Few seeds of *Cuminum cyminum*; Deir El Medineh; New Kingdom; identified by V. Täckholm (1950); donated by the French Institute of Oriental Archaeology with permission of the Antiquity Department (1933).
Dominant	● *Cuminum cyminum* Ca. 100 fruits.
Contam.	● *Coriandrum sativum* 1 fruit. ● *Hordeum vulgare* Chaff. ● *Lolium* cf. *temulentum* Spikelet. ● Cruciferae 1 fruit fragment (valve, cf. *Brassica*).

No.	4220
Cat. 3/146	A nut of *Moringa aptera*; Elephantine; probably Graeco-Roman period; identified by

V. Täckholm (1950); [part of text unreadable]; donated by the Antiquity Department (1937/38).

Dominant
● *Moringa peregrina*
1 whole fruit (17 x 15 mm).

No. 4222
Cat. 3/147 Fragment of *Acacia albida* fruit; Deir El Medineh; New Kingdom; identified by V. Täckholm (28-01-1950); donated by the Antiquity Department (1933).

Dominant
● *Faidherbia albida*
1 fruit.

No. 4223
Cat. 3/147 Floss of ushar fruits (*Calotropis procera*); Elephantine; probably Graeco-Roman period; identified by V. Täckholm (1951); donated by P.H.H. Bovier-Lapierre (1936).

Dominant
● *Calotropis procera*
Pappus (length: ca. 35 mm) and 1 seed.

No. 4225
Cat. 3/147 Quantity of sant seeds (*Acacia nilotica*) found in a big pottery vase (height: 21 cm); uncertain provenance; Graeco-Roman period; bought in Luxor (1933).

Dominant
● *Acacia* cf. *nilotica*
Thousands of seeds, relatively large specimens. No fruit fragments found.

Contam.
● *Carthamus tinctorius*
Fruit fragment.

No. 4226
Cat. 3/148 Quantity of sant seeds (*Acacia nilotica*) found in a big pottery vase (height: 23.5 cm); uncertain provenance; Graeco-Roman period; bought in Luxor (1933).

Dominant
● *Acacia* cf. *nilotica*
Thousands of seeds.

No. 4227
Cat. 3/148 Quantity of sant seeds (*Acacia nilotica*) found in a big pottery vase (height: 39 cm); uncertain provenance; Graeco-Roman period; bought in Luxor (1933).

Dominant
● *Acacia nilotica*
Thousands of seeds, relatively large specimens and a few fruit fragments.

Contam.
● *Carthamus tinctorius*
Fruit fragment (recent?).
● *Citrullus lanatus*
1 seed.

No. 4228
Cat. 3/148 Fragments of *Acacia nilotica* legumes; Karanis (Kôm Aushim); Roman period (30 BC- AD 640); donated by the Antiquity Department (1934); [29|158★|VIII].

Dominant
● *Acacia nilotica*
Fruits, mostly whole specimens, open and without seeds. Fruits with insect damage.

Contam.
● *Triticum turgidum* ssp. *durum*
Rachis fragment.
● Gramineae tribe Triticeae
Culm fragment.

No. 4229
Cat. 3/148 Fragments of sant legumes (*Acacia albida*); Elephantine; Roman period (30 BC – AD 640); P.H.H. Bovier-Lapierre; donated by the Antiquity Department (1934).

Dominant
● *Faidherbia albida*
2 fragments of fruit (width: 23 and 12 mm).

No. 4230
Cat. 3/148 Few sant seeds (*Acacia albida*); Saqqara; probably Late period; J.E. Quibell; donated by the Antiquity Department (1934).

Dominant
● *Acacia* cf. *nilotica*
Ca. 50 seeds.

No. 4231
Cat. 3/149 Few sant seeds (*Acacia albida*); Saqqara; probably Late period; J.E. Quibell; donated by the Antiquity Department (1934).

Dominant.
● *Acacia nilotica*
Few fruit fragments.

No. 4232
Cat. 3/149 Particles of sant legumes (*Acacia nilotica*); Karanis (Kôm Aushim); Roman period (30 BC-AD 640); donated by the Antiquity Department (1934); [29|B179L|D].

Dominant
● *Acacia nilotica*
Ca. 100-150 fruit fragments.

No. 4233
Cat. 3/149 Legumes of *Acacia albida*; uncertain provenance and period; bought in Luxor (1933).

Dominant
● *Faidherbia albida*
Dozens of fragmented fruits. No seeds present.

No. 4234
Cat. 3/149 Particles of *Acacia nilotica*; Karanis (Kôm Aushim); Roman period (30 BC-AD 640);

	donated by the Antiquity Department (1934); [28	C29A	Z III].		of tetra- and hexaploid wheats, have been observed.
Dominant	● *Acacia nilotica* Several hundreds of fruit fragments and loose seeds.	Contam.	● *Hordeum vulgare* ssp. *vulgare* (hulled) Spikelets and rachis fragments. ● Gramineae tribe Triticeae Single culm node (charred).
No. Cat. 3/154	4263 Few rhizomes of tiger-nuts (*Cyperus esculentus*); Mustagada (Badari); Prehistoric period; excavations of G. Brunton (1928); donated by the Antiquity Department (27-01-1942).	No. Cat. 3/159	4290 Red pottery vase (height: 23 cm) containing barley grains; probably Deir El Medineh; New Kingdom; bought in Luxor (1933).
Dominant	● *Cyperus esculentus* Ca. 100-200 tubers.	Dominant	● *Triticum turgidum* ssp. *dicoccon* Large sample of threshing remains. Only small sample checked.
No. Cat. 3/154	4264 Few capsules of *Linum usitatissimum*); Mustagada (Badari); Prehistoric period; excavations of G. Brunton (1928); donated by the Antiquity Department (1928).	Contam.	● *Hordeum vulgare* ssp. *vulgare* (hulled) Few spikelets.
		No. Cat. 3/159	4292 Small quantity of *Lolium temulentum* found with barley; granaries of the step pyramid at Saqqara (1927); IIIe Dynasty (3000 BC); donated by the Antiquity Department (1933).
Dominant	● *Linum usitatissimum* 44 whole fruits (some fragmented, a few charred).	Dominant.	● *Lolium temulentum* Hundreds of spikelets.
No. Cat. 3/155	4265 Remains of wheat spikelets; Mustagada (Badari); Badarian period; excavations of G. Brunton (1928); donated by the Antiquity Department (27-01-1942).	Contam.	● *Hordeum vulgare* ssp. *vulgare* (hulled) Few spikelets.
		No. Cat. 3/160	4294 Large quantity of wheat grains (*Triticum dicoccum*); northwest granaries of the step pyramid at Saqqara; IIIe Dynasty; excavation of J-Ph. Lauer; identified by E. Greiss (03-06-1939): wheat, barley and *Lolium*; donated by the Antiquity Department (1933).
Dominant	● *Triticum turgidum* ssp. *dicoccon* Dozens of spikelets, no upper spikelet with twisted internode present. Poor preservation.	Dominant.	● *Triticum turgidum* ssp. *durum* Thousands of spikelets, including upper spikelets with twisted internode.
No. Cat. 3/155	4266 Few wheat stems (burnt or carbonised); Mustagada (Badari); IVe Dynasty, excavations of G. Brunton (1928); donated by the Antiquity Department (27-01-1942).	Contam.	● *Hordeum vulgare* ssp. *vulgare* (hulled) Spikelets and rachis fragments. ● *Lolium temulentum* Spikelets.
Dominant	● Gramineae tribe Triticeae Ca. 13 burnt culm fragments (ca. 3-5 cm long).	No. Cat. 3/160	4295 Small quantity of carbonised wheat and barley grains; Maadi; Predynastic period; excavation of Dr. M. Amer Bey; donated by the Antiquity Department (1934).
No. Cat. 3/158	4287 Small quantity of barley grains; the pyramids of Zoser at Saqqara; IIIe Dynasty; identified by H. Helbaek (22-06-1955): chaff of wheat (*Triticum dicoccum*), small quantity of barley and one seed of *Vicia hirsutus*; donated by the Antiquity Department (1933).	Dominant.	● *Hordeum vulgare* ssp. *vulgare* (hulled) Hundreds of charred spikelets.
		Contam.	● *Triticum turgidum* ssp. *dicoccon* Few charred grain kernels.
Dominant	● *Triticum turgidum* ssp. *dicoccon* Only a few complete spikelets. Most spikelets have fallen apart, most probably due to damage by insects. Top spikelets with twisted rachis fragments, being indicative	No. Cat. 3/160	4296 Small quantity of barley grains; Deir El Medineh;

Dominant.	XVIIIᵉ Dynasty; excavations of the French Institute of Oriental Archaeology (1934). ● *Hordeum vulgare* ssp. *vulgare* (hulled) Hundreds of slender spikelets with rather long awns. Lateral spikelets conspicuously twisted.
No. Cat. 3/160	4297 Fragments of wheat (*Triticum dicoccum*); Karanis (Kôm Aushim); Roman period; Italian excavations, donated by the Antiquity Department (1934).
Dominant	● *Triticum turgidum* ssp. *durum* 1 ear consisting of 6 rachis nodes.
No. Cat. 3/163	4321 Capsules of linseeds; Saqqara; probably Late period; J.E. Quibell; donated by the Antiquity Department (1933); [11|3|15|21].
Dominant.	● *Linum usitatissimum* Hundreds of whole fruits.
Contam.	● *Medicago polymorpha* Many fruits. ● *Lolium* cf. *remotum* Whole inflorescences with ripe fruits.
No. Cat. 3/163	4322 Seeds of flax preserved in a modern glass frame; Thebes; XVIIIᵉ Dynasty; donated by the Antiquity Department (1933).
Dominant	● *Linum usitatissimum* Recent and 18 subfossil seeds in display slide made of glass.
No. Cat. 3/173	4387 Sample of date; Deir El Medineh; New Kingdom; donated by the French Institute of Oriental Archaeology (1933).
Dominant	● *Phoenix dactylifera* Several hundreds of fruits. Pieces of textile are still present on the surface of several fruits, indicating that they once were stored in a piece of textile.
No. Cat. 3/173	4388 Dates; tomb of Tut-Ankh-Amun at Thebes; New Kingdom; donated by the Egyptian Museum (1937).
Dominant	● *Phoenix dactylifera* Ca. 60-80 small (unripe?) fruits with some pieces of textile.
No. Cat. 3/174	4389 Pottery dish (diameter: 14.5 cm) containing dates; probably Thebes; New Kingdom; bought in Luxor (1932).
Dominant	● *Phoenix dactylifera* Ca. 100 fruits, including some seeds. Few fruits are charred.
No. Cat. 3/174	4390 Dates found in a polished pottery dish (diameter: 19 cm); Thebes; New Kingdom; bought in Luxor (1932).
Dominant	● *Phoenix dactylifera* Ca. 200 fruits, partly disintegrated (ca. 25-50 x 15-20 mm).
No. Cat. 3/174	4392 Pottery dish (diameter: 14.5 cm) containing dates; Deir El Medineh; New Kingdom; donated by the French Institute of Oriental Archaeology (1933).
Dominant	● *Phoenix dactylifera* Ca. 50 fruits, including some seeds.
Contam.	● *Vitis vinifera* 1 fruit.
Comment	See figure 18.12 in Darby *et al.*, 1977
No. Cat. 3/174	4393 Polished red pottery dish (diameter: 14.5 cm) containing dates; Deir El Medineh; New Kingdom (1400 BC); donated by the French Institute of Oriental Archaeology (1933).
Dominant	● *Phoenix dactylifera* 12 fruits (ca. 32 x 13 mm).
No. Cat. 3/175	4395 Kernels of dôm fruits (*Hyphaene thebaica*); Elephantine; Graeco-Roman period; donated by Antiquity Department (1934).
Dominant	● *Hyphaene thebaica* 29 whole fruits, 1 part of endocarp (cut mark); 1 fragment of seed.
No. Cat. 3/175	4396 Part of dôm bunch (*Hyphaene thebaica*) (length: 7-15 cm); Deir El Medineh; New Kingdom; donated by the French Institute of Oriental Archaeology (1933).
Dominant	● *Hyphaene thebaica* 2 long fruiting stalks.
No. Cat. 3/175	4397 7 fruitstones of *Medemia argun*; Elephantine; Graeco-Roman period; donated by P.H.H. Bovier-Lapierre (1935).
Dominant	● cf. *Medemia argun*

	6 endocarps (?) each with small hole at one end. In one specimen this hole has been enlarged (similar to No. 4398).
No.	4398
Cat. 3/175	4 fruitstones of *Medemia argun*; uncertain provenance and period, probably Graeco-Roman; period donated by the Egyptian Museum (1935).
Dominant	● cf. *Medemia argun* 4 specimens of unknown fruit/seed (same as No. 3275): small hole at one end and branched vein pattern present on surface (30 x 22 mm) (similar to No. 4397).
No.	4402
Cat. 3/176	Small basket (margounah; diameter: 13 cm; cover: 10 cm) of date palm leaves and halfa, containing raisins; Deir El Medineh; New Kingdom; donated by the French Institute of Oriental Archaeology (1933).
Dominant	● *Vitis vinifera* Ca. 30 fruits.
No.	4455
Cat. 3/180	2 bulbs of garlic, one in bad condition; tomb of Tut-Ankh-Amun at Thebes; XVIIIᵉ Dynasty; donated by the Egyptian Museum (22-10-1953).
Dominant	● *Allium sativum* 2 bulbs, each with part of the stem.
No.	4456
Cat. 3/180	3 fruits of juniper (*Juniperus communis*); tomb of Tut-Ankh-Amun at Thebes; XVIIIᵉ Dynasty; M. Drar; donated by the Egyptian Museum (22-10-1953).
Dominant	● *Juniperus* 3 fruits (diameter: 15 mm).
No.	4457
Cat. 3/181	*Vitis vinifera* and *Juniperus*; tomb of Tut-Ankh-Amun at Thebes; XVIIIᵉ Dynasty; donated by the Egyptian Museum (22-10-1953).
Dominant	● *Vitis vinifera* Ca. 15 fruits, the greater number desiccated.
Contam.	● *Juniperus* 1 fruit (9 x 9 mm).
.	● *Coriandrum sativum* 1 fruit.
No.	4461
Cat. 3/181	5 fruits of custard-apple (*Annona squamosa*) (3.9 cm; 3.2 cm; 3 cm, 2.7 cm and 3.3 cm); Tuna El Gebel (emperor Trojan); probably Roman period; identified by M. Drar, not by V. Täckholm; bought from Qena (Tadros Girgis) (27-04-1932).
Dominant	● *Annona squamosa* 4 complete fruits, 3 labelled: A, B and C. Length: 40 mm (A), 32 mm (B), 26 mm (C) and 23 mm.
Comment	Probably origin West Indies. Considered as recent.
No.	4462
Cat. ?	In the catalogue this number refers to an insect.
Dominant	● *Mimusops laurifolia* Ca. 30 fruits (18-27 x 21-32 mm) and dozens of seeds.
No.	4463
Cat. 3/182	Basket (margounah; 19 x 11 cm) made of fibres of date palm and halfa, containing some fruits of sycamore figs, dates, *Balanites*, *Ziziphus* and raisins; Deir El Medineh; New Kingdom; donated by the French Institute of Oriental Archaeology (1933).
Dominant	● *Phoenix dactylifera* ● *Ficus sycomorus* Accessory fruits (syconia), some with incisions. ● *Ficus carica* Accessory fruits (syconia). ● *Balanites aegyptiaca* ● *Cyperus esculentus* Small tubers. ● *Vitis vinifera* 3 fruits, each in small linen bag, and loose specimens. ● Prepared food
No.	4464
Cat. 3/182	Fruits of hazel (*Corylus avellana*) in a box; probably Graeco-Roman period; bought in Luxor (Mohareb Tadros) (27-04-1932).
Dominant	● *Corylus avellana* Rather flat specimens (ca. 18-25 mm long).
No.	4465
Cat. 3/182	Fragments of *Punica granatum*; uncertain provenance; Roman period; bought in Luxor (Mohareb Tadros) (27-04-1932).
Dominant	● *Punica granatum* 2 small fruits (24 x 22 mm) and 2 flattened fruits that could be pomegranate.

No.	4466		No.	4476
Cat. 3/182	One endocarp of peach (diameter: 3 cm); uncertain provenance; probably Roman period; bought in Luxor (Mohareb Tadros) (1933).		Cat. 3/184	4 fruits of *Ziziphus*; Deir El Medineh; New Kingdom; donated by the French Institute of Oriental Archaeology (1933).
Dominant	● *Pinus pinea* 1 seed and 3 cone scales.		Dominant	● *Ziziphus spina-christi* 4 whole fruits and 1 endocarp.
No.	4467		No.	4477
Cat. 3/183	3 endocarps of *Prunus* (3.1 cm, 2.2 cm and 2.2 cm); Elephantine; probably Roman period; donated by the Egyptian Museum (P.H.H. Bovier-Lapierre) (1936).		Cat. 3/184	Small pottery dish (diameter: 9.6 cm) containing *Coriandrum sativum* seeds; provenance and period not mentioned; identified by V. Täckholm (1950); donated by the French Institute of Oriental Archaeology (1933).
Dominant	● *Prunus persica* 4 complete endocarps (22 mm, 23mm, 31 mm and 32 mm).		Dominant	● *Coriandrum sativum* 100-150 fruits.
No.	4468		No.	4878
Cat. 3/183	4 endocarp fragments of *Amygdalus* (2.5 cm, 2.4 cm, 2.3 cm and 2 cm); Saqqara, probably Graeco-Roman period; donated by the Egyptian Museum (15-02-1936).		Cat. 4/45	Dish (diameter: 3.4 cm) containing half a fruit of *Juglans* regia; Tebtynis (Fayum); Graeco-Roman period; donated by the Italian excavations (1933).
Dominant	● *Amygdalus communis* Endocarp fragments of 2 fruits.		Dominant	● *Juglans regia* Half fruit (3.5 cm).
No.	4469		No.	4925
Cat. 3/183	2 fruits of *Juniperus* (diameter: 1 cm); uncertain provenance and period; bought in Luxor (Molatham) (24-12-1934).		Cat. 4/45	*Pinus halepensis*, it has small seeds, not snobar; Saqqara; from J.E. Quibell.
Dominant	● *Juniperus* 2 fruits (diameter: 15 mm).		Dominant	● *Pinus* Cone, of which most of the scales have been removed. Small part is burnt.
No.	4470		No.	4926
Cat. 3/183	*Corylus*, not complete (diameter: 1.5 cm); uncertain provenance and period; bought in Luxor (Molatham) (27-06-1934).		Cat. 4/56	2 cones of *Pinus pinea* (9 cm and 8 cm); Suez?; period uncertain; donated by the Egyptian Museum (J.E. Quibell) (1933).
Dominant	● *Corylus avellana* Gnawed fruit, rather 'fresh' appearance.		Dominant	● *Pinus* 2 cones, ca. 6-7 cm long, thick stalk.
No.	4474		No.	4927
Cat. 3/183	Quantity of fruits of *Ziziphus*; tomb of Tut-Ankh-Amun at Thebes; New Kingdom; donated by the Egyptian Museum (1936).		Cat. 4/56	3 endocarps of *Amygdalus persica* (2.6 cm, 2.7 cm and 2.1 cm); Saqqara; Late period (612 BC); bought in Luxor (Mohareb Tadros) (24-12-1934).
Dominant	● *Ziziphus spina-christi* Hundreds of whole fruits, although most specimens have lost their exocarp owing to preservation conditions.		Dominant	● *Prunus persica* 3 endocarps (28 x 21 mm, 29 x 20 mm and 22 x 17 mm).
No.	4475		No.	4928
Cat. 3/184	8 fruits of *Balanites*; Deir El Medineh; New Kingdom; donated by the French Institute of Oriental Archaeology (1933).		Cat. 4/57	2 endocarps of *Amygdalus persica* (2.5 cm and 2.4 cm); Saqqara; Late period (612 BC); bought in Luxor (Mohareb Tadros) (24-12-1934).
Dominant	● *Balanites aegyptiaca* 8 whole fruits, although exocarp is partly missing.		Dominant	● *Prunus persica* 2 endocarps (31 x 25 mm and 28 x 22 mm).
			Comment	See figure 18.19 in Darby *et al.*, 1977

No. 4929
Cat. 4/57 2 endocarps of *Amygdalus persica* (2.8 cm and 2.5 cm); uncertain provenance and period; bought in Luxor (Mohareb Tadros) (24-12-1934).
Dominant • *Prunus persica*
2 endocarps (27 x 19 mm and 26 x 20 mm).

No. 4930
Cat. 4/57 Endocarp of *Amygdalus persica* (2.3 cm); uncertain provenance and period; bought in Luxor (Mohareb Tadros) (24-12-1934).
Dominant • *Prunus persica*
1 endocarp (25 x 21 mm).

No. 4931
Cat. 4/57 Quantity of *Amygdalus communis* fruits; uncertain provenance and period, maybe Late period (612 BC); bought in Luxor (Mohareb Tadros) (24-12-1934).
Dominant • *Amygdalus communis*
100-150 endocarps.

No. 4932
Cat. 4/57 Quantity of *Amygdalus communis* fruits; uncertain provenance and period; maybe Late period (612 BC); bought in Luxor (Mohareb Tadros) (24-12-1934).
Dominant • *Amygdalus communis*
10 endocarps (17-24 x 15-18 mm).

No. 4933
Cat. 4/58 2 pomegranate fruits (4.6 cm and 4.1 cm) out of 6; one from Thebes tomb; New Kingdom (1500 BC); 6 fruits bought in Luxor (Mohareb Tadros) (27-02-1933).
Dominant • *Punica granatum*
2 fruits.

No. 4934
Cat. 4/58 Fruits of pomegranate; Deir El Medineh; New Kingdom, XVIIIe Dynasty (1555 BC); donated by the French Institute of Oriental Archaeology (1933).
Dominant • *Punica granatum*
8 fruits (48-33 x 38-29 mm).

No. 4935
Cat. 4/58 3 fruits of pomegranate (6 cm, 5.1 cm and 4.7 cm) out of 6; period uncertain; 6 fruits bought in Luxor (Mohareb Tadros) (27-02-1933).
Dominant • *Punica granatum*
3 fruits (64 x 62 mm; 50 x 50 mm; 55 x 55 mm).

No. 4936
Cat. 4/58 3 fruits of pomegranate, one is broken (4.3 cm, 3.9 cm and 3.2 cm); Medinet Habu; New Kingdom (1500 BC); [M.H 30 | MG8] (1933).
Dominant • *Punica granatum*
3 fruits, 1 specimen fragmented.

No. 4937
Cat. 4/58 Fruits of pomegranate, Deir El Medineh, New Kingdom (1555 BC); donated by the French Institute of Oriental Archaeology (1933).
Dominant • *Punica granatum*
8 fruits, medium sized.

No. 4938
Cat. 4/59 Fruits of pomegranate; Saqqara; probably Late period (612 BC); bought in Luxor (Mohareb Tadros) with other fruits (24-12-1934).
Dominant • *Punica granatum*
11 whole fruits (36-56 x 26-43 mm) and some peel fragments.

No. 4939
Cat. 4/59 3 fruits of pomegranate (5 cm, 2.3 cm and 2.2 cm); Karanis (Kôm Aushim); Roman period (30 BC – AD 640); 1933); [30 | C123 CF2 | F].
Dominant • *Punica granatum*
3 fruits (25 x 25 mm and 48 x 53 mm), smallest specimen without upper part (not measured).
Comment See figure 18.24 in Darby et al., 1977

No. 4953
Cat. 4/63 Lupinus seeds (*Lupinus termis* L.); Karanis (Kôm Aushim); Roman period (30 BC – AD 640); bought in Luxor (Mohareb Tadros) (09-06-1932).
Dominant • *Lupinus albus*
26 seeds (ca. 9 mm).

No. 4954
Cat. 4/63 Lupinus seeds (*Lupinus termis* L.); uncertain provenance; probably Roman period; donated by the Egyptian Museum (hall 53, case O); [7 | 33 | 27 | 22].
Dominant • *Lupinus albus*
26 seeds (ca. 7-9 mm).

No. 4955
Cat. 4/63 Lupinus seeds (*Lupinus termis* L.); uncertain provenance; Roman period (30 BC – AD 640); bought in Luxor (Mohareb Tadros) with other seeds (09-06-1932).
Dominant • *Lupinus albus*

	76 seeds (ca. 7-9 mm). Seeds look rather old, some seed-coats are partly eroded.	Contam.	● *Galium aparine* Several fruits. ● *Phalaris* Spikelet. ● *Triticum turgidum* ssp. *durum* Rachis fragment. ● *Hordeum vulgare* cf. ssp. *vulgare* (hulled) Spikelets.
No. Cat. 4/63	4956 Lupinus seeds (*Lupinus termis* L.); uncertain provenance; Roman period (30 BC – AD 640); bought in Luxor (Mohareb Tadros) with other seeds (09-06-1932).		
Dominant	● *Lupinus albus* Hundreds of seeds, without damage.	No. Cat. 4/65	4963 Leaves of *Ficus sycomorus*; period uncertain; donated by the Egyptian Museum (1933).
Comment	Probably depicted by Darby *et al.*, 1977 in figure 17.12.	Dominant	● *Ficus sycomorus* 3 whole leaves.
No. Cat. 4/64	4957 4 seeds of grass-pea (*Lathyrus sativus*) [crossed out]; Deir El Medineh; New Kingdom (1500 BC); excavations of the French Institute of Oriental Archaeology (1933).	No. Cat. 4/66	4966 Fruits of *Balanites aegyptiaca*; Deir El Medineh (Luxor); XVIIIᵉ Dynasty (1400 BC); excavations of the French Institute of Oriental Archaeology (1933).
Dominant	● *Lathyrus sativus* 4 seeds (ca. 7 mm long).	Dominant	● *Balanites aegyptiaca* 33 whole fruits.
No. Cat. 4/64	4958 Chick pea seeds (*Cicer arietinum*); Deir El Medineh; New Kingdom (1500 BC); excavations of the French Institute of Oriental Archaeology (1933).	No. Cat. 4/66	4967 Fruits of *Balanites aegyptiaca*; Deir El Medineh (Luxor); XVIIIᵉ Dynasty (1400 BC); excavations of the French Institute of Oriental Archaeology (B.C.M.J. Bruyère) (1933).
Dominant	● *Cicer arietinum* 9 seeds.	Dominant	● *Balanites aegyptiaca* 26 fruits, most of them still complete, 5 with insect hole.
No. Cat. 4/64	4959 Chick pea seeds (*Cicer arietinum*); Deir El Medineh; New Kingdom (1500 BC); excavations of the French Institute of Oriental Archaeology (1933).	No. Cat. 4/66	4971 Quantity of seeds of *Citrullus vulgaris*, Deir El Medineh (Luxor), XVIIIᵉ Dynasty (1555 BC); excavations of the French Institute of Oriental Archaeology (1933).
Dominant	● *Lathyrus sativus* Several hundreds of seeds, with damage by insects.	Dominant	● *Citrullus lanatus* Ca. 50 seeds.
Contam.	● *Triticum* Fragments of grain kernels. ● *Lolium* Part of spikelet.	Comment	See figure 18.8 in Darby *et al.*, 1977
No. Cat. 4/64	4960 A bean seed (*Vicia faba* L.); Karanis (Kôm Aushim); Roman period (30 BC – AD 640); (1937).	No. Cat. 4/67	4972 Endocarps of *Olea europaea*; Thebes (Luxor); New Kingdom, XVIIIᵉ Dynasty (1500 BC); bought in Luxor (Mohamed Mansour) (26-06-1934).
Dominant	● dung pellet of sheep/goat	Dominant	● *Olea europaea* Ca. 100 endocarps.
No. Cat. 4/65	4961 Quantity of Erivum lentils; Karanis (Kôm Aushim); Roman period; (1937); [CS$_2$|214 70-C X 37].	Contam.	● *Ziziphus spina-christi* 1 endocarp.
Dominant	● *Lens culinaris* Diameter: ca. 4 mm.	Comment	See figure 18.9 in Darby *et al.*, 1977 The label on the photograph erroneously mentions '4973', probably labels of No. 4972 and No. 4973 were changed.

No.	4973
Cat. 4/67	Endocarp of *Olea europaea*; Thebes (Luxor); New Kingdom, XVIIIᵉ Dynasty (1500 BC); bought in Luxor (Mohamed Mansour) (26-06-1934).
Dominant	● *Olea europaea* 1 whole fruit.

No.	4975
Cat. ?	–
Dominant	● *Balanites aegyptiaca* 50-100 endocarps (a few specimens with insect damage and gnawed holes)
Contam.	● *Ficus sycomorus* Accessory fruit (syconium).

No.	5003
Cat. 4/67	Buds of qaranful (clove tree); uncertain provenance and period; bought in Luxor (Mohamed Mansour) (14-04-1935).
Dominant	● *Syzygium aromaticum* 9 flower buds (blackish).
Contam.	● *Triticum turgidum* ssp. *dicoccon* Spikelet (brownish).

No.	5010
Cat. 4/75	Fruits of *Ziziphus spina-christi*; Thebes and Gebelein; Middle and New Kingdom; donated by L. Keimer (1933).
Dominant	● *Ziziphus spina-christi* 8 whole fruits and 1 endocarp.

No.	5011
Cat. 4/75	Fruits of *Ziziphus spina-christi*, Thebes and Gebelein, Middle and New Kingdom; donated by L. Keimer (1933).
Dominant	● *Ziziphus spina-christi* Ca. 30 whole fruits (some fragmented), including 1 specimen with 3 carpels.
Contam.	● *Olea europaea* 1 endocarp.

No.	5024
Cat. 4/78	*Ficus sycomorus* fruits (diameter of the large one: 2.3 cm); Step pyramid of Saqqara (1934); IIIᵉ Dynasty; donated by the Antiquity Department (1934); [C.G.65913].
Dominant	● *Ficus sycomorus* 24 accessory fruits (syconia), outside brown, inside blackish (obviously not charred).

No.	5034
Cat. 4/81	6 fruits of *Arceuthos drupaces* (diameter of the large one: 1.8 cm); El Asasif (Thebes); Middle Kingdom (2000 BC); donated by the Antiquity Department (1934).
Dominant	● *Juniperus* 1 fruit (diameter: 12 mm). ● *Juniperus drupacea* 5 fruits (diameter biggest specimen: 27 mm).

No.	5035
Cat. 4/81	9 fruits of *Juniperus* (diameter of the large one: 1 cm) in a pottery dish; Deir El Medineh (Luxor); XVIIIᵉ Dynasty (1500 BC); excavations of the French Institute of Oriental Archaeology (1933).
Dominant	● *Juniperus* 8 fruits.
Contam.	● *Phoenix dactylifera* 1 fruit.

No.	5039
Cat. 4/82	14 fruits of *Hyphaene thebaica* (4-7.5 x 3.5-5.5 cm); Thebes; New Kingdom (1500 BC); donated by the Egyptian Museum (1933).
Dominant	● *Hyphaene thebaica* 12 whole fruits and 2 endocarps, 1 specimen with cut mark.

No.	5100
Cat. 4/94	Seeds of *Hibiscus esculentus*; Fayum; Pharaonic period; excavations of the Antiquity Department.
Dominant	● *Pisum sativum* Ca. 100 seeds, slightly pappilose.

No.	5101
Cat. 4/94	*Raphanus sativum* seeds / *Hibiscus esculentus* [both names are mentioned]; Deir El Medineh; New Kingdom; excavations of the Antiquity Department.
Dominant	● *Raphanus sativum* Thousands of charred seeds.
Contam.	● *Triticum* Few grain kernels.

No.	5115
Cat. 4/96	Sample of *Lens esculentus*; Thebes; New Kingdom; bought from an antique dealer in Luxor.
Dominant	● *Lens culinaris* 60-80 seeds, no damage by insects.

The next series of samples is exposed in the VIP-hall.

No. 5210
Cat. 4/105 *Ziziphus spina-christi* fruits; Deir El Medineh; New Kingdom.
Dominant • *Ziziphus spina-christi*
3 whole fruits and 1 endocarp.

No. 5211
Cat. 4/105 *Cyperus esculentus*, Gebelein; Middle Kingdom (2000 BC).
Dominant • *Cyperus esculentus*
Ca. 50 tubers.
Contam. • *Triticum turgidum* ssp. *dicoccon*
1 spikelet.

No. 5212
Cat. 4/105 *Citrullus lanatus* seeds.
Dominant • *Ceratonia siliqua*
Ca. 25 seeds, badly preserved.

No. 5213
Cat. 4/105 Raisins; Deir El Medineh; New Kingdom.
Dominant • *Vitis vinifera*
3 fruits.

No. 5214
Cat. 4/105 3 fruits of pomegranate; Deir El Medineh; New Kingdom.
Dominant • *Punica granatum*
3 whole fruits.

No. 5222
Cat. 4/106 2 seeds of date palm; Saqqara; Old Kingdom.
Dominant • *Phoenix dactylifera*
2 seeds.

No. 5223
Cat. 4/107 2 seeds of date palm; Saqqara; Old Kingdom.
Dominant • *Phoenix dactylifera*
2 fruits.

No. 5224
Cat. 4/107 4 seeds of date palm; Saqqara; Old Kingdom.
Dominant • *Phoenix dactylifera*
4 seeds.

No. 5225
Cat. 4/107 Fruits of *Ficus sycomorus*; store of Zoser pyramid at Saqqara; Old Kingdom, IIIe Dynasty.
Dominant • *Ficus sycomorus*
Ca. 10 accessory fruits (syconia).

No. 5226
Cat. 4/107 One fruit of *Ficus carica*; Deir El Medineh; New Kingdom.
Dominant • *Ficus* cf. *sycomorus*
1 accessory fruit (syconium).

No. 5227
Cat. 4/107 2 fruits of *Ficus carica*; Deir El Medineh; New Kingdom.
Dominant • *Ficus sycomorus*
2 accessory fruits, 1 specimen with small hole.

No. 5228
Cat. 4/107 *Citrullus lanatus* seeds; Deir El Medineh; New Kingdom; excavations of the French Institute of Oriental Archaeology.
Dominant • *Citrullus lanatus*
Ca. 50 seeds.

No. 5229
Cat. 4/108 Seeds of *Hibiscus esculentus*; Fayum.
Dominant • *Pisum sativum*
Ca. 100 seeds.

No. 5230
Cat. 4/108 Longitudinal section in the fruit of *Hibiscus esculentus*; bought from an antique dealer in Luxor.
Dominant • animal, probably part of a jaw of a fish.

No. 5231
Cat. 4/108 Seeds of *Raphanus sativus*; Deir El Medineh; New Kingdom (1400 BC).
Dominant • *Raphanus sativus*
Hundreds of seeds, mostly charred.
Contam. • *Triticum turgidum* ssp. *dicoccon*
1 grain kernel.
Contam. • *Hordeum vulgare* (hulled)
1 spikelet.

No. 5232
Cat. 4/108 Seeds of *Lens esculentus* / *Cicer arietinum* [both names are mentioned]; Deir El Medineh; New Kingdom (1400 BC).
Dominant • *Lathyrus sativus*
Ca. 100-200 seeds.

No. 5233
Cat. 4/108 Seeds of *Lens esculentus*; Deir El Medineh; New Kingdom (1400 BC).
Dominant • *Lathyrus sativus*
11 seeds.

No. 5234
Cat. 4/108 5 seeds of *Lupinus termis*; Thebes; Late period.
Dominant • *Lupinus albus*
5 seeds, no damage by insects.

No. 5235
Cat. 4/109 Seeds of *Lupinus termis*; Roman period; bought in Luxor.
Dominant • *Lupinus albus*
Ca. 100 seeds, no damage by insects.

No. 5236
Cat. 4/109 5 fruits of *Balanites aegyptiaca* (one endocarp and the large specimen damaged); Deir El Medineh; New Kingdom (1400 BC).
Dominant • *Balanites aegyptiaca*
4 whole fruits and 1 endocarp.

No. 5237
Cat. 4/109 Fruits of *Carthamus tinctorius*; Karanis (Kôm Aushim); Roman period.
Dominant • *Carthamus tinctorius*
Ca. hundreds of fruits.

No. 5238
Cat. 4/109 Grains of barley; Deir El Medineh; New Kingdom.
Dominant • *Hordeum vulgare* ssp. *vulgare* (hulled)
Ca. 100 spikelets.
Contam. • *Triticum turgidum* ssp. *dicoccon*
Spikelets.

No. 5239
Cat. 4/109 Quantity of barley; store of Step pyramid at Saqqara (02-1927); Old Kingdom, IIIe Dynasty.
Dominant • *Hordeum vulgare* (hulled)
Chaff.
Contam. • *Triticum turgidum* ssp. *dicoccon*
Rachis fragment (spikelet fork).
• *Scorpiurus muricatus*
Fruits.

No. 5240
Cat. 4/109 Grains of wheat (*Triticum dicoccum*); Badari; Prehistoric period; [31|573].
Dominant • *Hordeum vulgare* ssp. *vulgare* (hulled)
Badly preserved chaff and only a few whole spikelets.

No. 5241
Cat. 4/110 Grains of wheat; store of Step pyramid at Saqqara; Old Kingdom, IIIe Dynasty.
Dominant • *Triticum turgidum* ssp. *dicoccon*
Hundreds of spikelets, top spikelets with twisted internodes.
Contam. • *Hordeum vulgare* ssp. *vulgare* (hulled)
Spikelets.
• *Lotus*
Fruit.

No. 5242
Cat. 4/110 *Triticum pyramidale*; Deir El Medineh; New Kingdom.
Dominant • *Triticum aestivum/durum*
Ca. 30 grain kernels.

No. 5243
Cat. 4/111 Flax fruits; Saqqara; Late period.
Dominant • *Linum usitatissimum*
Ca. 50 fruits.
Contam. • *Lolium remotum*
Spikelets.
• *Medicago polymorpha*
Fruits.

No. 5244
Cat. 4/111 Flax fruits; Thebes; New Kingdom.
Dominant • *Linum usitatissimum*
3 fruits and a few seeds.

No. 5245
Cat. 4/111 *Lactuca sativa* seeds; Karanis (Kôm Aushim); Roman period.
Dominant • Cruciferae
Hundreds of seeds.

No. 5246
Cat. 4/111 *Ricinus communis* seeds; Karanis (Kôm Aushim); Roman period; (separate label: *Balanites aegyptiaca*; Deir El Medineh; New Kingdom).
Dominant • *Ricinus communis*
24 seeds.

No. 5247
Cat. 4/111 *Olea europaea* endocarps; Roman period; bought from an antique dealer in Luxor.
Dominant • *Olea europaea*
30 endocarps.

No. 5249
Cat. 4/111 Grains of wheat; Badari; Prehistoric period (5000 BC).
Dominant • *Triticum turgidum* ssp. *dicoccon*
Ca. 50 grain kernels.
Contam. • *Hordeum vulgare* ssp. *vulgare* (hulled)
1 spikelet.
• *Lolium temulentum*
1 spikelet.

No.	5250
Cat. 4/111	Sample of barley grains; store of the Step pyramid at Saqqara; Old Kingdom, IIIe Dynasty.
Dominant	● *Triticum aestivum/durum* Ca. 30 grain kernels.

No.	5251
Cat. 4/111	Grains of wheat inside their spikelets.
Dominant	● *Hordeum vulgare* ssp. *vulgare* (hulled) 14 spikelets.

No.	5256
Cat. 4/112	Grains found inside 10 pottery containers; Deir El Medineh; excavations of the French Institute of Oriental Archaeology.
Dominant	● *Hordeum vulgare* Clay objects in which chaff has been stored/mixed.

No.	5257
Cat. 4/112	2 fruits of *Hyphaene thebaica*; Deir El Medineh; New Kingdom
Dominant	● *Hyphaene thebaica* 2 whole fruits.

Samples stored in the laboratory. These samples have no official numbers.

No.	-
Cat. ?	-
Dominant	● *Papaver somniferum* Whole fruit with some incisions on the outside.

No.	-
Cat. ?	-
Dominant	● *Hordeum vulgare* ssp. *vulgare* (hulled) Spikelets. ● *Triticum turgidum* ssp. *dicoccon* Spikelets. ● *Lolium temulentum* Spikelets.

No.	-
Cat. ?	-
Dominant	● *Hordeum vulgare* ssp. *vulgare* (hulled) Spikelets. ● *Triticum turgidum* ssp. *dicoccon* Spikelets, top-spikelet with twisted internode.

No.	-
Cat. ?	-
Dominant	● *Triticum turgidum* ssp. *dicoccon* Ca. 50-100 grain kernels.

No.	-
Cat. ?	-
Dominant	● *Lathyrus sativus* Ca. 50 seeds.

No.	-
Cat. ?	Gebelein
Dominant	● *Mimusops laurifolia* 12 seeds.

No.	-
Cat. ?	*Triticum dicoccum* Schübl.; identified by E. Åberg (Uppsala, surelen) (14-11-1949); [4 | 3 | 27 | 29].
Dominant	● *Triticum turgidum* ssp. *dicoccon* Ca. 50-100 grain kernels.
Contam.	● *Lotus* Fruit fragment.

No.	-
Cat. ?	*Triticum dicoccum* Schübl.; identified by E. Åberg (Uppsala), (14-11-1949); [4 | 3 | 27 | 29].
Dominant	● *Triticum turgidum* ssp. *dicoccon*

No.	-
Cat. ?	*Triticum monococcum*; identified by E. Åberg (Uppsala), (14-11-1949); [4 | 3 | 27 | 29].
Dominant	● *Triticum turgidum* ssp. *dicoccon* Ca. 100 ears and spikelets (2-seeded), including top-spikelets with twisted internode.

No.	-
Cat. ?	Deir El Medineh (Luxor), VXIIIe Dynasty.
Dominant	● cf. *Papaver somniferum* Fruit seems immature which may explain why small seeds do not match those of *Papaver somniferum*.

Table 1: List of sites from where samples were collected. Uncertain provenance and plant remains bought from antique dealers have been placed between brackets. Samples without official numbers have been omitted.

Sites	Sample numbers	N
Abu-Sir	3367	1
Assiut	3228, 4205	2
Badari (Upper Egypt)	1213, 5240, 5249	3
Cheikh Abd El Qurnah (Thebes)	0393, 3356, 4083	3
Deir El Bahari	0377, 0392	2
Deir El Medineh (Luxor)	0081, (0371) 1218, 1350, 1401, 1407, 1414, 1416, 1417, 1421, 1426, 1427, 1432, 1439, 1440, 1441, 1543, 1545, 2662-2664, 2686, 2688, 2689, 2785, 2786, 3216, (3218) 3219, 3220, 3227, 3229, 3265, 3267-3269, 3271, 3276, 3280, (3281), (3370), 3371, (4170), 4171-4173, 4176, 4180, 4183, 4196, 4203, 4211, 4219, 4222, (4290), 4296, 4387, 4392, 4393, 4396, 4402, 4463, 4475, 4476, 4934, 4937, 4957-4959, 4966, 4967, 4971, 5035, 5101, 5210, 5213, 5214, (5226), 5227, 5228, 5231-5233, 5236, 5238, 5242, (5246), 5256, 5257	90
Denderah	4079	1
Dra Abu El Naga (Thebes)	0400, 0401	2
Elephantine	2105, 3212, 4195, 4220, 4223, 4229, 4395, 4397, 4467	9
El Asasif (Thebes)	5034	1
El Rizeikat (near Luxor)	(1442), (1445)	2
El Sheik Ibada (Antinoöpolis)	2791, 2792	2
Fayum	3279, (4212), 5100, 5229	4
Gebelein (Upper Egypt)	(0371), 0372, 0378-0380, 0382-0386, 0389, 0394-0397, 0399, 0403, 1214, 1216, 1312-1314, 1316, 3205, 3206, 3214, 3215, 3372, 4096, 5010, 5011, 5211	32
Hawara (Fayum)	0408	1
Karanis (Kôm Aushim)	0271-0277, 0279-0289, 0292-0295, 0460, 1047, 1425, 1429, 1433, (1436), 1444, 1968, 2054-2057, 2693, 2694, (2696), 2699, (3272), 3289, 3294-3298, (3299), 4175, 4177, 4191, 4194, 4197, 4228, 4232, 4234, 4297, 4939, (4953), 4960, 4961, 5237, 5245, (5246)	62
Kuseih and Meir	2249, 3362, 4086, 4087	4
Luxor (bought!)	1355-1359, 1361, 1363, 1365-1368, 1370, 1372-1375, 1404, 1411, 1423, 1434, 1436, 1442, 1445, 1541, 1542, 1544, 1966, 2103, 2104, 2250-2252, 2273, 2666, 2667, 2687, 2695, 2698, 2700, 3218, 3222-3226, 3273-3275, 3278, 3281, 3282, 3357, 3370, 3378, 3379, 4059, 4060, 4167-4170, 4178, 4179, 4181, 4185, 4187-4190, 4192, 4193, 4200, 4201, 4210, 4212, 4213, 4216, 4225-4227, 4233, 4290, 4389, 4390, 4464-4466, 4469, 4470, 4927-4933, 4935, 4938, 4953, 4955, 4956, 4972, 4973, 5003, 5115, 5230, 5235, 5247	108
Maadi	4295	1
Medinet Habu	3208, 3209, 4199, 4936	4
Merimde (west of Bani Salama)	1431, 4088, 4089	3
Mustagada (Badari)	4263, 4264, 4265, 4266	4
Qena (bought!)	4461	1
Qurna (Thebes)	3376	1
Saqqara	0373, 0374, 0404, 0405, 0939, 0940, 1219, 1402, 1412, 1430, 1437, 1546, 2106, 2665, 3207, 3210, 3211, 3284, 3288, 3290, 3293, (3299), 3300, 3301, 4085, 4174, 4204, 4230, 4231, 4287, 4292, 4294, 4321, 4468, 4925, (4927), (4928), (4938), 5024, 5222-5225, 5239, 5241, 5243, 5250	47
Suez	(4926)	1
Tebtynis (Fayum)	(2696), 2697, 3366, 4878	4
Thebes	0406, (1356), 1438, (2103), (2104), (3218), 3270, 3361, 3374, 3375, (4181), 4184, (4188-4190), (4193), 4198, 4206, 4207, (4213), 4322, 4388, (4389), (4390), 4455, 4456, 4457, 4474, (4933), (4972), (4973), 5010, 5011, 5039, (5115), 5234, 5244	37
Tuna El Gebel	1576, 3292 , (4461)	3

Table 2: List of plants and samples. Non-botanical remains, lichens, resins and prepared food are excluded. Uncertain identifications both to the level of genus or species are placed between brackets.

Plant name	Sample numbers	N
Acacia	1361, 4187	2
Acacia nilotica	1367, 1576, (4225), (4226), 4227, 4228, (4230), 4231, 4232, 4234	10
Allium cepa	1442, (2686), 2687, 4181, 4185	5
Allium cepa var. *aggregatum*	4182	1
Allium sativum	0276, 1444, 1445, 4181, 4183, 4184, 4185, 4192, 4455	9
Ambrosia maritima	0406	1
Amygdalus communis	0408, 1576, 3294, 4468, 4931, 4932	6
Anastatica hierochuntica	4210	1
Annona squamosa	4461	1
Anthemis	0287	1
Asphodelus tenuifolius	4213	1
Balanites aegyptiaca	0939, 1375, 1426, 2786, 3216-3218, 3220, 3267, 3281, 3371- 3373, 4463, 4475, 4966, 4967, 4975, 5236	19
Beta vulgaris	0286, 1968	2
Brassica	0406	1
Brassica napus	(4188), (4189)	2
Brassica nigra	2273	1
Calotropis procera	4205, 4223	2
Carthamus tinctorius	0272-A/B, 0281, 0285, 1437, 2693-2697, 4177, 4178, 4225, 4227, 5237	14
Ceratonia siliqua	4216, 5212	2
Cicer arietinum	0403, 1434, 2251, 4169, 4179, 4958	6
Citrullus colocynthis	1432, 4176, 4190	3
Citrullus lanatus	1439, 1441, 2252, 4097, 4187, 4193, 4198, 4227, 4971, 5228	10
Citrus medica	(4199), (4200)	2
Cocculus pendulus	(0397)	1
Convolvulus arvensis	1968, 2056, 2057	3
Cordia myxa	1541, 1542, 1544-1546, 2250	6
Coriandrum sativum	1432, 4206, 4219, 4457, 4477	5
Corylus avellana	0280, 4464, 4470	3
Cruciferae	0283, 1438, 2700, 4095, 4219, 5245	6
Cucumis sativus	1440, (4197)	2
Cucurbitaceae	4201	1
Cuminum cyminum	0406, 1432, 4219	3
Cupressus	0395	1
Cyperaceae	3358	1
Cyperus	(1576)	1
Cyperus articulatus	1312	1
Cyperus esculentus	0383, 0394, 0401, 1213-1219, 1350, 2249, 2791, 3205, 3206, 3214, 3215, 3362, 3364, 3368, 3369, 4095, 4263, 4463, 5211	25
Daucus carota	1216, 3206, (3214)	3
Echium rauwolfii	1432	1
Emex spinosa	4206	1
Faidherbia albida	4222, 4229, 4233	3
Ficus carica	1414, 3267, 4463	3
Ficus sycomorus	0371, 0380, 0399, 1316, 1368, 1411, 1412, 1576, (2250), 2688, 2689, 2785, (2792), 3222, 3223, (3224), 3225, 3226, 3228, 3229, 3267, 3364, 4085, 4463, 4975, 5024, 5225, (5226), 5227	30
Galium aparine	0274, 1968, 2056, 2057, 4961	5
Galium tricornutum	0286	1

Plant name	Sample numbers	N
Gramineae	2696, 3358	2
Gramineae tribe Triticeae	0281, 0285, 1358, 1365, 1968, 2056, 2694, 3267, 3273, 3357, 3360, 4207, 4228, 4266, 4287	15
Grewia	4206, 4207	2
Hordeum vulgare	4219, 5256	2
Hordeum vulgare (hulled)	0277, 0286, 0287, 0382, 0401, 1432, 1438, 2693, 3214, 3215, 3267, 3379, 4089, 4203, 5231, 5239	16
Hordeum vulgare ssp. *distichum* (hulled)	3361	1
Hordeum vulgare ssp. *vulgare* (hulled)	0135, 0274, 0275, (0284), (0285), 0385, 0386, 0403, 0409, 1047, 1215, 1217, 1358, 1430, 1439, 1968, 2054-2057, 2667, 3358-3360, 3362, 3364, 3369, 4086, 4087, 4095, 4198, 4287, 4290, 4292, 4294-4296, 4961, 5238, 5240, 5241, 5249, 5251	43
Hyoscyamus muticus	4204	1
Hyphaene thebaica	1342, 1372, 1373, 1416, 1417, 2103, 2104, 2105, 2106, 3267, 3270-3276, 3374-3376, 4395, 4396, 5039, 5257	24
Juglans regia	0279, 4878	2
Juniperus	0377, 0392, 3219, 3293, 3370, 4207, 4456, 4457, 4469, 5034, 5035	11
Juniperus drupacea	5034	1
Juniperus oxycedrus	(1423)	1
Lagenaria siceraria	0130, 0290, 4187, 4194-4196	6
Lathyrus hirsutus	1432, 4172, 4206	3
Lathyrus sativus	(0385), (1217), 1432, 3265, 3360, 4171-4173, 4176, 4957, 4959, 5232, 5233	13
Leguminosae	0406, 2251	2
Lens culinaris	0284, 0351, 0406, 3360, 3364, 4089, 4198, 4961, 5115	9
Linum usitatissimum	0289, 1359, 1438, 3214, 3215, 4059, 4060, 4083, 4095, 4096, 4168, 4264, 4321, 4322, 5243, 5244	16
Lolium	4959	1
Lolium remotum	(4321), 5243	2
Lolium temulentum	(0406), 3214, 3267, 4089, (4219), 4292, 4294, 5249	8
Lotus	0274, 3362, 5241	3
Lupinus albus	0282, 1436, 2699, 3281, 3384, 4084, 4167, 4170, 4953-4956, 5234, 5235	14
Lupinus digitatus	2791	1
Malva nicaeensis/sylvestris	0274, 1968	2
Malva sylvestris	0287	1
Maerua crassifolia	0286, 0385, 0398	3
Medemia argun	0372, 0374, (3275), (3376), (4397), (4398)	6
Medicago polymorpha	(0275), 4321, 5243	3
Medicago sativa	0136, 4079	2
Mimusops laurifolia	1337, 1357, 1401, 1402, 1404, 2662-2667, 3235, 3369, 4462	14
Moringa peregrina	4220	1
Myrtus communis	4212	1
Nelumbo nucifera	3349, 3350	2
Nigella sativa	0406	1
Olea europaea	0285, 0404, 0460, 1425, 1541, 2251, 2693, 3210, 3212, 3281, 4179, 4180, 4972, 4973, 5011, 5247	16
Papaver somniferum	–	1
Phalaris	0406, 4961	2
Phalaris paradoxa	0274	1
Phoenix dactylifera	0273, 0285, 0381, 0404, 1363, 1966, 2687, 3265, 3267, 3273, 3278-3282, 3284, 4207, 4382, 4387-4390, 4392, 4393, 4463, 5035, 5222-5224	29
Pinus	0288, 3290, 4925, 4926	4
Pinus pinea	0379, 0383, 0384, 0404, 3287-3289, 3291, 3292, 3372, 4466	11

Plant name	Sample numbers	N
Pistacia	(1355)	1
Pisum sativum	3360, 4173, 4176, 5100, 5229	5
Prunus persica	3295-3299, 4467, 4927-4930	10
Pseudoconyza viscosa	0295	1
Punica granatum	0294, 1421, 3300-3302, 3364, 4465, 4933-4939, 5214	15
Raphanus raphanistrum	0274, 0286, 1968, 2056	4
Raphanus sativum	5101, 5231	2
Ricinus communis	0271, 2698, 5246	3
Rumex dentatus	(0405), 4191	2
Scorpiurus muricatus	5239	1
Sesamum indicum	4206	1
Silene coeli-rosa	0407	1
Sinapis	(0286)	1
Sinapis arvensis	(0274), (1968), (2056)	3
Sorghum	1313	1
Sorghum bicolor	3357	1
Sorghum halepense	1365, 3356, 4198, (4207)	4
Syzygium aromaticum	5003	1
Tamarix	2055	1
Trifolium	0281, 2273, 2694, 4191	4
Trifolium alexandrinum	(0287)	1
Trigonella foeno-graecum	(0406)	1
Triticum	2273, 3358, 4959, 5101	4
Triticum aestivum	0135, 1358, 2054, 3379	4
Triticum aestivum/durum	0081, 0286, 0393, 5242, 5250	5
Triticum turgidum ssp. *dicoccon*	0373, 0382, 0394, 0396, 0401, 0406, (1214), 1216, 1356, 1427, 1430-1432, 3206, 3214, 3265, 3267, 3356, 3357, 3359- 3362, 3364, 3367, 3368, 3378, 4086-4089, 4095, 4198, 4265, 4287, 4290, 4295, 5003, 5211, 5231, 5238, 5239, 5241, 5249	44
Triticum turgidum ssp. *durum*	0135, 0272-B, 0274, 0275, 0277, 0281, 0285, 1047, 1365, 1429, 1968, 2054, 2056, 2693, 3356-3358, 3366, 3369, 4095, 4228, 4294, 4297, 4961	24
Vaccaria hispanica	(0286), (2273)	2
Vicia	(3206)	1
Vicia ervilia	(0284), 2251, 4168	3
Vicia faba	3358, 4174	2
Vicia faba var. *minor*	0393, 1433, 4175	3
Vitis vinifera	0378, 0393, 0396, 0401, 1370, 1407, 1432, 3215, 3227, 3265, 3267-3269, 3273, 3358, 4206, 4392, 4402, 4457, 4463, 5213	21
Zilla spinosa	4095	1
Ziziphus spina-christi	0383, 0389, 0406, 0940, 1423, 2792, 3207, 3208-3212, 3227, 3370, 4206, 4207, 4474, 4476, 4972, 5010, 5011, 5210	22
Zygophyllum coccineum	(2057)	1